T0318596

Herbicides

Emerging Issues in Analytical Chemistry

Series Editor
Brian F. Thomas

Co-published by Elsevier and RTI Press, the *Emerging Issues in Analytical Chemistry* series highlights contemporary challenges in health, environmental, and forensic sciences being addressed by novel analytical chemistry approaches, methods, or instrumentation. Each volume is available as an e-book, on Elsevier's ScienceDirect, and in print. Series editor Dr. Brian F. Thomas continuously identifies volume authors and topics; areas of current interest include identification of tobacco product content prompted by regulations of the Family Tobacco Control Act, constituents and use characteristics of e-cigarettes and vaporizers, analysis of the synthetic cannabinoids and cathinones proliferating on the illicit market, medication compliance and prescription pain killer use and diversion, and environmental exposure to chemicals such as phthalates, endocrine disrupters, and flame retardants. Novel analytical methods and approaches are also highlighted, such as ultraperformance convergence chromatography, ion mobility, in silico chemoinformatics, and metallomics. By highlighting analytical innovations and new information, this series advances our understanding of chemicals, exposures, and societal consequences.

2016 *The Analytical Chemistry of Cannabis: Quality Assessment, Assurance, and Regulation of Medicinal Marijuana and Cannabinoid Preparations*
Brian F. Thomas and Mahmoud ElSohly

2016 *Exercise, Sport, and Bioanalytical Chemistry: Principles and Practice*
Anthony C. Hackney

2016 *Analytical Chemistry for Assessing Medication Adherence*
Sangeeta Tanna and Graham Lawson

2017 *Sustainable Shale Oil and Gas: Analytical Chemistry, Geochemistry, and Biochemistry Methods*
Vikram Rao and Rob Knight

2017 *Analytical Assessment of e-Cigarettes: From Contents to Chemical and Particle Exposure Profiles*
Konstantinos E. Farsalinos, I. Gene Gillman, Stephen S. Hecht, Riccardo Polosa, and Jonathan Thornburg

2018 *Doping, Performance-Enhancing Drugs, and Hormones in Sport: Mechanisms of Action and Methods of Detection*
Anthony C. Hackney

2018 *Inhaled Pharmaceutical Product Development Perspectives: Challenges and Opportunities*
Anthony J. Hickey

2018 *Detection of Drugs and Their Metabolites in Oral Fluid*
Robert White

2020 *Smokeless Tobacco Products*
Wallace Pickworth

2020 *Analytical Methods for Biomass Characterization and Conversion*
David Dayton

2020 *Particulates Matter: Impact, Measurement, and Remediation of Airborne Pollutants*
Vikram Rao and William Vizuete

Herbicides

Chemistry, Efficacy, Toxicology, and Environmental Impacts

Edited by

Robin Mesnage

King's College London, London, United Kingdom

Johann G. Zaller

Institute of Zoology, University of Natural Resources and Life Sciences (BOKU), Vienna, Austria

Elsevier
Radarweg 29, PO Box 211, 1000 AE Amsterdam, Netherlands
The Boulevard, Langford Lane, Kidlington, Oxford OX5 1GB, United Kingdom
50 Hampshire Street, 5th Floor, Cambridge, MA 02139, United States

Library of Congress Cataloging-in-Publication Data
A catalog record for this book is available from the Library of Congress

British Library Cataloguing-in-Publication Data
A catalogue record for this book is available from the British Library

ISBN: 978-0-12-823674-1

For information on all Elsevier publications
visit our website at https://www.elsevier.com/books-and-journals

Publisher: Susan Dennis
Acquisitions Editor: Kathryn Eryilmaz
Editorial Project Manager: Mariana Kuhl
Production Project Manager: Sruthi Satheesh
Cover Designer: Margaret Reid

Typeset by SPi Global, India

Contents

Contributors

Michael Antoniou King's College London, London, United Kingdom

Charles M. Benbrook Heartland Health Research Alliance, Brookfield, WI, United States

Rachel Benbrook Heartland Health Research Alliance, Brookfield, WI, United States

Carsten A. Brühl iES Landau, Institute for Environmental Sciences, University of Koblenz-Landau, Landau, Germany

Mirco Bundschuh Universität Koblenz-Landau, Landau, Germany

Nicolas de Sadeleer Université Saint-Louis, Brussels, Belgium

Sylvain Dulaurent University Hospital of Limoges, Limoges, France

Souleiman El Balkhi University Hospital of Limoges, Limoges, France

Robin Mesnage King's College London, London, United Kingdom

John Peterson Myers Environmental Health Sciences, Charlottesville, VA, United States

Sophie Oster Universität Koblenz-Landau, Landau, Germany

Franck Saint-Marcoux University Hospital of Limoges, Limoges, France

András Székács Agro-Environmental Research Centre, Institute of Environmental Sciences, Hungarian University of Agriculture and Life Sciences, Budapest, Hungary

Laura N. Vandenberg University of Massachusetts Amherst, Amherst, MA, United States

Johann G. Zaller Institute of Zoology, University of Natural Resources and Life Sciences (BOKU), Vienna, Austria

Foreword

Despite Rachel Carson's epic warnings in *Silent Spring*,[1] total pesticide use, including insecticides, herbicides, fungicides, and other "cides" around the globe, has increased fivefold since the book was published in 1962.[2] The herbicidal properties of glyphosate, the active ingredient in Monsanto's blockbuster Roundup, were unknown to Rachel Carson. She died before the 1970s when Monsanto's John E. Franz discovered them.[3] Since the very beginning of glyphosate use, Monsanto has claimed that it is safe for people.[4]

Since we published *Our Stolen Future* in 1996,[5] the use of glyphosate-based herbicides has grown at least 10-fold (through 2014) and more than 100-fold since the late 1970s.[6] It is now the most widely used herbicide in the world. In spite of Monsanto's ongoing claims of safety, over the last 10 years hard data has emerged from independent scientific research documenting glyphosate toxicity to vertebrates.

And then came the lawsuits about Roundup and cancer, followed by documents, guilty verdicts, and large punitive damages. Whoops, maybe glyphosate is not so safe after all. Wait until you read Carey Gillam's Grishamesque book, *The Monsanto Documents*, out in Spring 2021.

I offer that context to set up the following conclusion: It's about time a comprehensive, scholarly book like this one is being published, written by trusted scientific experts without problematic conflicts of interest and taking an unvarnished, deep look at herbicides. I wish it had been available decades ago.

When Dr. Robin Mesnage asked me in July 2020 to write a foreword, he framed the request with this:

> *We decided to write this book because there is a need for a one-volume comprehensive overview of herbicide chemistry, use, toxicity, and environmental effects by selected internationally recognized experts. Information presented will inform the discussion on the use of herbicides in modern agricultural systems and their potential nontarget effects in human populations and various ecosystems.*

That resonated with me because I knew it was true. There is a need for this book. From 2014 through 2017, I worked with Robin and several authors of this book's chapters on two reviews of glyphosate safety.[7, 8] Based on that experience with Robin and his colleagues, I was confident of the quality the book would attain.

But why is this book truly needed? Scientists working in this area need to understand all dimensions of the playing field. One dimension is the excellent science presented by the authors. Use the book to guide your scientific explorations of herbicides.

But researchers also need to be aware of the corruption that has plagued herbicide science. That's the purpose of this foreword. They need to develop a nose for what's real and what's not. They need to be prepared to detect and counteract "manufactured doubt," the phenomenon that the chemical and herbicide industries employ to undermine scientific evidence of harm, so clearly documented in Monsanto's own internal memos and described in Gillam's book.

Fortunately, there is a rich literature on manufactured doubt, including books by Dan Fagin and Marianne Lavelle,[9] Gerald Markowitz and David Rosner,[10] Naomi Oreskes and Richard Conway,[11] and David Michaels.[12, 13] The body of work presented in this volume–trustworthy, accurate, and up to date–is an effective and much-needed counterpoint to industry. This book isn't manufactured doubt. It's distilled reality.

Manufacturing doubt has become standard practice in the chemical industry. Large revenue streams from the sale of successful chemicals can be applied to the toolkits of doubt practitioners. Consider these recent examples of industries using manufactured doubt to defend against complaints of harm: Syngenta and atrazine,[14] Johnson and Johnson and asbestos in talc,[15] Volkswagen's diesel scandal,[13] and Monsanto and glyphosate[4]. And read Michaels[13] for a litany of more.

A whole new trophic level of scientific "research" has been created: "product defense firms," whose business model is to deliver science that defends the interested party's products by creating enough uncertainty in the minds of regulators that they, the regulators, can tell the opposing parties "come back when a scientific consensus is reached." Adept product defense firms powered by the monetary value of keeping the product on the market can delay that "come back" moment for decades.

The investigative reporter Paul Thacker, writing in *Environmental Science and Technology*,[16] revealed an elaborate plan developed by the Weinberg Group, a product defense firm, to help DuPont withstand a growing public health scandal swirling around its perfluorooctanoic acid (PFOA) plant in West Virginia. This scandal ultimately escalated into the 2019 feature film *Dark Waters*, starring Mark Ruffalo. There is a lesson there for any company contemplating manufactured doubt. If you lose, you lose very big. The reputation of DuPont is forever sullied.

There are more cases in the wings, probably many. The next one will be a lawsuit against Syngenta for decades of misrepresenting, according to the plaintiffs, the dangers of the herbicide paraquat. The case is set to be tried in Illinois in early April 2021. Depositions are largely complete. The plaintiffs' lawyers also claim that documents obtained in discovery lay bare a profound disregard by Syngenta for human life and suffering since the 1960s, when paraquat came on the market. Around the world, thousands have died. Those who lived were at great risk of Parkinson's disease as they aged. The science is clear, they say. Perhaps the plaintiffs' lawyers' claims will be rejected by the court. By the time this book is published, the trial could be over and you will already know the outcome.

Are these and other transgressions just the doings of a few bad apples? I think not. I think it's about a bad barrel. The system guiding pesticide regulations, including herbicides, is broken. Globally, it needs deep structural reform. The reality is that manufacturing doubt works, and it's worked on multiple issues going back to tobacco and lead. It's gotten more and more sophisticated at execution. They are rewarded for doing it, and the system allows it. Perhaps manufacturing doubt is too soft a phrase. They are betraying science. They are abusing the law. Their delays hurt and even kill people.

I have faced manufactured doubt ever since I started in the field of endocrine disruption 30 years ago. I sometimes wonder what might we have been able to accomplish for public health were it not so rampant. This issue became much more poignant as we learned in 2020 that comorbidities such as hypertension, heart disease, and diabetes, among others, were exacerbating the seriousness of Covid-19,[17, 18] and that some of them had endocrine disruption as a possible cause.[19] Without manufactured doubt, would we have made more progress in regulating endocrine-disrupting chemicals? I am certain the answer is yes. What fraction of Covid-19 deaths could have been avoided with better endocrine-disrupting chemical regulations? When the next pandemic hits (as it will), will we have made more progress in reducing the comorbidities? I hope so.

So how is the system broken? There are multiple factors.

First, the ties between the regulated community (pesticide manufacturers) and the regulators are too close. A revolving door of regulators going to work for the regulated after a stint in government is too common for it to be healthy. It works the other way, too, with the regulated joining government and bringing with them the agenda they were pushing while in the private sector. There may never have been a more blatant and deeper example than during the tenure of Donald J. Trump as President, but it didn't start there. And it's global, not just in the United States.

I have a personal story to report, something I experienced directly. In 2012, before Trump, I was chairing a writing process on low dose/low concentration effects of endocrine-disrupting compounds. It led to a paper[20] (now cited more than 2200 times in the scientific literature). The conclusion of the paper was that using high-dose testing to anticipate low-dose impacts was scientifically inappropriate and based upon misguided assumptions from the 16th century, literally. Vandenberg et al.[20] was presaged by Myers et al.[21]; when the latter had no impact, I set out to organize a much more ambitious effort, which became the Vandenberg paper. As Vandenberg et al. headed toward publication, my coauthors and I met with scientists from the US Food and Drug Administration (FDA), specifically the FDA's Center for Food Safety and Applied Nutrition (CFSAN), to discuss the implications of our conclusions for regulatory science. Basically, we had concluded that the standard practice of testing at high doses and extrapolating downward to a "safe dose" was unacceptable because different things happen at low doses. Sometimes those low-dose effects are just the opposite of what happens at high doses. Any standard regulatory testing regime would miss the low-dose impacts, and the estimated safe dose would be wildly misleading, that is, too high.

I was seated at a table with one of the FDA's top food safety scientists. She explained that we were wrong because they never saw effects like that. I countered, "You don't see that because you don't test at those doses." She said (more or less), "Well, that's true." Fifteen minutes later I heard her take the same line of argument with one of my colleagues. "We don't see those effects." And then a year later she had jumped the FDA ship and began working for a product defense firm earning more than 10 times her FDA salary, plus bonuses.

So deep structural reform starts with dramatically slowing down the revolving door of regulators swapping jobs with the regulated. A very fast win would be a complete cleanout of the "experts" in CFSAN.

But it can't stop there. Every meeting of a regulator with a stakeholder needs to be recorded, with the MP4 file saved in a publicly accessible database. And by stakeholder, I mean everybody, on all sides of the table. I would prefer regulators to have 24-h 7-day body cams, but some of my civil liberty friends think that would be going too far. My parry is that the schmoozing that takes place out of business hours can be all too influential as decisions are made. What do you want: chemical safety, or curtailment of personal private communications with people with products to defend?

Another change: We need at least civil if not criminal penalties for failure to disclose conflicts of interest in scientific papers that are published relevant to regulatory decisions. A recent exposé by *Le Monde*[22] and *Environmental Health News* about hidden conflicts of interest by supposed experts on endocrine disruption had a salutary impact on negotiations leading to the European Commission's new (October 2020) Chemical Strategy for Sustainability. Investigative reporting by *Le Monde* revealed that the authors of an editorial thrown into the debate over how the strategy should approach endocrine disruption had massive undisclosed conflicts. Unfortunately, nondisclosure is all too common, and will likely remain so without penalties.

David Michaels in *The Triumph of Doubt* writes that not only should manufacturers "pay for toxicity studies and risk determination (i.e., polluter pays), but the studies should be conducted by independent scientists who design, execute, and interpret the results without interference." Moreover, "top corporate executives should sign off on the test results just as they must affirm the accuracy of their financial accounting under the threat of criminal sanctions for wrongdoing."

Deep structural reform also means changes to how science is applied to regulation. The effects of mixtures must be acknowledged and incorporated into regulations. So too, the challenge of low-dose/low-concentration effects that cause impacts not predictable from the typical high-dose experiments used by regulatory science. The tests used must be altered from traditional toxicology assays to ones that are clearly relevant to human diseases, which today's are not. They must incorporate 21st century biology instead of assays designed in the mid-20th century, or earlier. They must explicitly pay attention to epidemiological findings relevant to disease causation. And instead of beginning with a process that uses inappropriate criteria to eliminate the majority of rigorous academic studies from consideration in risk assessment, they must embrace these studies openly and honestly. As good as it sounds, the absence of good laboratory practice (GLP) cannot be used as an exclusionary criterion for academic data. GLP was about eliminating fraud, not about good science.[23] Rigorous peer review will ensure that good science is incorporated into decision making.

The false dichotomy between active and inert ingredients has no scientific basis in the regulation of pesticides. "Inert" ingredients can potentiate the impact of the active ingredient, and even add effects of their own. Tests of pesticides, including herbicides, should evaluate the toxicity of each product sold. This would lead to more meaningful scientific results and also slow down the proliferation of formulations that, because of their sheer number, defy epidemiological testing. "No safety data (especially on endocrine disruption), no market" should be applied to formulations, not just to the active ingredient.

This brief discussion is insufficient to consider all the structural reforms necessary to make pesticide regulation work for public health and the environment. I have touched on a few items that I think are exceptionally important. David Michaels in the final chapter of *The Triumph of Doubt* covers more. And I welcome informed suggestions from readers of this foreword about other important interventions. My email address is jpmyers@ehsciences.org.

Thank you, Robin, for taking the initiative to pull this book together. I think you put a bit too much hope behind machine learning as a future solution; the text comparing machine learning to animal experiments unfortunately made the comparison with regulatory tests as opposed to testing informed by 21st century science, giving an unfair advantage to machine learning. But overall, these chapters will help improve how we establish what is safe, and what is not, in the world of herbicides.

Statement on possible conflict of interest. I am a founder and board member of Sudoc LLC, a chemical company that produces catalysts capable of amplifying the oxidation efficacy of hydrogen peroxide. The catalysts biomimic the action of natural peroxidase enzymes in human cells to oxidize microbes and chemical contaminants. They are the intellectual creation of Terrence J. Collins, Teresa Heinz Professor of Green Chemistry at Carnegie Mellon University. To eliminate the potential conflict, I have donated all my shares in the LLC to an irrevocable grantor trust. I cannot benefit economically from distributions to my shares. Instead, distributions will be used to fund charitable activities that advance the fields of environmental health and safe, sustainable chemistry. On the website, you will also note that the company has established a Leadership Council on Endocrine Disruption to guide our testing of the safety of our chemicals. Our tests for safety will be far more stringent than those of the US Environmental Protection Agency, the FDA, or their counterparts around the world. They will be based on 21st century science. And the process will be transparent.

Pete Myers
Co-author, *Our Stolen Future*
Environmental Health Sciences, Bozeman, MT, United States
Carnegie Mellon University, Pittsburgh, PA, United States

References

1. Carson R. *Silent spring*. Houghton Mifflin; 1962.
2. Cribb J. *Earth DeTox*. Cambridge University Press; 2021.
3. Chemical and Engineering News. Monsanto's John E. Franz Wins 1990 Perkin Medal. *Chem Eng News* 1990;**68**(11):23–30. https://doi.org/10.1021/cen-v068n011.p029.
4. Gillam C. *The Monsanto papers*. Island Press; 2021.
5. Colborn T, Dumanoski D, Myers JP. *Our stolen future: are we threatening our fertility, intelligence, and survival? A scientific detective story*. Dutton; 1996.

6. Benbrook C. Trends in the use of glyphosate herbicide in the US and globally. *Environ Sci Eur* 2016;**28**, 3.
7. Myers JP, Antoniou MN, Blumberg B, Carroll L, Colborn T, Everett LG, et al. Concerns over use of glyphosate-based herbicides and risks associated with exposures: a consensus statement. *Environ Health* 2016;**15**:19–32.
8. Vandenberg LN, Blumberg B, Antoniou MV, Benbrook CM, Carroll L, Colborn T, et al. Is it time to reassess current safety standards for glyphosate-based herbicides. *J Epidemiol Community Health* 2017. https://doi.org/10.1136/jech 2016-208463.
9. Fagin D, Lavelle M. *Toxic deception. How the chemical industry manipulates science, bends the law and endangers your health.* Birch Lane Press Book; 1996.
10. Markowitz G, Rosner D. *Deceit and denial. The deadly politics of industrial pollution.* University of California Press; 2002.
11. Oreskes N, Conway R. *Merchants of doubt. How a handful of scientists obscured the truth on issues from tobacco smoke to global warming.* Bloomsbury Press; 2010.
12. Michaels D. *Doubt is their product: how industry's assault on science threatens your health.* Oxford University Press; 2008.
13. Michaels D. *The triumph of doubt. Dark money and the science of deception.* Oxford University Press; 2020.
14. Howard C. *Special Report: syngenta's campaign to protect atrazine, discredit critics*; 2013. Environmental Health News https://www.ehn.org/special-report-syngentas-campaign-to-protect-atrazine-discredit-critics-2646375953.html (downloaded 12 December 2020).
15. Girion L. *Johnson and Johnson knew for decades that asbestos lurked in its baby powder*; 2018. A Reuters Investigation. https://www.reuters.com/investigates/special-report/johnsonandjohnson-cancer/ (downloaded 12 December 2020).
16. Thacker P. The Weinberg proposal. *Environ Sci Technol* 2006;**40**(6):915–38.
17. Birnbaum L, Cohen A, Myers JP. *Endocrine disrupting chemicals and Covid-19*; 2020. Webinar. https://bit.ly/Covid19EDC (downloaded 12 December 2019).
18. Grandjean P, Timmermann CAG, Kruse M, Nielsen F, Vinholt PJ, Boding L, et al. Severity of COVID-19 at elevated exposure to perfluorinated alkylates. *medRxiv* 2020. https://www.ncbi.nlm.nih.gov/pmc/articles/PMC7605584/ (downloaded 12 December 2014).
19. Gore AC, Chappell VA, Fenton SE, Flaws JA, Nadal A, Prins GS, et al. EDC-2: the endocrine society's second scientific statement on endocrine-disrupting chemicals. *Endocr Rev* 2015;**36**(6). https://doi.org/10.1210/er.2015-1010. E1-E150.
20. Vandenberg LN, Colborn T, Hayes TB, Heindel JJ, Jacobs DR, Lee DH, et al. Hormones and endocrine disrupting chemicals: low-dose effects and nonmonotonic dose response. *Endocr Rev* 2012;**33**:378–455. http://bit.ly/A25AWs. https://doi.org/10.1210/er.2011-1050.
21. Myers JP, Zoeller RT, vom Saal FS. A clash of old and new scientific concepts in toxicity, with important implications for public health. *Environ Health Perspect* 2009;**117**:1652–5. http://bit.ly/Ljwb37.
22. Foucart S, Horel S. *Perturbateurs endocriniens: ces experts contestés qui jouent les semeurs de doute*; 2020. Le Monde. https://bit.ly/19editorsLeMonde (downloaded 12 December 2020). English translation (with permission) in Environmental Health News https://bit.ly/19editors (downloaded 12 December 2020).
23. Myers JP, vom Saal FS, Akingbemi BT, Belcher S, Colborn T, Chahoud I, et al. Why public health agencies cannot depend upon 'good laboratory practices' as a criterion for selecting data: the case of bisphenol A. *Environ Health Perspect* 2009;**117**:309–15.

Preface

Herbicides constitute the biggest portion of global pesticide use. They have become an integral tool for weed management. About 90% of global use is for industrialized agriculture to support the intensification of crop production, but nontrivial amounts go into urban environments and private gardens for cosmetic purposes and even in nature conservation areas to kill invasive neophytes. Evidence of adverse impacts on human health and ecosystems is mounting, but there is no book that summarizes the scientific knowledge in a concise and multidisciplinary way. Such a book is especially needed, as the active ingredient of the world's most often used herbicide–glyphosate–continues to fuel scientific and societal debate, including multibillion-dollar lawsuits to compensate for alleged health damage. This book has gathered a team of internationally recognized experts to shed light on a complex topic and present the state of the science, including chemistry, efficacy, analysis, toxicology, environmental impact, and legal aspects.

We are living in an information society where material about herbicides can easily be accessed on the internet. However, it has become increasingly difficult to differentiate between evidence-based knowledge and deliberate misinformation, which seems to be spreading faster than the truth because it is specifically crafted to induce emotion over reason. It is often cleverly exploited by advocate groups to spread doubt about scientific evidence. As academic scientists working on herbicide toxicity, we have been exposed to the strategies developed by agrochemical companies to lobby for their products and the strategies of other, nongovernmental advocacy groups to undermine the companies' interests. This is why we decided to write the book.

The objective is to provide an overview while acknowledging the precautionary principle, that is, the responsibility to protect the public and environment from unnecessary exposure when there is a plausible risk.

Chapter 1 sets the scene with a brief history of herbicide use from ancient to modern times. It describe how herbicides and nonchemical weed management methods address the challenges of modern food production, and gives an overview of the different factors influencing the herbicide market and the development of new products. The multidisciplinary nature of the debate is emphasized, as agriculture is intertwined with economy, public health, and environmental protection. Herbicides are not a one-size-fits-all solution, and their use should not be disconnected from the need to understand weed-crop ecology. This is illustrated in Chapter 2, on the changes in herbicide use patterns that have occurred in recent decades on soybeans with the rise of glyphosate-resistant weeds after the rapid adoption of glyphosate-resistant crops. Changes in use patterns are triggered by multiple factors such as tillage systems, plant genetics, relative treatment costs, regulatory actions, and the introduction of new pesticides.

Herbicide-based management systems rely on a constant rate of innovation to avoid weed resistance. The steady creation of new mechanisms of action is needed. Chapter 3 describes the mode of action of dozens of active ingredients. Success is also dependent on the physicochemical properties of the spray mixture, so adjuvant technologies and coformulants have become cornerstones of chemical weed control. Chapter 4 describes how the solubility, volatilization, adherence, penetration, rainfastness, foaming, and drift of herbicides can be controlled by the inclusion of coformulants. Coformulant toxicity has been at the heart of the controversy on glyphosate's human health effects.

Many controversies arise from the results of studies reporting the presence of herbicide residues in human bodily fluids. Challenges in the measurement of herbicide analytes in food and human samples are discussed in Chapter 5. Glyphosate is taken as an example for exploring the bottlenecks in the realization of biomonitoring studies with chromatography methods coupled with tandem mass spectrometry.

Human health effects have been debated for decades. All herbicides carry risks to nontarget organisms. Exposure to spray dilutions is implicated in the development of cancer, diabetes, infertility, and neurodegenerative disorders in agricultural workers. Whether current exposure of the broader public to environmental levels of residues in air or food causes adverse effects is less clear. Safety profiles of major herbicides are presented in Chapter 6.

Contrary to a widespread assumption, herbicides not only kill weeds but have direct and indirect impacts on a wide variety of nontarget organisms and the function of ecosystems. Direct effects, of course, concern weed diversity, but also nontarget crop plants, for instance via spray drift. Chapter 7 addresses effects on invertebrates, vertebrates, and microorganisms inhabiting terrestrial ecosystems at the species and population levels. Chapter 8 covers indirect effects of altering overall biodiversity and food web interactions that further affect ecosystem functions and services. This includes the consequences of spray drift and erosion by wind and water, the promotion of crop diseases, and the influence on natural biocontrol processes. Herbicides are often used to control aquatic weeds or are leached into water bodies; in an invited close-up, Sophie Oster and Mirco Bundschuh provide a short overview of the effects on aquatic food webs.

The literature reviews of the effects on human health (Chapter 6), other nontarget organisms (Chapter 7), and overall biodiversity and food web interactions (Chapter 8) reveal an inescapable time gap between the introduction of a new product and the detection of side effects. Chapter 9, on toxicity testing, suggests a way forward by moving away from animal assays toward high-throughput screening of human samples with computational approaches. In a close-up, Laura Vandenberg focuses on endocrine disruption, a specific kind of toxicity that is governed by the principles of endocrinology, and for which current toxicity paradigms such as the classical dose-response relationship do not apply.

Every herbicide carries risk. Safe use requires the definition of an acceptable level. Weighing the risk-benefit ratio involves toxicological, economic, social, and environmental considerations. This cannot be fully understood without a basic knowledge of the regulatory system and legal framework. Chapter 10 examines the procedures that regulate approval under European Union law for glyphosate as an active substance, and the authorization of pesticides containing it, in light of the precautionary principle.

Chapter 11 attempts to set the contents in the practical world to promote the safe use of herbicides in the future, and to offer insight toward sustainable weed management that supports resilient agricultural systems while securing global food security. We hope that this book will engage specialists in the research community and inform persons responsible for legislative, funding, and public health matters in the community at large.

Acknowledgments

We are grateful to all authors for their extraordinary contributions and their engagement to work together and meet the rather strict deadlines. The highly multidisciplinary nature of this book in evaluating the application, usage, chemistry, toxicity, ecotoxicology, and legislation of herbicides makes us hope that it will become a major reference for scientists, regulators, and health practitioners, and that it will be useful for advanced undergraduate and graduate courses.

We thank the Elsevier publications team for supporting this project, and Anna Wetterberg, Senior Manager of RTI Press. We are especially appreciative of Gerald T. Pollard for his editorial refinement of the text and his unremitting commitment to seeing this book through to completion.

Although we carefully edited all contributions, we would like to emphasize that the responsibility for the final content of the chapters rests with the individual authors.

Herbicides: Brief history, agricultural use, and potential alternatives for weed control

Robin Mesnage[a], *András Székács*[b], *and Johann G. Zaller*[c]
[a]King's College London, London, United Kingdom, [b]Agro-Environmental Research Centre, Institute of Environmental Sciences, Hungarian University of Agriculture and Life Sciences, Budapest, Hungary, [c]Institute of Zoology, University of Natural Resources and Life Sciences (BOKU), Vienna, Austria

Chapter outline

Herbicides. https://doi.org/10.1016/B978-0-12-823674-1.00002-X

What is a herbicide?

Herbicides are agrochemicals applied to prevent or interrupt normal plant growth and development. They are increasingly used to manage weeds in agriculture, but they are also used by railway companies, landscapers, greenskeepers, sports field managers, municipalities, and private gardeners. For instance, the active ingredient glyphosate is applied on 39% of the arable land in Germany.[1] Herbicides save the labor and energy of mechanical weeding with plows or harrows. They target physiological processes in weeds that can therefore be controlled by herbicide applications instead of labor-intensive weeding. Finding a specific biochemical target in the weed without causing phytotoxicity to the crop is difficult (see Chapter 3 for details). Thus, weed control is often the most difficult task in plant protection management by agrochemicals. In organic (ecological) farming or other systems that restrict the use of synthetic herbicides, weeds are controlled by mechanical or so-called cultural means such as mowing, mulching, crop rotation, cover crops, adjustments of crop density, flaming, and soil solarization.

Herbicides are distinguished by time of application as pre-emergent or postemergent. Pre-emergent agents, mostly applied to the soil, prevent germination and the early growth of weed seeds. Consequently, they have to be used prior to planting, or they would inhibit crop germination as well. They are "total" (nonselective or broadband) agents that exert phytotoxicity to all vegetation. In contrast, postemergent agents are applied on the emerged crop; therefore, they must have specificity toward weeds. Specific herbicides target either monocotyledons (grass-like weeds such as Bermuda grass, quackgrass, wild fescue) or dicotyledons (herbaceous weeds such as ragweed, thistles, plantains). A potential toxicity to the crop can be suppressed by herbicide safener (antidote) additives that trigger the enzymatic decomposition of the herbicide in the crop.[2] Postemergence agents can be applied to both the soil and the standing crop. Herbicides are also used preharvest, which in certain crops (e.g., soybeans, lentils, cotton) simplifies harvest by desiccating green plant parts. Some cereals such as wheat are frequently desiccated with glyphosate-based herbicides in Northern Europe. This application has a particularly significant environmental impact, is prone to leave residues in the harvested produce, and is therefore restricted in many countries.

Herbicides are also classified by their absorbance characteristics. Contact agents absorb to the plant surface (mostly the leaf epidermis), do not become translocated from there, and exert phytotoxicity only on the plant tissue that they contact. Consequently, their weed control effect is limited to their time on the plant surface, and is less effective against perennial weeds that can re-emerge from their roots, tubers, or rhizomes. Important contact herbicide active ingredients are glufosinate-ammonium and paraquat. Systemic herbicides penetrate into the plant tissue to some distance from the point of contact, and therefore exert wider effects. Systemic agents are taken up by the weed and translocated in it. Translocation may occur upward with the xylem to the leaves, shoots, and flowers (acropetal) and/or downward with the phloem to the roots (basipetal). In consequence, basipetally and acropetally translocated agents are used in foliar and soil applications. The phytotoxic effects of systemic herbicides are mostly (but not always) slower than those of contact herbicides and are more effective against perennials. Efficacy often differs significantly between monocotyledonous and dicotyledonous plants because of physiological differences—for example, lower absorbance on the narrow, erect leaves of monocots. Important systemic herbicides are glyphosate, atrazine, 2,4-D, and dicamba.

What is a weed? It depends

Weeds are ubiquitous, common, and bothersome plants that have been described in terms of their habitat or their behavior. Generally, a weed is an undesirable plant, or to put it simply, a plant present at the wrong place at the wrong time. Plants growing in agricultural fields that have not been planted intentionally may not be considered undesirable if they are not inconvenient. Definitions comprise poetic descriptions, didactic terms, and agronomic, control-oriented aspects.[3] The Weed Science Society of America defines it as "Any plant that is objectionable or interferes with the activities or welfare of man," and weed control as "The process of reducing weed growth and/or infestation to an acceptable level."[4] Nevertheless, many herbicide-intensive management measures aim at eradication from the field rather than tolerating a certain level that is not affecting yield, which would be according to the principle of integrated pest management.

So, defining a weed is somewhat subjective and context-specific. Certain plant species can clearly be regarded as weeds (e.g., ragweed), and some crops can also be regarded as weeds (e.g., sunflowers emerging in a cereal field) (Fig. 1.1).

The ecological role of weeds can be seen in very different ways, depending on one's perspective. Most commonly, they are perceived as unwanted intruders that compete for limited resources, reduce crop yields, and force the use of large amounts of human labor and technology to prevent even greater crop losses. In developing countries, farmers may spend 25–120 days hand-weeding a hectare of cropland, yet still lose a quarter of the potential yield to weed competition.[5] In the United States, where farmers spend $6 billion annually on herbicides, tillage, and cultivation for weed control, crop losses due to weed infestation may exceed $4 billion per year.[6] At the other end of the spectrum, weeds can be viewed as valuable agroecosystem components that provide services complementing those obtained from crops. In India and Mexico, farmers consume the "weed" species *Amaranthus*, *Brassica*, and *Chenopodium* as nutritious foods before crop species are ready to harvest.

The rate of weed growth depends upon environmental conditions. Disruption of the framework of application by wet fields, windy spray conditions, or time and labor constraints could result in poor weed control and yield loss.

Crops can coexist with weeds for some time without yield loss (the so-called critical period). There are two such periods, defined by when weeds emerge: early season and late season. The time over which weed control efforts must be maintained before a crop can effectively compete with late-emerging weeds and prevent crop yield loss is

Fig. 1.1 What is a weed? The sunflower is a cash crop, but when emerging in a sorghum field where it was cultivated in the previous year, it is considered a weed because it is undesired there. Ragweed is not used in agriculture and is always considered a weed.
(Image: J. Zaller.)

only a few weeks for tall crops such as maize because fields cannot be treated with tractors when crops are too high.

A very brief history of weed control

Prehistoric

The concept of a weed had little meaning for hunter-gatherers. The first weeds, proto-weeds, were undesirable wild plants that grew in small cultivated plots of the first farming communities. The oldest evidence for them was found in a 23,000-year-old hunter-gatherer sedentary camp on the southwestern shore of the Sea of Galilee,[7] a large quantity of well-known species such as threehorn bedstraw (*Galium tricornutum*). This suggested that nonedible plants were growing with cereals collected for human consumption. In the Neolithic site of Atlit-Yam in Israel, weeds were even found along with the grain pest beetle (*Sitophilus granarius*), suggesting that stored crops were invaded by pests.[8]

Some proto-weeds were probably part of the normal diet. A large proportion of plants found in the first sedentary camps were apparently collected for their medicinal properties or for crafting clothes, for example, the purple nutsedge (*Cyperus rotundus*), one of the most prolific and tenacious.[9] A study of dental calculus removed from ancient teeth excavated in the central Sudanese site of Al Khiday revealed that *C. rotundus* was regularly ingested and might have contributed to the unexpectedly low level of caries found in these populations.[10] Many plants identified in Paleolithic and Neolithic sites have medicinal but no nutritional properties.[11]

Even in modern agriculture, a plant can be an invasive weed or a nutritious staple. *Amaranthus palmeri* (pigweed) is one of the most problematic weeds in the United States, where it became resistant to glyphosate.[12] On the other hand, it is commonly eaten by native peoples in several countries and is a promising crop to feed a booming population.[13] Knotweed (*Polygonum* spp.) is another; the leaves are rich in proanthocyanidins, a family of compounds with potent antioxidant activity and possibly beneficial health effects.[14]

Weeds of former times were important in developing modern cereal crops. The Neolithic cereal variety emmer (*Triticum dicoccum*) and the sticky and protein-rich tetraploid grains of the so-called macaroni wheats (*T. durum*) were developed from hybridization with their own weeds, the quack grasses (*Agropyron* species) and weed grasses (*Aegelops* species).[15] The original cross would still have been diploid, as wild emmer is, but chromosome doubling commonly occurring in plants produced the domesticated emmer, one of the most common cultivated wheats. These hybridizations did not result from human design. Cereals and weeds just happened to be growing intermixed. Back then, several types of grain reached cultivation as weeds in the main crop. Rye (*Secale cereale*) originally grew as a weed in fields of soft wheat (*T. aestivum*) or einkorn (*T. monococcum*), and harvesters could not separate it from the crop. In regions with mild winters, wheat constituted the greatest number of plants. Where winters were severe, rye surpassed wheat. As farmers migrated northward carrying mixed seeds, rye became the principal crop. Similarly, in the British Isles, oats (*Avena sativa*) started as a weed growing in fields where emmer was cultivated and ended up as a separate crop in the Scottish Highlands.

Greece and Rome

Greek and Roman literature describes undesirable plants and methods of control. Democritus (460–370 BC) first reported using lupine-flower soaked in hemlock juice.[16] Salted water and human feces were also used.[17] Religious practices and folk magic were very common. Theophrastus's large botanical treatise *Enquiry into Plants* (c. 350–287 BC) and Pliny the Elder's *Natural History* (AD 77) are among the most important.[17] The Latin for weed is *herba inutilis*. Removal was done by hand or plowing, or their growth was prevented by preparations based on amurca (olive oil lees). Amurca, an abundant byproduct of olive oil production, may be considered one of the first herbicides. Cato the Elder's manual on running a farm *De Agri Cultura* (c. 150 BC) describes the use of amurca to make a threshing floor, prevent weed growth, keep weevils and mice from harming grain, keep caterpillars off the vines, and protect clothing from moths.

These advances in science and technology were lost during the early Middle Ages, and so was weed control, though some Arabic literature survived. Most of the knowledge about farming practices during Roman times was collected during the 10th century in Constantinople in the 20-volume book series *Geoponica*.[17] The Persian Islamic scholar Ibn Qutaybah's encyclopedia described mixtures of duck excrement and salt to kill plants.[18] With the invention of the printing press in the 15th century, the Greek and Roman knowledge reappeared, notably in the collection *Scriptores rei rusticae* of Palladius. Salt was extensively described as a herbicide from the 16th century.[18]

The age of chemistry

The first successful modern chemical weed control was probably the use of a 6% copper sulfate solution to kill charlock selectively in cereals during the 19th century.[19] This gave rise to a vast number of agrochemicals. The first commercial herbicide was also an inorganic compound, sodium arsenite, that, due to its broad toxicity, showed antibacterial and zoocidal activity as well. Eventually, the first organic herbicide, sodium dinitrocresylate (Sinox), was developed in France in 1896.

Rachel Carson's *Silent Spring* (1962)[20] describes the devastating effect of insecticides, especially DDT, on the environment and human health. Not so well known is that she also criticized the heavy use of herbicides outside of agricultural fields. She describes "one of the most tragic examples of our unthinking bludgeoning of the landscape" in the sagebrush lands of the American West, with vast campaigns to destroy sagebrush and substitute grass. Sagebrush is the common name of several woody and herbaceous species of the genus *Artemisia*. She noted that the program had been under way for several years with the involvement of government agencies and industry. In the late 1950s, besides about 4 million acres of rangeland being sprayed each year, some 50 million acres were routinely treated for "brush control." An estimated 75 million acres of mesquite lands in the Southwest were sprayed, and an unknown but very large timber-producing acreage was aerially sprayed to "weed out" the hardwoods from the more spray-resistant conifers. Treatment of agricultural lands doubled in the decade following 1949, totalling 53 million acres in 1959. The combined acreage of US private lands, parks, and golf courses already being treated in the late 1950s must have reached an astronomical figure. The most widely used herbicides were 2,4-D, 2,4,5-T, and related compounds (Fig. 1.2). Herbicide use patterns have changed radically in recent decades. In 1968, atrazine and 2,4-D made up more than half of the amount applied. Forty years later, glyphosate alone had a share of 50%.

The Green Revolution

The Green Revolution (sometimes called the Third Agricultural Revolution) was the adoption of high-yielding crop varieties, monocultures, chemical fertilizers, agrochemicals, irrigation, and heavy machinery to boost yield. Revolution cultivars created during the 1960s helped feed the world and prevent starvation in developing countries. They changed the architecture of wheat and rice plants and led to higher

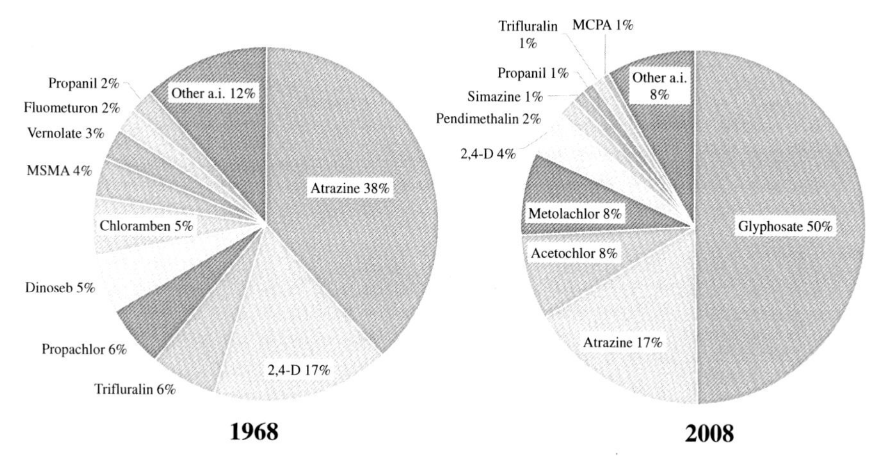

Fig. 1.2 Amounts of herbicide active ingredients applied to 21 selected crops in 1968 and 2008 in the United States.
Data from Economic Research Service with USDA and proprietary data. Fernandez-Cornejo J, Nehring RF, Osteen C, Wechsler S, Martin A, Vialou A. Pesticide use in US agriculture: 21 selected crops, 1960–2008. *USDA-ERS economic information bulletin 124*; 2014.

grain yields when appropriate production needs (fertilizer and irrigation) were provided. They also changed microclimatic conditions in the field and thus weed establishment.[21] The main factor was a selection for shorter straw length that made cereals less likely to lodge with higher rates of fertilizer input. The effect of short straw length was a changed light environment providing more light at the soil surface, which stimulates germination and the growth of weed seedlings. In the changed habitat, new weeds emerged. The connotation of the adjective "green" in the term Green Revolution can be misleading nowadays because it is mostly used to qualify practices that are environmentally preferable. When "Green Revolution" was coined, it was not aimed to reflect a transition to ecological practices.

With the Green Revolution, agriculture in many countries shifted from a mix of crops on a farm to extensive monocultures. Wheat, soybeans, maize, and cotton dominate the landscape in parts of the United States. Monocultural environments create ecological changes that determine what weeds will succeed. Each agricultural practice can influence the density and survival of weed species in a cropped field. This includes tillage, irrigation, nutrient supply (fertilization), pH, date of planting, growing period, shading period, seed dispersal at harvest, and weed control strategy.[21]

Agriculture is one of the largest environmental interactions of humans. It has shaped landscapes and ecosystems for thousands of years. It is one of the sectors of human activity with the highest source of pressure on the environment, such as transport, forestry, households, tourism, and recreation. Its impact clearly indicates the need for a "greener revolution," an environmentally sustainable one.[22] In practical terms, this means using existing knowledge to reduce agriculture's inevitable environmental effects while increasing its productivity. Regarding herbicides, one solution is the greater use of integrated weed and other pest management programs application specificity (only where weeds grow, not to entire fields; also termed precision agriculture); site- and time-appropriate amounts and irrigation; cover crops to reduce erosion, especially on fallow land; interrow areas and buffer strips between fields; and the appropriate use of more productive cultivars to increase yield and reduce fertilizer, water, and pesticide runoff to nonagricultural areas.[22]

Yield will continue to be an important measure of the success of agricultural programs. However, it cannot be the only measure if we are to have a greener revolution—what Gordon Conway[23] calls a Doubly Green Revolution—to feed all in the 21st century. If yields do not increase, more land will be required, and it is estimated that a near doubling of cropped land is necessary just to maintain current food consumption levels to feed the projected 8+ billion population by 2050. However, it is important to note that this calculation does not consider that current yields are often used as livestock feed to support a meat-dominated diet in industrialized countries that is not always healthy, that a great proportion of agricultural land is used for industry purposes (e.g., production of ethanol, biogas, sweeteners), and that a substantial amount of agricultural yield is wasted.[24] Agriculture-dependent gains and losses in the value of such ecosystem goods and services as potable water, biodiversity, carbon storage, pest control, pollination, fisheries, and recreation[22] should be included as well, not just yield gains.

Herbicides in current weed management

Weeds cause an estimated crop yield loss of 43% worldwide.[25] Control in practice is not simple. The effect on crop yield is sigmoidal, not linear with weed infestation. At very low weed density, there is none. As density increases, the effect is barely discernible.[21] As density continues to increase, yield drops quickly but never goes to zero; even very high density does not eliminate all crop plants. An effective broad-spectrum herbicide usually does not control all weeds all the time, and farmers resort to multiple applications to get the results they desire. The level of control desired by farmers is often greater than that needed to prevent yield loss.[26]

Herbicides decrease the manual labor required to control weeds, thus freeing producers for other tasks. However, they can also cause unintentional problems such as off-target movement to sensitive crops (spray drift) and small amounts persisting in the soil and disrupting future crops (see Chapter 8 for details). Also, weed populations can shift in response to herbicides and other production practices, and the establishment of new weeds can lead to crop loss.

Because producers expect that fields free of weeds will increase productivity, they will probably continue to use herbicides as a primary method.[26] Good weed control depends on cultural knowledge—what a good farmer or plant grower knows. This is different from the scientific knowledge that led to herbicide development and its successful use. Scientific knowledge that tells us what *can* be done and cultural knowledge that tells us what *should* be done are both essential.[21] Good weed control is not always associated with herbicide use.

Farmers need to be aware of biological time constraints that affect weed control and crop yield. If herbicides are applied too early, too many weeds emerge after treatment; if they are applied too late, there is a risk for herbicide-induced injuries to crop plants. Thus, the relative impact of weeds on crop yield depends on the time of weed emergence, the rate of weed growth, the diversity of weed communities, and the time of weed removal.[27]

The adoption of strategies has much to do with who provides the information that influences farmers' weed management decisions. US Department of Agriculture surveys have shown that many farmers obtain advice from pesticide manufacturers, not independent agriculture extension personnel.[28, 29] The herbicide business is very competitive, and the market rewards those who sell the most chemicals. This push serves the short-term profit motives of industry but not necessarily the long-term interests of farmers, consumers, and the environment.

Herbicide-tolerant genetically modified (GM) crops, weeds, and herbicides

Herbicide-tolerant crops—particularly those resistant to glyphosate, dicamba, and 2,4-D—have changed the face of weed management.[30] These herbicides are non-selective: they do not distinguish between weeds and crop plants. GM corn, soybeans, and other crops tolerate them, so farmers can apply them over the entire field without fear of injury to the crop.

Past reliance on glyphosate and mistakes made in the stewardship of the glyphosate-resistant cropping system have led directly to the current weed-resistance problems. One response of seed and herbicide producers has been to combine resistance traits, as in the GM soybean that tolerates dicamba-, 2,4-D, isoxaflutole, glyphosate, and glufosinate.

Weed adaptation to herbicides is an area of active research. The fear is that as genetically engineered crops are released into the environment, their genes will spread into weed and pathogen populations and increase crop loss. Data prove that genes can be transferred between some crops and their weedy wild relatives, for example, sorghum and radish.[31, 32] In such cases, it is likely that a herbicide-resistance gene will move from a crop to a weed and make the weed impossible to control with the specific herbicide or related ones.

Integrated weed management

The use of diverse weed management practices to reduce the selection pressure for herbicide-resistant weed evolution is essential to preserve the utility of new traits. The use of herbicides with differing modes of action, ideally in combination as mixtures but also in rotation as part of a weed management program, may slow the evolution of resistance in some cases. Increased selection pressure from the effects of some herbicide mixtures may lead to more cases of metabolic herbicide resistance. The most effective long-term approach is integrated weed management (IWM), which accounts for the ecological complexity of the cropping system.[21] Given the challenges, IWM will likely play a critical role in enhancing future food security for a growing global population.

Although their use continues to increase, herbicides represent only a part of the chemical input into most systems. An underestimated aspect is how key biological time constraints such as the periodicity of weed emergence, the rate of crop growth and development, and critical periods of weed control can influence the outcome of IWM systems.[27] A key component lies in the ability of the crop producer to align individual time and labor management with existing biological time constraints.

Herbicides versus nonchemical methods

Combinations of mechanical, cultural, biological, and other nonchemical methods have increased the scope of available techniques for many crops. Among those are the reintroduction of crop rotation, cultivation designed to minimize erosion, timing of planting, high-density planting, cover cropping, crop row orientation, allelopathy, specific irrigation, intercropping, and biological control.[33–37] Field experiments running for more than 20 years have shown that herbicide-free management can produce yields equivalent to those with conventional herbicide use.[38, 39] These findings demonstrate that extreme reliance on herbicides is unsound and unsustainable and has harmful environmental consequences.

Weed control in organic agriculture

There is no simple standard solution for weed control in organic agriculture.[40] While a farmer engaged in industrialized agriculture may rely on herbicides that can be applied with short notice to cure a field from an ongoing weed infestation, the organic farmer needs to maintain a long-term perspective while taking preventive measures to avoid yield loss. Direct and cultural methods need to be integrated in organic agriculture, with a long-term goal of preventing weed-induced yield losses while keeping costs down.

In permanent grassland with fewer options for mechanical control such as tillage, some plant species can become dominant. The presence of dock (*Rumex crispus* and *R. obtusifolius*) is taken to indicate mismanagement: high soil nitrogen and potassium levels, slurry and farmyard manure application, sward disturbance, improper cutting frequency, improper grazing, improper plowing, or soil compaction.[41] On farms, *Rumex*-infested grasslands are treated with herbicides (e.g., glyphosate, thifensulfuron, fluroxypyr, mecoprop-p). On organic farms, cultural measures dominate. To prevent damage to the turf, nonchemical measures for dock are often limited to frequent pulling and cutting. However, field experiments with an endangered, very frugal sheep variety that feeds exclusively on dock (*Ovis aries* cv. East Prussian Skudden) resulted in a significant suppression of dock in sheep-grazed pastures compared to mown grasslands.[42]

Weed management should consider not only measures in the crop field but also the treatment of harvest products or manure containing weed seeds to prevent the redistribution of weed species. Experiments comparing vermicomposting (with compost worms and not allowing high heat) and conventional composting (with temperatures up to 60°C) stopped *Rumex* germination after 4 months while *Rumex* seeds stored at room temperature still had a germination rate of 89%.[43] Including such integrated strategies at the farm and landscape level could substantially reduce weed pressure in the field.

Tillage

The primary alternative to herbicides is tillage. A study found yield declines in wheat under direct drilling with herbicide use compared to conventional tillage.[44] The conclusion was that the declines were not related to direct herbicide effects but rather to root-inhibitory pseudomonads prevalent in undisturbed, herbicide-treated soil as compared with tilled soil. This is an example of the difficulty in distinguishing direct herbicide effects from indirect knock-on effects altering yield-relevant processes.

Conservation tillage is widely used to reduce erosion because it leaves at least 30% of the crop residue on the soil surface. Often, herbicides are used to prepare a weedless seed bed for the following crop. However, very few studies compare the relative effects of this method of soil cultivation and herbicide input on yield and biodiversity conservation.[45]

Tillage with moldboard plows (also called inversion tillage because soil is turned) is still a popular and very effective weed control method in organic farming. The disadvantage is high fuel consumption and potential detrimental effects on the soil biota. Reduced tillage, which avoids soil inversion, produces good weed control with decreased fuel demand and less impact on soil life. In an experiment in Germany, loosening the soil to 30 cm was as suppressive for weeds in rye (*S. cereale*) as deep moldboard plowing; however, it resulted in a higher weed biomass in barley (*Hordeum vulgare*).[46] The year-long application of reduced tillage resulted in a lower yield than moldboard plowing in barley fields but a similar yield in rye fields.[47] This highlights the importance of developing crop-specific weed control strategies and including the cultural knowledge of farmers to adapt techniques to a given situation.

Crop rotation including cover crops

Crop diversification is an important cultural measure for weed management in organic systems. A six-field rotation scheme with red clover (*Trifolium pratense*) as the cover crop after the harvest showed good suppression of perennial weeds.[48] In contrast, rotation with a higher proportion of cereals in the crop was less successful in suppressing weeds. As certain crops favor specific weed species, it is important to implement a rotation that suppresses the weed populations that have been expanding during a former cropping season. A proper rotation adapted to the actual farm situation is the core of organic farming and should become more important for industrialized farming that aims to reduce herbicide (and overall pesticide) use. The early identification of an upcoming weed problem is necessary, and a wide range of control measures may be combined to keep weed populations at an acceptable level. Whatever crop is employed, soil bed preparation and management should be directed toward the rapid establishment and maximum competitive ability of the crop under cultivation. Direct control measures need to be employed against weeds as early as possible to prevent them from competing with the crop, but also in a later phase to prevent them from replenishing their seed bank. Perennial weeds especially should be prevented, but if they do occur, revising a planned rotation by, for example, establishing a perennial ley may contribute to a long-term solution.

The use of plants with strong weed-suppressing ability as cover crops is well-suited in such a holistic approach, as they provide other agroecosystem services.[49, 50] Living cover crops suppress the development of weed populations through niche pre-emption, and their residues suppress or retard weed emergence and growth by both allelopathic and physical effects. Compared to herbicides, a cover crop needs careful follow-up throughout the intended period to maximize its agroecosystem services and minimize disservices. The important characteristics to consider before selecting a weed-suppressing cover crop include high initial growth rate and rapid establishment of a close, dense canopy.

Toward ecologically based weed management systems

Herbicides helped weed scientists to define weeds as the enemy, and only in the 1990s did this begin to change. The overuse of herbicides led them to neglect other weed management strategies, notably those of organic agriculture.[21] Herbicide use has masked the importance of weed prevention and the need to understand weed-crop ecology. The shift toward ecologically based systems is occurring for several reasons[21]:

- Weeds highly susceptible to available herbicides have been replaced by species more difficult to control.
- Herbicide resistance has developed in many weed species, some of which are resistant to several herbicides. Multiple resistance to agents from chemical families with different modes of action has occurred.
- There are weed problems in monocultural agriculture that cannot be solved easily with the present management techniques.
- New weed problems have appeared in reduced and minimum tillage systems.
- Economic factors have forced the consideration of alternative control methods.
- There is increased awareness of the environmental costs of herbicides.

The European Union (EU) recently launched the European Green Deal program to reduce the use of pesticides in the EU by 50% by 2030, and support agroecology to reduce excess fertilization, increase ecological farming, and reverse biodiversity loss. These ambitious goals were set by recognizing the urgent need to reduce dependency on pesticides and antimicrobials, to reduce excess fertilization, to increase organic farming, to improve animal welfare, and to reverse biodiversity loss. Nonchemical methods of weed control facilitate compliance with these objectives.

Dynamics of the market and development of new herbicides

The global herbicide market was estimated at $27 billion in 2016 and is projected to reach $39 billion by 2022.[51] The situation in high-income areas such as the EU is different from that in the developing world. In the EU, total herbicide sales remained constant between 2011 and 2016 and could even decrease because the share of organic farming is steadily increasing[52] in response to consumer demand. However, organic agriculture generally requires more area to produce a similar amount of food.[53]

As the most recent development, the European Green Deal policy[54] aims to implement radical reductions in pesticide use and, in parallel, to support agroecology. Rationalizing with the fact that pesticides contribute to soil, water, and air pollution, the European Community (EC) declared it would take steps to halve the use of highly hazardous pesticides by 2030. To promote ecological agriculture as environmentally friendly, the EC will help to increase its profile, with the goal of converting 25% of agricultural land by 2030.

To keep up with growing food demand, agriculture is intensifying in the developing world, where pesticide use is steadily increasing. This is the case in Latin America where Brazil and Argentina have become leaders in world corn and soybean production and export. According to the Phillips McDougall consultancy, besides climatic, political, and demographic forces, the following factors are known to affect agrochemical research and development.[55]

Rising cost of pesticide registration and requirements of regulatory bodies

The costs associated with pesticide development have steadily increased from $152 million in 1995 to an average of $286 million for 2010–14.[56] These costs are attributed mainly to registration ($33 million), research ($107 million), and development ($146 million for field trials, toxicology, chemistry). Such sums deter smaller companies from entering the market.

The size and number of companies involved

The pesticide market is concentrated. There were 10 major companies at the end of the 1990s, 6 in 2009 (Syngenta, Bayer, BASF, Dow Chemical, DuPont, and Monsanto),[57] and 4 in 2017 because Bayer acquired Monsanto, ChemChina acquired Syngenta, and Dow merged with DuPont to form Corteva, and BASF.[55]

Company portfolio management

Difficulties in developing new active ingredients explain why companies are spending a relatively large proportion of their revenue on lobbying to maintain the commercial approval of existing products.[58]

The number of active herbicide ingredients registered for use in the EU has decreased in the last two decades.

This underwent two major revision periods. Between 1993 and 2009, 343 products were approved by the EC and 797 were banned or withdrawn out of 1217 assessed.[59] After this revision, authorization has been governed by EU Regulation 1107/2009, entered into force in 2011. This gives the EC and the member states shared competence in evaluations and decisions, with active ingredients being approved at the European level (with the vote of the member states) and formulated products at the member state level. The evaluation process has to clarify whether the agent poses "unacceptable risks" to users, consumers, and the environment. Among 13 newly approved pesticides between 2011 and 2019, only two are herbicides, florpyrauxifen-benzyl and halauxifen-methyl. At the time of writing (October 24, 2020), 473 agents, including 112 herbicides, were approved by the EC.

In the past three decades, pesticide producers have also developed a seed segment or acquired seed companies,[60] further increasing market concentration. This was driven mostly by the introduction of herbicide-tolerant GM crops.

Development of GM crops tolerant to herbicides

The major companies have reoriented their strategy to combine the development of herbicides and GM herbicide-tolerant crops. While the pesticide market grew at a 3.8% annual rate from 2001 to 2016, the GM seed market grew three times faster, at 13.3%.[57] In 2014, 12 traits were commercialized for glyphosate tolerance as well as 8 for glufosinate tolerance, and traits conferring tolerances to 2,4-D, sulfonylurea, dicamba, isoxaflutole, oxynil, HPPD, and imidazolinone were either being developed or in the regulatory pipeline.[61] Glyphosate-based products, the major pesticides used worldwide, were associated with the cultivation of 80% of agricultural GM organisms.[61]

Weed resistance

Reliance on a single herbicide has favored the appearance of resistant weeds. Since the introduction of the Roundup Ready GM cropping system, the United States has reported 14 species that have evolved resistance to glyphosate in 32 of the 50 states that forewarned growers that the use of glyphosate alone was unsustainable.[62] Globally, 220 species (130 dicots and 90 monocots) have evolved resistance to 21 of the 25 known herbicide sites of action.[62]

New formulations by improving coformulant technology

Agrochemical companies can gain competitive advantage and market share by improving the efficiency of a known herbicide. Coformulants can be added to improve solubility, volatilization, adherence, penetration, rainfastness, foaming, and drift of spray mixtures (see Chapter 4 for more details). Upon expiration of its glyphosate patent, Monsanto introduced new formulations to diversify its portfolio and compete with other companies that had started selling generic glyphosate products.[63]

Emergence of new technologies

In response to the emergence of glyphosate-resistant weeds, biotech companies have developed GM crops with combined tolerance to glyphosate and other herbicides. All major players have now developed stacked herbicide-tolerant GM crops to fight glyphosate-resistant weeds: Roundup-Ready Xtend from Monsanto (glyphosate + dicamba), Enlist Duo from Dow AgroSciences (2,4-D + glyphosate), Balance GT from Bayer CropScience (isoxaflutole + glyphosate), and Optimum GAT technology from DuPont (acetolactate synthase (ALS) inhibitor + glyphosate). Syngenta and Bayer CropScience even codeveloped the MGI system conferring tolerance to three herbicides (mesotrione + glufosinate + isoxaflutole).

More weed scientists are calling for increased emphasis on the study of weed ecology and population biology to design management strategies that reduce herbicide use or make it obsolete.[21]

Herbicide use patterns

Pesticides are used more intensively on fruits and vegetables than on arable crops. However, the proportion of herbicides in overall pesticide use varies among crop species. In the United States, herbicides are applied to 92%–97% of acreage in corn, cotton, soybeans, and citrus; 87% in potatoes; 75% in vegetables; and 67% in apples and other fruits.[26] Herbicide use is least extensive on winter wheat, at 56%.

Fig. 1.3 shows the share of herbicides of the total pesticide use on various crops in Switzerland between 2009 and 2014. It was >70% for maize, pulses, sugar beet, oilseed rape, and other cereals; 10%–50% for potatoes, winter wheat, and winter barley; and below 10% for the perennial stone fruits, grapevines, and pome fruits.

In the United Kingdom, wheat was the most cultivated arable crop in 2018, accounting for 42% of the agricultural area. On average, wheat received three

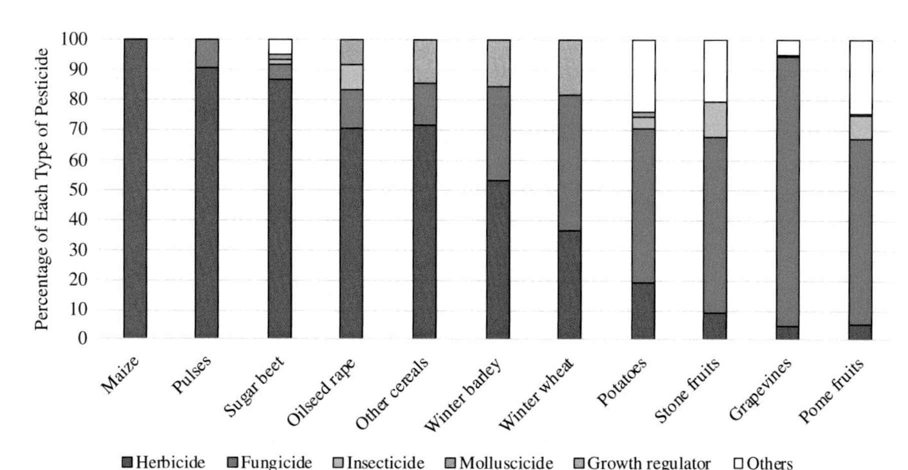

Fig. 1.3 Percentage of each major pesticide active substance used on crops in Switzerland between 2009 and 2014. Data provided by Agroscope.[64]

herbicide, three fungicide, two growth regulator, and one insecticide spray rounds during the cropping season. Herbicides were applied mostly during spring and fall for general weed control and specifically to eliminate blackgrass. Glyphosate was the most applied herbicide and desiccant (Fig. 1.4), followed by diflufenican/flufenacet, iodosulfuron-methyl-sodium/mesosulfuron-methyl, and fluroxypyr.

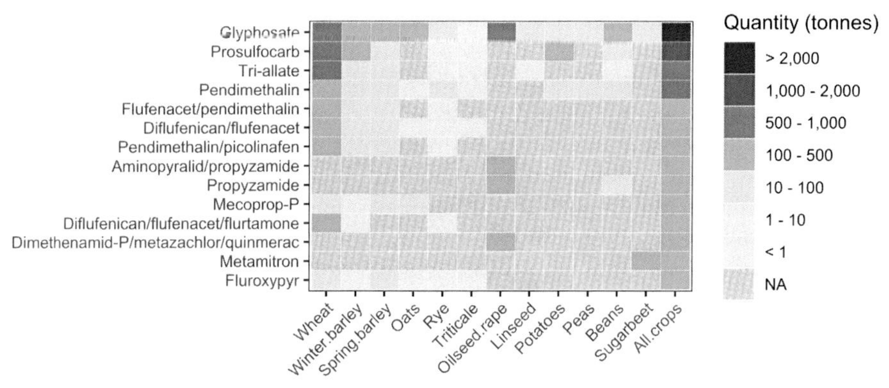

Fig. 1.4 Herbicides used on arable crops in the United Kingdom, 2018. Heatmap shows the combination of active ingredients.[65] Color scale indicates volume applied (NA = none).

An aspect of herbicide usage not considered in this book is nonagricultural and noncommercial, which can be substantial. Use on amenity areas (roads, pavements, parks, gardens, sports grounds, and other areas administrated by public authorities) accounts for 0.2%–2.7% of the annual total in seven European countries,[66] most of it herbicides. In Great Britain,[67] glyphosate was most commonly used for weed control in the amenity sector, representing 51.8% of the total volume of pesticide applied (Table 1.1), followed by ferrous sulfate (16.0%), MCPA (10.0%), and 2,4-D (7.9%).

Table 1.1 Herbicide use by sector in the United Kingdom in 2012.[67]

Sector	Total area (ha)	Total weight applied (kg)	Glyphosate (kg)
All crops	**24,626,379**	**7,783,665**	**1,540,628**
Cereals	16,020,961	4,186,410	843,371
Oilseeds	3,469,082	1,371,972	321,981
All amenity	**504,395**	**671,370**	**339,911**
Golf	75,796	60,322	8554
Industrial	2986	5238	3498
Infrastructure	193,991	273,797	217,815
Public authorities	90,290	245,878	106,081
Residential	12,822	9347	2546
Turf	128,510	76,788	1417

Industrial uses include applications for electrical and communication systems, pipes carrying petroleum, natural gases, water, buildings and protective fencing, and environs. Infrastructure includes aquatic banks, water bodies, roads, and railways. Public authorities are control operations contracted for commercial facilities, educational institutions, cemeteries, parks, hospitals, and airports, among others. Residential is mostly lawn care and landscaping.
Data from the UK Department for Environment, Food & Rural Affairs.

In Switzerland, the use of glyphosate in horticulture and by professional gardeners accounted for 7% of glyphosate sales.[68] In Germany, nonprofessional use made up 6.8% (3041 tons) of herbicide use in 2018.[69] In the United States in 2012, agriculture accounted for 91% of herbicide use, home and garden 5%, and industrial and governmental 4%.[70]

Conclusions

From the first agricultural settlements in Mesopotamia to the modern biotechnology-based agricultural systems, weed control has always been a cornerstone of food production. Although herbicide use was recorded in Greek and Roman times, it accelerated massively after World War II to allow large-scale monocultures, which were widely adopted to intensify food production during the Green Revolution.

The efficacy of weed control is usually based on the proportion of the weed population killed. However, crop yield might be adequately maintained by weed injury with much lower herbicide use rather than weed eradication in the field with all the ecological consequences. Weed resistance against herbicides is an increasing problem, and severe impacts can occur when there are no alternative herbicides to control the resistant genotypes or when available alternatives are too expensive.

Herbicides (and other pesticides) should not automatically be given the highest priority when managing agricultural crops. Rather, these measures should be evaluated in conjunction with available alternative management practices, not only with respect to efficacy, cost, and ease of implementation but also with respect to long-term sustainability, environmental impact, and human health.[26]

References

1. Dickeduisberg M, Steinmann H-H, Theuvsen L. Erhebungen zum Einsatz von Glyphosat im deutschen Ackerbau. In: *Nr 434 (2012): Tagungsband 25 Deutsche Arbeitsbesprechung über Fragen der Unkrautbiologie und -bekämpfung/Neue Entwicklungen in der chemischen Unkrautkontrolle/Recent developments in chemical weed control*; 2012.
2. Jablonkai I. Herbicide safeners: effective tools to improve herbicide selectivity. In: *Herbicides-current research and case studies in use*. IntechOpen; 2013.
3. Zimdahl RL. *Weed science: a plea for thought—revisited*. Dordrecht, NL: Springer; 2012.
4. WSSA. Terms, definitions, and abbreviations. *Weed Sci* 1985;**33**(S1):12–5.
5. Liebman M. Weed management: a need for ecological approaches. In: Mohler CL, Staver CP, Liebman M, editors. *Ecological management of agricultural weeds*. Cambridge: Cambridge University Press; 2001. p. 1–39.
6. Bridges D, Anderson R. Crop losses due to weeds in the United States. In: Bridges D, editor. *Crop losses due to weeds in the United States—1992*. Champaign, IL, USA: Weed Science Society of America; 1992. p. 1–74.
7. Snir A, Nadel D, Groman-Yaroslavski I, et al. The origin of cultivation and proto-weeds, long before neolithic farming. *PLoS One* 2015;**10**(7), e0131422.
8. Hartmann Shenkman A, Kislev ME, Galili E, Melamed Y, Weiss E. Invading a new niche: obligatory weeds at Neolithic Atlit-Yam, Israel. *Veg Hist Archaeobot* 2015;**24**(1):9–18.

9. Marble SC, Prior SA, Runion GB, Torbert HA. Control of yellow and purple nutsedge in elevated CO_2 environments with glyphosate and halosulfuron. *Front Plant Sci* 2015;**6**:1.
10. Buckley S, Usai D, Jakob T, Radini A, Hardy K. Dental calculus reveals unique insights into food items, cooking and plant processing in prehistoric Central Sudan. *PLoS One* 2014;**9**(7), e100808.
11. Hardy K. Paleomedicine and the use of plant secondary compounds in the Paleolithic and Early Neolithic. *Evol Anthropol* 2019;**28**(2):60–71.
12. Vieira BC, Samuelson SL, Alves GS, Gaines TA, Werle R, Kruger GR. Distribution of glyphosate-resistant *Amaranthus* spp. in Nebraska. *Pest Manag Sci* 2018;**74**(10):2316–24.
13. Cheng A. Review: shaping a sustainable food future by rediscovering long-forgotten ancient grains. *Plant Sci* 2018;**269**:136–42.
14. Bensa M, Glavnik V, Vovk I. Leaves of invasive plants-Japanese, bohemian and giant knotweed-the promising new source of flavan-3-ols and proanthocyanidins. *Plants (Basel)* 2020;**9**(1):118.
15. Dethier VG. *Man's plague? Insects and agriculture*. Princeton, NJ, USA: The Darwin Press; 1976.
16. Zadoks JC. *Crop protection in medieval agriculture: studies in pre-modern organic agriculture*. Sidestone Press; 2013.
17. Smith AE, Secoy DM. Forerunners of pesticides in classical Greece and Rome. *J Agric Food Chem* 1975;**23**(6):1050–5.
18. Smith AE, Secoy DM. Early chemical control of weeds in Europe. *Weed Sci* 1976;**24**(6):594–7.
19. Butler O. Chemical, physical, and biological properties of Bordeaux mixtures. *Ind Eng Chem* 1923;**15**(10):1039–41.
20. Carson R. *Silent spring*. Houghton Mifflin: USA; 1962.
21. Zimdahl RL. *Fundamentals of weed science*. Elsevier Science; 2007.
22. Tilman D, Fargione J, Wolff B, et al. Forecasting agriculturally driven global environmental change. *Science* 2001;**292**(5515):281–4.
23. Conway G. *The doubly green revolution*. London: Penguin Books; 1997.
24. Zaller JG. *Daily poison. Pesticides—an underestimated danger*. Basel, Switzerland: Springer International Publishing; 2020.
25. Oerke EC. Crop losses due to pests. *J Agric Sci Cambr* 2006;**144**(1):31–43.
26. National Research Council. *The future role of pesticides in US agriculture*. Washington, DC, USA: The National Academies Press; 2000.
27. Gunsolus J, Buhler DA. Risk management perspective on integrated weed management. *J Crop Prod* 1999;**2**:167–87.
28. Jin S, Bluemling B, APJ M. Information, trust and pesticide overuse: Interactions between retailers and cotton farmers in China. *NJAS-Wagen J Life Sic* 2015;**72–73**:23–32.
29. Mohammadrezaei M, Hayati D. The role of agricultural extension services in integrated pest management adoption by Iranian pistachio growers. *Int J Agric Ext* 2015;**3**:47–56.
30. Gage KL, Krausz RF, Walters SA. Emerging challenges for weed management in herbicide-resistant crops. *Agriculture* 2019;**9**(8):180.
31. Klinger T, Arriola PE, Ellstrand NC. Crop-weed hybridization in radish (*Raphanus sativus*): effects of distance and population size. *Am J Bot* 1992;**79**(12):1431–5.
32. Krimsky S, Wrubel RP. *Agricultural biotechnology and the environment: science, policy, and social issues*. Champaign, IL, USA: University of Illinois Press; 1996.
33. Cramer C, Bowman G, Brusko M, Cicero K, Hofstetter B, Shirley C, editors. *Controlling weeds with fewer chemicals*. Emmaus, PA, USA: Rodale Institute; 1991.

34. Mohammadi GR. Alternative weed control methods: a review. In: Soloneski S, Larramendy ML, editors. *Weed and pest control. Conventional and new challenges.* London, UK: InTechOpen; 2013 [chapter 6].
35. Coolong T. Using irrigation to manage weeds: a focus on drip irrigation. In: Soloneski S, Larramendy ML, editors. *Weed and pest control. Conventional and new challenges.* London, UK: InTechOpen; 2013 [chapter 7].
36. Bond W, Grundy AC. Non-chemical weed management in organic farming systems. *Weed Res* 2001;**41**(5):383–405.
37. Mennan H, Jabran K, Zandstra BH, Pala F. Non-chemical weed management in vegetables by using cover crops: a review. *Agronomy* 2020;**10**(2):257.
38. Lotter D, Seidel R, Liebhardt W. The performance of organic and conventional cropping systems in an extreme climate year. *Am J Altern Agric* 2003;**18**:146–54.
39. Delate K, Cambardella C, Chase C, Turnbull R. A review of long-term organic comparison trials in the U.S. *Sustain Agric Res* 2015;**4**(3):5–14.
40. Lundkvist A, Verwijst T. Weed biology and weed management in organic farming. In: Nokkoul R, editor. *Research in organic farming.* London, UK: IntechOpen; 2011. https://doi.org/10.5772/31757 [chapter 10].
41. Zaller JG. Ecology and non-chemical control of *Rumex crispus* and *R. obtusifolius* (Polygonaceae): a review. *Weed Res* 2004;**44**(6):414–32.
42. Zaller JG. Sheep grazing vs. cutting: regeneration and soil nutrient exploitation of the grassland weed *Rumex obtusifolius. BioControl* 2006;**51**:837–50.
43. Zaller JG. Seed germination of the weed *Rumex obtusifolius* after on-farm conventional, biodynamic and vermicomposting of cattle manure. *Ann Appl Biol* 2007;**151**:245–9.
44. Simpfendorfer S, Kirkegaard J, Heenan D, Wong P. Reduced early growth of direct drilled wheat in southern New South Wales—role of root inhibitory pseudomonads. *Crop Pasture Sci* 2002;**53**:323–31.
45. Barré K, Le Viol I, Julliard R, Chiron F, Kerbiriou C. Tillage and herbicide reduction mitigate the gap between conventional and organic farming effects on foraging activity of insectivorous bats. *Ecol Evol* 2018;**8**(3):1496–506.
46. Vakali C, Zaller JG, Köpke U. Reduced tillage effects on soil properties and growth of cereals and associated weeds under organic farming. *Soil Tillage Res* 2011;**111**:133–41.
47. Vakali C, Zaller JG, Köpke U. Reduced tillage in temperate organic farming: effects on soil nutrients, nutrient content and yield of barley, rye and associated weeds. *Renewable Agric Food Syst* 2015;**30**:270–9.
48. Zarina L, Gerowitt B, Melander B, Salonen J, Krawczuk R, Verwijst T. Crop diversification for weed management in organic arable cropping systems. *Environ Technol Resour Proc Int Sci Pract Conf* 2015;**2**:333.
49. Euteneuer P, Wagentristl H, Steinkellner S, Scheibreithner C, Zaller JG. Earthworms affect decomposition of soil-borne plant pathogen Sclerotinia sclerotiorum in a cover crop field experiment. *Appl Soil Ecol* 2019;**138**:88–93.
50. Lemessa F, Wakjira M. Cover crops as a means of ecological weed management in agroecosystems. *J Crop Sci Biotechnol* 2015;**18**(2):123–35.
51. Marketsandmarkets. *Herbicides market by type (glyphosate, 2, 4-D, diquat), crop type (cereals & grains, oilseeds & pulses, fruits & vegetables), mode of action (non-selective, selective), and region—global forecast to 2022*; 2020. Available at: https://wwwmarketsandmarketscom/Market-Reports/herbicides-357html.
52. EEA. *Environmental indicator report—pesticide sales*; 2018. Available at: https://www.eea.europa.eu/airs/2018/environment-and-health/pesticides-sales.

53. European Commission. Organic farming in the EU—a fast growing sector. In: *EU agricultural markets briefs No 13*; 2019.
54. EC. *The European green deal: communication from the commission to the European Parliament.* The European Council, The Council, The European Economic and Social Committee and the Committee of the Regions; 2019. COM(2019) 640 final https://eur-lex.europa.eu/resource.html?uri=cellar:b828d165-1c22-11ea-8c1f-01aa75ed71a1.0002.02/DOC_1&format=PDF.
55. Phillips MWA. Agrochemical industry development, trends in R&D and the impact of regulation. *Pest Manag Sci* 2020;**76**:3348–56.
56. McDougall P. Agrochemical research and development: the cost of new product discovery, development and registration. In: *A consultancy study for CropLife International, CropLife America and the European Crop Protection Association United Kingdom*; 2016.
57. Nishimoto R. Global trends in the crop protection industry. *J Pestic Sci* 2019;**44**(3):141–7.
58. Heisey P, Schimmelpfennig D. Regulation and the structure of biotechnology industries. In: Just RE, Alston JM, Zilberman D, editors. *Regulating agricultural biotechnology: economics and policy*. Boston, MA: Springer US; 2006. p. 421–36.
59. Anton A, Fekete G, Darvas B, Székács A. Environmental risk of chemical agriculture. In: Gruiz K, Meggyes T, Fenyvesi E, editors. *Engineering tools for environmental risk management. Vol. 1. Environmental deterioration and contamination—problems and their management.* Boca Raton, FL, USA: CRC Press; 2014. p. 93–112.
60. Bonny S. Corporate concentration and technological change in the global seed industry. *Sustainability* 2017;**9**(9):1632.
61. Parisi C, Tillie P, Rodriguez-Cerezo E. The global pipeline of GM crops out to 2020. *Nat Biotechnol* 2016;**34**(1):31–6.
62. Heap I. Global perspective of herbicide-resistant weeds. *Pest Manag Sci* 2014; **70**(9):1306–15.
63. Perry ED, Hennessy DA, Moschini G. Product concentration and usage: behavioral effects in the glyphosate market. *J Econ Behav Organ* 2019;**158**:543–59.
64. BLW. *Agrarbericht 2016*; 2016 https://2016agrarberichtch/de/umwelt/wasser/einsatz-von-pflanzenschutzmittelnSchweizerischeEidgenossenschaftBundesamtfürLandwirtschaft BLW.
65. FERA. *Pesticide usage survey reports*; 2018 https://secureferadefragovuk/pusstats/surveys/indexcfm.
66. Kristoffersen P, Rask AM, Grundy AC, et al. A review of pesticide policies and regulations for urban amenity areas in seven European countries. *Weed Res* 2008;**48**(3):201–14.
67. FERA. *Pesticide usage survey reports*; 2013 https://secureferadefragovuk/pusstats/surveys/indexcfm.
68. Hanke I, Wittmer I, Bischofberger S, Stamm C, Singer H. Relevance of urban glyphosate use for surface water quality. *Chemosphere* 2010;**81**(3):422–9.
69. BVL. *Absatz an Pflanzenschutzmitteln in der Bundesrepublik Deutschland.* Berlin, Germany: German Federal Agency for Consumer Protection and Food Safety; 2019. www.bvl.bund.de/psmstatistiken.
70. EPA U. *Pesticides industry sales and usage 2008–2012 estimates*; 2017. Available at: https://www.epa.gov/pesticides.

A minimum data set for tracking changes in pesticide use

2

Charles M. Benbrook and Rachel Benbrook
Heartland Health Research Alliance, Brookfield, WI, United States

Chapter outline

A frequently asked but deceptively simple question often arises about pesticide use on a given farm or crop: Is pesticide use going up, down, or staying about the same?

Where substantial changes in pesticide use are occurring, it is also important to understand the factors driving change. These might include more or fewer hectares planted, a change in the crop mix, a higher or lower percentage of hectares treated, or higher or lower rates of application and/or number of applications. Or, it might arise from a shift to other pesticides applied at a higher or lower rate and/or lessened or greater reliance on nonpesticidal strategies and integrated pest management (IPM).

Questions about whether pesticide use is changing and why arise for a variety of reasons. Rising use typically increases farmer costs and cuts into profit margins. It generally raises the risk of adverse environmental and/or public health outcomes. It can accelerate the emergence and spread of organisms resistant to applied pesticides. If the need to spray more continues year after year for long enough, farming systems become unsustainable.

Lessened reliance on and use of pesticides, on the other hand, are typically brought about and can only be sustained by incrementally more effective prevention-based biointensive IPM systems (bioIPM).[1–3] Fewer pesticide applications and fewer pounds/kilograms of active ingredient applied reduce the impacts on nontarget organisms and provide space for beneficial organisms and biodiversity to flourish. Such

[illegible]. https://doi.org/10.1016/B978-0-12-823674-1.0000[illegible]

systems reduce the odds of significant crop loss in years when conditions undermine the efficacy of control measures, leading to spikes in pest populations and the risk of economically meaningful loss of crop yield and/or quality.

Introduction

There are many ways to track changes in pesticide use:

- Pounds/kilograms of active ingredient applied per unit area.
- Number of applications made on a given acre/hectare in a crop year.
- Rate of application of a pesticide on a given crop.
- Number of different active ingredients required to control target pests and bring a crop to harvest.
- Expenditures on pesticide products and their application.

Tracking changes in pesticide use requires multiple metrics because no single metric will provide a reliable lens through which to evaluate pesticide use. For this reason, a Minimum Data Set (MDS) approach is recommended to answer questions dependent on an empirical accounting of pesticide use.

The metrics in an MDS will vary as a function of why pesticide use is of interest. Use trends have real-world consequences in three important ways. The first is in determining whether pest management systems are stable, improving, or in decline relative to the efficacy of control. Second, in the world of risk assessment, use is an essential variable in tracking actual and potential adverse effects on public health or the environment. Third, changes in pesticide use are key factors when identifying pest-management-system efficacy and effects on farm-level productivity and profitability, and hence economic sustainability and food security.

Tools and metrics are available to analyze each of the above real-world impacts, but with highly variable—and generally modest—precision.

To determine whether pest management systems are stable, it is not enough to just track changes in pounds/kilograms applied or the number of treatments made on a given field. This is because pesticides vary greatly in two ways that directly affect measures of use and efficacy: (1) rates of application, and (2) breadth of control, duration of control, and degree of control achieved by a given spray application.

One example: To replace a soil fumigant applied at 100 kg/hectare that controls several soil-borne insects and pathogens and some weeds, farmers sticking with a predominantly pesticide-based system may need to apply a dozen other pesticides at rates between 0.1 and 1.5 kg/hectare. Some products may need to be applied multiple times. Still, metrics based on kilograms applied will show a large reduction in use from 100 kg to fewer than 5. Does that mean the farmer has reduced reliance on pesticides? Not necessarily.

To accurately determine whether a change in a pest management system shifts reliance from pesticides to other control measures, it is necessary to quantify and then track three dimensions of pesticide efficacy: (1) how many target pests will a given application affect, (2) how long will susceptible target pests be controlled, and (3) the degree and reliability of control. Such efficacy-focused evaluations are available to one degree or another for many crop-pest-geographic regions but they vary significantly, even from farm to farm, as a function of many factors (e.g., soil type and health, irrigation methods, cropping patterns/rotations, cultural practices, plant genetics). While analytically challenging, each farmer's best guess at answers to these questions plays an important role in driving the selection of which pesticides they apply, when, and how.

Another key factor at the farm level is the cost of pesticide products plus their application. To the degree possible, farmers weigh efficacy versus cost when deciding which pesticides to use. Heavy weight tends to be placed on each farmer's recent, personal experience with how well the pesticides they applied last year worked. In cases where adequate control has been achieved in recent years at acceptable and/or comparable costs to known alternatives, farmers are generally reluctant to try something new, especially if alternatives are unproven in the area and/or more costly.

But if farmers are concerned about slipping efficacy, perhaps driven by the emergence and spread of resistant organisms or a new pest taking hold, they are more open to incorporating new pesticides into their control strategies. Then, tradeoffs between expected efficacy and cost must be evaluated to the extent possible and decisions made. In working through such decisions, farmers always must contend with some degree of uncertainty on the efficacy side, coupled with generally hard numbers on the cost side.

From the farmer's perspective—and society's—the most important metric is whether, year to year, a farmer needs more, less, or about the same pesticide-delivered "control power" to bring a crop to harvest. Control power encompasses the impacts of pesticides on target pests, whether through "kill" power (acute toxicity), effects on reproduction (insect pheromones, herbicide plant growth regulators targeting flowering or reproduction), metabolic disruption (through altering energy metabolism or blocking enzyme-controlled metabolic pathways), physical barriers (applying horticultural oils to protect fresh fruit from viral and bacterial pathogens), or biological controls (microbial biocontrol agents).

While a wide range of factors contribute to control power, tracking changes over time is reasonably straightforward and generally entails taking into account the efficacy of pesticide products in just a few families of chemistry. In many cases, at least qualitatively, the differences in efficacy across pesticide alternatives are known and widely accepted. For example, the spread of target pests newly resistant to a previously applied product is typically tracked and known, and can reduce the control power of pesticides. Shifts in the complex of pests in a given area and cropping system are often driven by pesticide control power, whereby populations of well-controlled pests decline over time and other, less susceptible pests fill ecological voids.

Pest management challenges become more difficult for farmers and their advisors as a function of the scope and scale of change needed or desired in relative reliance on pesticides for control versus nonchemical preventive practices. The challenges are more complex in areas where high-value fruits and vegetables are being grown in hot, humid, and wet areas where multiple weeds, insects, and plant pathogens can cause major crop damage within days, leading to sizable economic loss.

Evaluating the environmental and public health effects of pesticide use patterns entails another set of analytical challenges that often are well beyond the reach of current science. The serious, generally well-funded efforts over the last 20 years to sort out the effects of neonicotinoid insecticides on pollinators, or glyphosate-based herbicides (GBHs, including Roundup) on weed management and cancer rates, are contemporary examples.[4–8] But all assessments of the environmental and public health risks stemming from pesticide use must begin with accurate data on the extent of use of any pesticide or family of chemistry.

Data on the extent of use in a region over a defined period have a major and direct role in identifying levels of exposure as well as who or what organisms are exposed, in which environments, and at what life stages. Only then can the tricky challenge be tackled—estimating pesticide-induced adverse effects on organisms, populations, and ecosystems.

In the next section, 18 metrics are outlined that can be drawn upon to provide a robust assessment of changes in pesticide use. They fall into two categories:

- Changes in land area treated (acres/hectares) planted in a given crop, percentage of acres/hectares treated with a given pesticide, and number of acre-treatments (see the caption of Fig. 2.1 for information on how to convert from acre-treatments to hectare-treatments).
- Changes in the intensity of pesticide use on a given acre/hectare (rate of application, number of applications, rate per crop year).

Fig. 2.1 illustrates the way multiple metrics can be integrated to assess changes in the use of a single pesticide, pesticides targeting a type of pest (herbicides/weeds, insecticides/insects, fungicides/plant diseases, others), or all pesticides. Applications of the metrics are discussed, along with methods to access sources of use data for crops in the United States. The penultimate section of the chapter, "Glyphosate-based herbicide

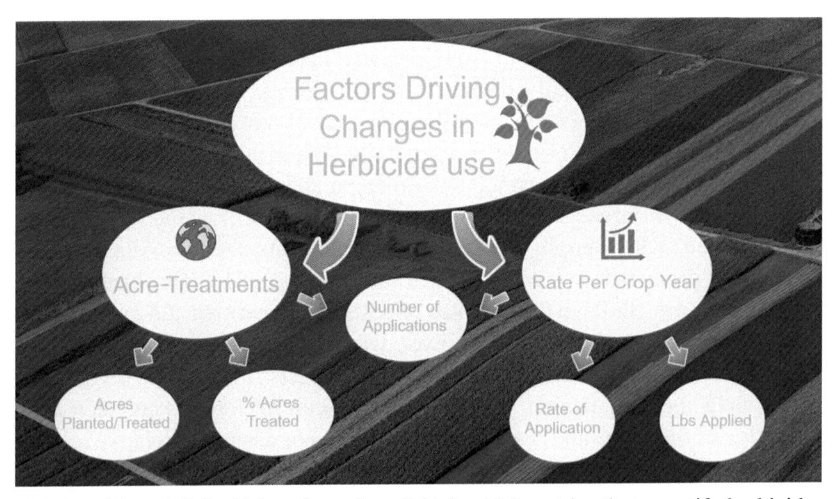

Fig. 2.1 Land-based (left side) and rate-based (right side) metrics that quantify herbicide use. Note that acre-treatments is a calculated variable that will not be converted into SI units for the purpose of this chapter. Multiply acre-treatment values by a conversion factor of 0.40686 to convert to hectare-treatments.

use on soybeans in the United States," is a case study in how the metrics can be applied to one important crop (soybeans) and type of pest (weeds).

Pesticide use metrics

Pesticide impacts on the environment and human health are a function of how widely and often a pesticide is applied (acre-treatments in Fig. 2.1) in addition to the rate at which it is applied (rate per crop year). A series of land-based metrics is needed to quantify the proportion of a crop treated and the number of times a given hectare is treated in a growing season. Rate-based metrics are dependent on the one-time rate of application on a crop coupled with the number of applications. The first seven metrics in Fig. 2.2 address land- and rate-based indicators.

PESTICIDE USE METRICS

Metrics by Active Ingredient:

1. Percent of Acres Treated
2. Number of Acres Treated
3. Acre Treatments
4. Rate per Application
5. Number of Application
6. Rate per Year
7. Pounds Applied

Metrics by High or Low Dose Chemistry:

8. Number Low-Dose Chemistry
9. Reliance Low-Dose Chemistry
10. Number High-Dose Chemistry
11. Reliance High-Dose Chemistry
12. Weighted Average Rate of Application

Metrics by Pesticide Type (Herbicide, Insecticide, Fungicide, Other):

13. Number of Active Ingredients
14. Number of Active Ingredients per Acre
15. Number of Acre-Treatments
16. Acre-Treatments per Acre
17. Pounds Applied
18. Pounds Applied per Acre

Fig. 2.2 The 18 pesticide use metrics, shown here in US units for consistency with the online database at http://www.hygeia-analytics.com/puds. Where possible without creating confusion, herbicide use values have been converted to SI units throughout this chapter.

Additional metrics are needed to gain deeper insight into how changes in one-time rates of application influence trends in overall use (metrics 8–12). This is because since the early 1980s, the pesticide industry has focused on identifying new modes of action that are highly specific to target pests and are usually effective at low per-hectare rates of application. Such active ingredients (AIs) typically target specific metabolic, biochemical, physiological, developmental, or reproductive processes in target pests. Several are applied at very low rates (e.g., 0.01 pound or less per acre or 0.0112 kg/hectare). Low-rate AIs bring down the overall pounds/kilograms per acre/hectare in a region or on a crop. Shifting cropland treated from one or a few relatively high-dose pesticides to low-dose ones reduces use measured in kilograms of AIs applied, but would appear as intensification of use relative to the metric number of acre/hectare-treatments required to achieve control.

Metrics 13–18 focus on the number and type of individual pesticides used on a given crop. The effects of annual use on farm production costs, the environment, the emergence and spread of resistant pests, and public health are driven by the total number, weight, environmental fate, and toxicity of the AIs applied. These metrics are needed to take into account how many different pesticides are required to bring a crop to harvest in a production cycle. They are tracked by three major categories: weeds (herbicides), insects (insecticides), and plant diseases (fungicides), referred to herein as H/I/F/O, with the "O" encompassing all other types of pesticides (e.g., desiccants, rodenticides, fumigants). The number and type of pesticides applied can be quantified per acre/hectare, on a field, across all fields on a farm, and/or in a county, state, or region, nationally or globally.

The 18 metrics constitute a robust MDS. When applied to a specified crop and region, the MDS metrics will yield multiple insights useful in answering the questions of whether and how pesticide use has changed. Note that acres are used in this metric list to be consistent with the online database, but units have been converted to SI in the tables, figures, and text of this chapter where possible.

1. *Percent of acres treated* is the percentage of the total acreage of a given crop that is treated one or more times with a given pesticide in an annual crop production cycle.
2. *Number of acres treated* is the number of acres of the crop treated one or more times with a given pesticide, calculated as the (percent acres treated) multiplied by (total crop acreage).

Those two metrics are needed to quantify the area treated with a given pesticide, but they fail to capture when a given acre/hectare is treated more than once in a season and do not account for differences in rates of application.

3. *Acre-treatments* takes into account the acres planted to a given crop in a region, the percent of acres treated one or more times, and the average number of applications made with each active ingredient applied. It is calculated as:

$$\text{Acre} - \text{treatments} = (\text{acres planted}) \times (\text{percent acres treated}) \times (\text{number of applications})$$

For example, if one-half of a farmer's soybean acreage is treated twice with a pesticide$_x$ and the rest once, the number of acre treatments on this farm with pesticide$_x$ would be 1.5 multiplied by the acreage of soybeans.

4. *One-time rate of application* is the average amount of active ingredient applied per acre/hectare on a given crop in a field, farm, region, or country.
5. *Number of applications* is the average number made with a specific active ingredient on a crop in a production cycle (usually a calendar year).
6. *Rate per crop year* is the average total rate of application per production cycle, and is calculated as (average one-time rate of application) multiplied by (average number of applications).
7. *Pounds applied* is the total pounds/kilograms of AIs per production cycle on a given crop. It is calculated for an individual pesticide by multiplying (acre treatments) by (average rate of application per crop year), or alternatively by (acres treated one or more times) multiplied by (rate per crop year).

The pounds of AIs applied per unit area is the most widely reported pesticide use metric. But application rates per acre/hectare vary by 100-fold or more. For this reason, a series of metrics that capture reliance on high- versus low-dose AIs is needed (focus below is on rates in pounds per acre).

8. *Number low-dose chemistry* is the number of AIs applied at a rate of $\leq$0.1 pound/acre.
9. *Reliance on low-dose chemistry* is the percent of acres treated with an AI applied at a rate of $\leq$0.1 pound/acre.
10. *Number high-dose chemistry* is the number of AIs applied at a rate of $\geq$1.0 pound/acre.
11. *Reliance on high-dose chemistry* is the percent of acres treated with an AI applied at a rate of $\geq$1.0 pound/acre.
12. *Weighted average rate of application* is the average rate per acre-treatment across all pesticides applied within a category (herbicides, insecticides, fungicides, and other), weighted by each pesticide's share of total acre-treatments.

Metric 12 is the best way to track the overall impact of changes in average rates of application. This indicator, coupled with trends in the average number of acre-treatments on any given acre, provides the most reliable empirical assessment of overall changes in the intensity of pesticide use on a given crop.

The last six metrics allow tracking changes in the use of a particular pesticide or category.

13. *Number of H/I/F/O applied* is the average number of different pesticide AIs used on a given crop in a specific region and year combination (e.g., number of different herbicide AIs applied on soybeans in Iowa in 2015).
14. *Number of H/I/F/O applied per acre* is the average number of AIs applied one or more times per acre. It is calculated as the sum of the percent of acres treated with each AI falling within a category (e.g., herbicides), divided by 100.
15. *Number of H/I/F/O acre-treatments* takes into account the number of applications per acre and the percent of acres treated. It quantifies the total number of acre-treatments with H/I/F/O AIs (e.g., a field treated twice with the same AI counts as one herbicide applied but two acre-treatments).
16. *H/I/F/O acre-treatments per acre* is the total number of acre-treatments divided by the number of acres planted.
17. *Pounds of H/I/F/O applied* is the sum of AI pounds applied across all agents within a type of pesticide in a given crop year/region combination.
18. *Pounds H/I/F/O applied per acre* is the total pounds applied divided by total acres planted.

Applying the metrics

GBHs are by far the most heavily used pesticides in history as measured by pounds/kilograms of AI applied, both in the United States and globally.[9] The introduction of glyphosate-tolerant crops in 1996 allowed for postemergence, over-the-top applications on soybeans and cotton, and, soon thereafter, on corn, alfalfa, and sugar beets. This resulted in dramatic increases in the land area and total pounds/kilograms of glyphosate and GBHs sprayed beginning in 1998 and peaking around 2015.

Overuse triggered the emergence of weed phenotypes that required higher rates of GBH application and/or multiple applications to achieve acceptable levels of control. Incremental increases in glyphosate-imposed selection pressure on weed populations eventually triggered the emergence of resistant weeds. Continued use then assured the spread of resistant weed seeds, and more and more serious resistant-weed challenges for farmers. As a direct result, herbicide use intensity in US soybean production has risen dramatically in recent years, increases that are obvious when changes are tracked with the 18 pesticide use metrics.

Sources of pesticide use data for US crops

Use data provide the foundation for tracking changes. The primary public sources on US crops are:

- The US Department of Agriculture (USDA) National Agricultural Statistics Service (NASS) pesticide use data by crop, chemical, and state.[10]
- California Pesticide Use Reporting (PUR) data by crop, chemical, product, and county.[11]
- Periodic reports issued by the US Environmental Protection Agency (EPA).[12]
- The US Geological Survey (USGS) database of pesticide use by crop, chemical, and county.[13]

Since 1990, USDA pesticide use surveys have been reasonably complete for major field, fruit, vegetable, and grain crops. Funding shortfalls have, in recent years, reduced the number of crops surveyed annually, thereby creating longer gaps between years with survey data. From the early 1970s to 1990, the USDA conducted three surveys that were limited to major crops and regions. By drawing on these data for the crops that account for the most pesticide use in the last half-century in the United States, long-term use trends can be tracked through the Pesticide Use Data System (PUDS) (https://hygeia-analytics.com/puds).

Data are compiled and reported by USDA for H/I/F/O by AI and crop, and at the national and state levels. States accounting for about 85% or more of national crop acreage in a given year are surveyed by NASS. Each crop-year table in an NASS report, accessible via the PUDS, includes the percentage of surveyed acres treated, the average one-time rate of application, the average number of applications, the rate per crop-year (average rate multiplied by average number of applications), and the pounds of AI applied. Over many years, NASS has conducted its surveys with a consistent methodology, producing the highest quality publicly accessible data on trends in use by US farmers. Since 1990, NASS has surveyed major grain and row crops on a nearly annual basis. Vegetable crops are surveyed in even years, fruit crops in odd years.

High-quality detailed data have been collected by the California Department of Food and Agriculture since the 1980s via the PUR system.[11] The availability of detailed, location-specific use data in California is a major reason why epidemiologists have been able to conduct sophisticated studies of the linkage between pesticide use and exposure as well as adverse health outcomes (e.g., see von Ehrenstein et al.[14]).

Use data periodically released by the EPA are valuable because they include information on nonagricultural use, and because they draw upon both USDA surveys and proprietary data provided to the agency by pesticide registrants and private firms.

The USGS data are available for only a few major row crops at the county level, and are drawn largely from USDA pesticide-use reports.

In addition to publicly accessible sources, several consulting firms have compiled proprietary datasets. While private survey data are often more detailed and complete than government data, gaining access to private survey data can cost hundreds of thousands of dollars per year, and is hence not an option for most analysts.

Changes in glyphosate-based herbicide use in the United States

Two data sources are drawn upon for tracking trends in glyphosate use in this chapter. The first is the detailed tables from the 2016 paper "Trends in glyphosate herbicide use in the United States and globally."[9] The second is the EPA's periodic ranking of the most heavily applied pesticides in the US agricultural sector (i.e., Grube et al.[15]).

Table 2.1 shows trends in agricultural and nonagricultural use of GBHs from 1974 through 2014. Of the estimated 635,000 kg of AI applied in 1974, 363,000 were on farms and ranches and 272,000 were outside agriculture, for a 57% to 43% split.

The data show some striking changes. From 1974 to 2001, Monsanto's Roundup brands were the only GBHs sold in the United States. By 1982, the total kilograms applied had risen to 3.5 million, 5.6-fold higher than in 1974. The share accounted for by agricultural use rose to 64% in 1982 and nonfarm use fell to 36%.

After 1982, growth in sales and kilograms applied continued at a slower but still significant pace through 1990, a year when the total kilograms applied reached 5.8

Table 2.1 Glyphosate active ingredient use in the United States, 1974–2014.

	1974	1982	1990	1995	2000	2005	2010	2012	2014
Glyphosate use (1000 kg)	635	3538	5761	18,144	44,679	81,506	118,298	118,753	125,384
Agricultural	363	2268	3357	12,474	35,720	71,441	106,963	107,192	113,356
Nonagricultural	272	1270	2404	5670	8958	10,065	11,335	11,562	12,029
Share agricultural	**57.2%**	**64.1%**	**58.3%**	**68.8%**	**79.9%**	**87.7%**	**90.4%**	**90.3%**	**90.4%**
Share nonagricultural	**42.8%**	**35.9%**	**41.7%**	**31.3%**	**20.0%**	**12.3%**	**9.58%**	**9.74%**	**9.59%**

Notes: Data in thousands of kilograms of glyphosate active ingredient.
Table from Benbrook C. Trends in glyphosate herbicide use in the United States and globally. *Environ Sci Eur* 2016; **28**(1):3; data from the National Agricultural Statistical Service pesticide use data and the Environmental Protection Agency pesticide industry and use reports.

million, 1.6-fold higher than in 1982. Sales and kilograms applied grew faster for nonagricultural use, shifting the agricultural to nonagricultural shares to 58% and 42%.

From 1990 to 1995, agricultural sales and kilograms applied more than tripled and nonagricultural use more than doubled. Total kilograms applied rose to 18 million, up 3.1-fold in just 5 years.

Beginning in the 1996–2000 period, agricultural glyphosate use began its decade-long run of spectacular, unprecedented growth, as farmers rapidly adopted the Roundup Ready seed and herbicide system. In 2000, farmers and ranchers applied 36 million kg, an absolute increase of 23 million since 1995, or a 2.9-fold increase. The rate of growth in nonagriculture sales and pounds applied was much lower, rising only by about 50% over 5 years. The agricultural to nonagricultural split moved to 80% and 20%.

Just the increase in the kilograms applied by farmers and ranchers in the 5 years from 1996 to 2000—23 million kg—exceeded by a wide margin the total growth in the preceding 25+ years that GBHs had been on the market.

While growth in the early years of the adoption of Roundup Ready seeds was unprecedented by any measure, it pales in comparison to the magnitude of growth in absolute agricultural sales and GBH kilograms applied from 2001 to 2005 and 2006 to 2010. The last US patents that prevented other pesticide companies from entering the GBH market lapsed in 2000, so since around 2001 it has been necessary to focus on total use of GBHs, not just Roundup brands.

From 2001 to 2005, agricultural and ranch use of GBHs nearly doubled from 36 million kg to 71 million kg. Meanwhile, nonagricultural use increased by just over 1 million kg, or just 12%. As a result, there was another sizable shift of agricultural relative to nonagricultural uses: 88% to 12%.

Rapid growth in GBH sales and kilograms applied persisted for 5 more years; for agricultural and ranch applications, this amounted to 107 million kg in 2010, an absolute increase of 35 million. This shifted the agricultural to nonagricultural shares to 90% and 10%, a split that has remained largely unchanged to 2020.

Monsanto successfully petitioned the EPA to approve tolerances and labels covering most agricultural and nonagricultural uses by 1982, a year when the total glyphosate use was 3.5 million kg. By 2014, as Table 2.1 shows, the total kilograms applied had risen to 125 million, a 35-fold increase in 22 years. No pesticide in history comes close to matching the sustained upward trajectory in glyphosate and GBH sales and pounds applied in the United States and globally.

EPA rankings of the most heavily applied pesticides

A series of reports issued by the EPA on the sale and use of pesticides in the United States provides a second data series and perspective on the growth in glyphosate. The reports list by year and rank order (1–25) the most heavily applied AIs on farms and ranches, along with the EPA's best estimate of total agricultural pesticide applications. They draw heavily on the USDA's periodic surveys, as does the trend analysis. The rankings were computed by the EPA every 2 years from 1993 through 2009 as well as in 1987 and 2012. Table 2.2 records the rapid rise and staying power of glyphosate relative to other widely used pesticides.

Table 2.2 Glyphosate's rise to the top, Parts I and II.

Part I—EPA rankings of the top pesticides by year in the US agricultural sector, 1987–2001

("Range" is in million kilograms active ingredient applied) (see "Notes")

		1987		1993		1995		1997		1999		2001	
Active ingredient	**Type**	**Rank**	**Range**	**Rank**	**Range**	**Rank**	**Range**	**Rank**	**Range**	**Rank**	**Range**	**Rank**	**Range**
Glyphosate	H	17	3–4	11	7–9	7	11–14	5	15–17	2	30–33	1	39–41
Atrazine	H	1	32–34	1	32–34	1	31–33	1	34–37	1	34–36	2	34–36
Metam sodium	Fum	15	2–4	8	11–14	3	22–24	3	24–26	3	27–29	3	26–28
Metolachlor-S	H									12	7–9	9	9–11
Acetochlor	H					11	10–12	7	14–16	4	14–16	4	14–16
Dichloropropene	Fum	4	14–16	6	14–16	5	17–20	6	15–17	11	8–9	8	9–11
2,4-D	H	5	13–15	7	11–14	6	14–16	8	13–15	6	13–15	5	13–15
Trifluralin	H	6	11–14	9	9–11	10	10–13	10	10–11	9	8–10	12	5–7
Propanil	H	13	3–4	15	3–5	17	3–5	22	3–4	18	3–5	17	3–4
Dicamba	H	23	2–3	16	3–5	18	3–5	16	3–5	22	3–4	24	2–3

Part II—Glyphosate stays at the top. EPA rankings of the top pesticides by year in the US agricultural sector, 2001–12

		2001		2003		2005		2007		2009		2012	
Active ingredient	**Type**	**Rank**	**Range**	**Rank**	**Range**	**Rank**	**Range**	**Rank**	**Range**	**Rank**	**Range**	**Rank**	**Range**
Glyphosate	H	1	39–41	1	58–60	1	67–76	1	77–86	1	95–104	1	122–132
Atrazine	H	2	34–36	2	34–36	2	30–34	2	32–36	2	13–31	2	29–34
Metam sodium	Fum	3	26–28	3	20–23	3	16–21	3	22–26	3	14–18	6	14–18
Metolachlor-S	H	9	9–11	6	13–15	5	11–16	4	12–17	6	11–15	3	15–20
Acetochlor	H	4	14–16	5	14–16	6	11–15	5	11–16	7	10–15	7	13–17
Dichloropropene	Fum	8	9–11	7	9–11	4	13–17	6	11–15	4	12–17	4	15–19

Continued

Table 2.2 Continued

Part II—Glyphosate stays at the top. EPA rankings of the top pesticides by year in the US agricultural sector, 2001–12													
Active ingredient	Type	2001		2003		2005		2007		2009		2012	
		Rank	Range	Rank	Range	Rank	Range	Rank	Range	Rank	Range	Rank	Range
2,4-D	H	5	13–15	4	14–16	7	10–14	7	10–15	5	11–15	5	14–18
Trifluralin	H	12	5–7	11	4–5	14	3–5	17	2–4	18	1–3	19	1–3
Propanil	H	17	3–4	18	2–3	18	1–3	18	1–3	17	1–3	17	1–3
Dicamba	H	24	2–3			22	0.5–2	—	0.5–2	25	0.5–2	18	1–3

Notes: "H" is herbicide; "Fum" is fumigant. (—) indicates that the pesticide did not make the 25 most commonly used pesticides ranking in the given year. Data does not include sulfur and petroleum oil.

Source: Data from US Environmental Protection Agency Pesticide Industry Sales and Usage Reports, accessible at https://www.epa.gov/pesticides/pesticides-industry-sales-and-usage-2006-and-2007-market-estimates.

In 1987, the first year EPA issued its ranking, glyphosate was 17th, at 3–4 million kg applied. No. 1 was the corn herbicide atrazine, 32–34 million, or 10 kg of atrazine for every kilogram of glyphosate. Six years later, glyphosate was No. 11, at 7–9 million, and atrazine remained No. 1, at 32–34 million. Glyphosate entered its rapid growth phase in the late 1990s and rose steadily to No. 1 by 2001, at 39–41 million, and atrazine fell to No. 2, at 34–36 million.

In the next 11 years, the amount of glyphosate applied almost tripled while atrazine declined modestly. In 2018, glyphosate exceeded 136 million—5 kg for every kilogram of atrazine.

Glyphosate-based herbicide use on soybeans in the United States: A case study

Dramatic changes have occurred in herbicide use on soybeans in the United States since the early 1990s. This case study describes trends in herbicide use metrics from 1991, a few years before the introduction of herbicide-tolerant crops, through 2018, the most recent year for which use data are available from the USDA. Units relied on in this section include acres/hectares, kilograms of active ingredient applied, and combinations of these units of measure.

Land area treated metrics

Table 2.3 covers the change in hectares of soybeans planted from 1991 through 2018 and the percent of hectares treated one or more times with any herbicide, GBH or non-GBH. The dramatic upward shift in reliance on GBHs stands out, as does the near doubling of hectares planted over this 27-year period. These data are displayed graphically in Fig. 2.3.

The total soybean hectares planted increased by ~50% from 1991 to 2018 while hectares treated with a GBH increased 27-fold, from 1.2 million in 1991 to a high

Table 2.3 Change in soybean hectares planted and herbicide use in the United States, 1991–2018.

	1991	1994	1997	2000	2003	2006	2009	2012	2015	2018
Soybean hectares planted	23.9	24.9	28.3	30.1	29.7	30.6	31.3	31.2	33.5	36.1
Percent soybean hectares treated										
All herbicides	206%	258%	252%	185%	155%	142%	179%	216%	274%	343%
Glyphosate-based herbicides	5%	15%	28%	62%	86%	96%	97%	98%	97%	87%
Nonglyphosate herbicides	201%	243%	224%	123%	69%	46%	82%	118%	177%	256%

Notes: Data downloaded from the Pesticide Use Dataset (https://hygeia-analytics.com/puds) and converted into hectares. Percentages above 100% indicate that hectarage was treated with a given herbicide more than once in the growing season.

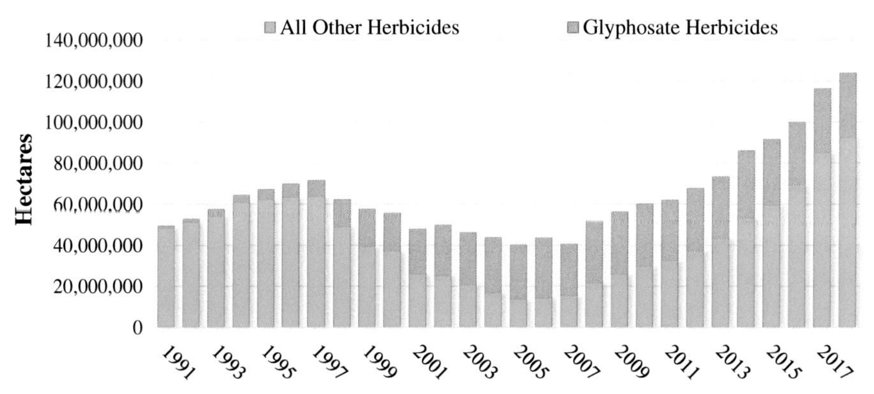

Fig. 2.3 US soybean hectares treated with glyphosate and all other herbicides, 1991–2018.

of 33 million in 2014. So, in terms of the total number of soybean hectares sprayed with a GBH one or more times, peak glyphosate use at 98% of total hectares planted occurred in 2012, then fell to 87% by 2018. Other indicators of peak glyphosate use occurred in other years, as evident in the following tables and figures.

In the early years of Roundup Ready technology (seeds genetically engineered to tolerate postemergence, over-the-top applications of a GBH), many farmers achieved acceptable to excellent weed control with a single application. But repeated use year after year triggered changes in weed communities and a loss of efficacy against some species, forcing farmers to make a second application and, on many farms, incrementally increase the rate applied per hectare.

Table 2.4 adds another metric—number of applications—to the appraisal of changes in soybean herbicide use. It reports the total hectares planted, the percent of hectares treated with a GBH one or more times, the percent of hectares treated with

Table 2.4 Change in soybean hectares planted and herbicide use in the United States, including acre-treatments, 1991–2018

	1991	1994	1997	2000	2003	2006	2009	2012	2015	2018
Soybean hectares planted	23.9	24.9	28.3	30.1	29.7	30.6	31.3	31.2	33.5	36.1
Percent soybean acres/hectares treated										
Glyphosate-based herbicides	5%	15%	28%	62%	86%	96%	97%	98%	97%	87%
Nonglyphosate herbicides	201%	243%	224%	123%	69%	46%	82%	118%	177%	256%
Millions of herbicide acre-treatments[a]										
Glyphosate-based herbicides	2.96	9.24	25.5	59.9	91.4	123.5	126.0	124.9	114.5	113.0
Nonglyphosate herbicides	120.3	154.2	162.1	92.5	51.8	36.0	65.8	96.9	153.9	251.7

[a] Multiply acre-treatments by the conversion factor of 0.40686 to convert them to hectare-treatments.

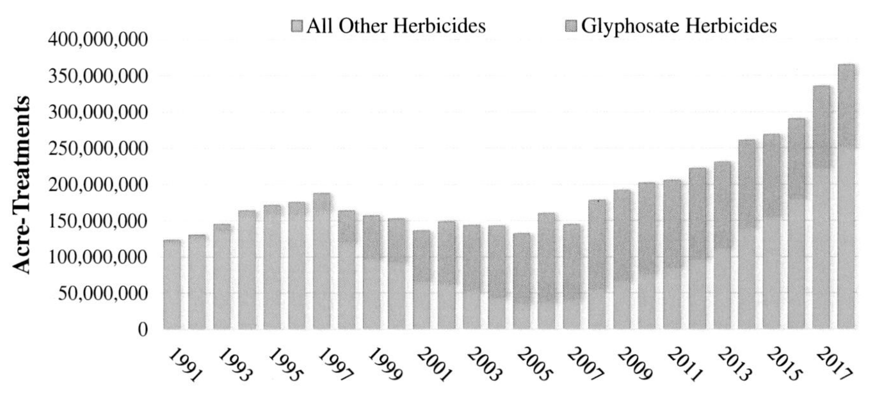

Fig. 2.4 Acre-treatments of glyphosate and all other herbicides on US soybeans, 1991–2018.

herbicides other than glyphosate, glyphosate acre-treatments and other herbicide acre-treatments, and glyphosate and other herbicide acre-treatments as a percent of total acre-treatments. Recall that multiplying acre-treatment values by a conversion factor of 0.40686 will give you the hectare-treatments.

The acre-treatments data are plotted in Fig. 2.4, with GBHs versus other herbicides. The average number of GBH applications rose from 1 in 1991 to 1.7 in 2006 and then fell back to 0.81 in 2018 (Table 2.5). This was associated with a rise in acre-treatments from 2.96 million to 113 million (1991 to 2018), or 38-fold (Table 2.4). The change in acre-treatments is one of the most important metrics when tracking changes in farmer reliance on a GBH or any other pesticide. Peak GBH acre-treatments as a percent of total herbicide acre-treatments occurred in 2006 at 77% and fell to 31% in 2018.

Table 2.5 Average application rate, number of applications, and rate per crop year, glyphosate and all other herbicides, 1991–2018.

	1991	1994	1997	2000	2003	2006	2009	2012	2015	2018
Average one-time application rate (kg/hectare)										
Glyphosate herbicides	0.897	0.583	0.684	0.762	0.821	0.890	0.963	1.039	1.092	1.004
All other herbicides	0.626	0.474	0.465	0.426	0.415	0.380	0.314	0.287	0.327	0.406
Average number of applications										
Glyphosate herbicides	1.00	1.00	1.30	1.30	1.45	1.70	1.68	1.65	1.43	0.810
All other herbicides	1.01	1.03	1.03	1.02	1.03	1.05	1.04	1.06	1.05	1.10
Average rate per crop year (kg/hectare)										
Glyphosate herbicides	0.897	0.583	0.889	0.991	1.19	1.51	1.61	1.72	1.56	0.81
All other herbicides	0.631	0.489	0.480	0.433	0.426	0.398	0.326	0.304	0.345	0.448

Application rate per acre metrics

The pounds/kilograms of any pesticide applied is a function of acre-treatments and the average rate of application on those acre-treatments. Table 2.5 reports the average one-time rate of application of glyphosate and all other herbicides, the average number of applications of glyphosate and all other herbicides, and the rate per crop year for glyphosate and all other herbicides. The rate of GBH application during this 27-year period rose 12% while that of other herbicides fell 35%. Even more startling, the peak glyphosate rate per crop year was in 2012, when the rate per year for GBHs was 91% higher than in 1991 while the rate per crop year for other herbicides fell 52% over the same period.

Table 2.6 combines the data on land area treated and rate per application to show trends over time in the amount of AI applied. Glyphosate use rose 41-fold. Other herbicides declined rapidly from 1997 through 2006, but then rose 751% by 2018. Glyphosate's share of the total herbicide kilograms applied peaked in 2006 at 89%, and fell to 52% in 2018. That share is likely to keep falling for several years as a result of slipping efficacy and the spread of glyphosate-resistant weeds.

Factors driving change in soybean herbicide use

The meteoric rise of GBH use on soybeans in the United States drove most of the changes in herbicide use metrics through about 2010. Since then, two additional trends have exerted significant influence on use patterns: an increase in the number of active ingredients required to achieve acceptable control (metric 13 in Fig. 2.2), and changes in the weighted average of rates of application (metric 12).

The spread of glyphosate-resistant weeds has required farmers to apply other herbicides, increase rates of application, or use nonchemical tactics. Roughly in order of impact when adopted, but not of prevalence, on US row crop farms, these tactics include diversified crop rotation, added tillage, preventing weeds that escape control from going to seed, planting cover crops, and changing planting and harvest timing. The surest way to reduce reliance on herbicides is to incorporate multiple nonchemical strategies into integrated weed management. This approach has appropriately been labeled the use of "many little hammers" by Iowa State University researchers.[16, 17]

Table 2.6 Kilograms applied, glyphosate and all other herbicides, 1991–2018.

	1991	1994	1997	2000	2003	2006	2009	2012	2015	2018
Millions of kilograms applied										
Glyphosate herbicides	1.11	2.18	7.2	19.8	30.6	43.4	48	52	52	46
All other herbicides	30.5	29.6	30.5	15.9	8.7	5.5	8.4	11.2	20.4	41.3
Percentage of total kilograms applied										
Glyphosate herbicides	4%	7%	19%	55%	78%	89%	85%	82%	72%	52%
All other herbicides	96%	93%	81%	45%	22%	11%	15%	18%	28%	48%

As Roundup Ready technology rapidly gained soybean sector market share from 1996 through the early 2000s, the rise in reliance on GBHs drove the decline in the use of nearly all other herbicides registered for use on soybeans. Ease of use and efficacy, achieved at prices comparable to herbicide-based options, drove the rapid rise in the use of GBHs on Roundup Ready soybeans.

Table 2.7 gives percent change indices in core glyphosate-soybean use metrics through 2018, with 1996 as the base year (when glyphosate-tolerant crops were introduced). The greater the change from the baseline, the larger a parameter's effect on the total pounds applied. By 2018, the parameters with the greatest to least effect on glyphosate use were acres/hectares treated (248%), average one-time rate of application (42%), acres/hectares planted (39%), and number of applications (32%).

Conclusions and research challenges

The answer to the question "Is pesticide use going up, down, or staying the same?" will vary depending on the metric chosen and how the metric is applied. An MDS of metrics is recommended as a starting point. The 18 metrics outlined in this chapter provide different and useful insights into what is driving changes in pesticide use for a given crop across time and space. Additional data and more refined models and research strategies will be required to deepen understanding of the factors driving changes in pesticide use and the consequences of such changes.

From the farmer's perspective, the two key factors are the degree and reliability of control and the cost of weed management systems. On conventional row-crop farms in

Table 2.7 Percent change of core soybean metrics from 1996 baseline, 1996–2018.

	1996	1999	2002	2005	2008	2011	2014	2017	2018
Soybean acres/ hectares planted	0%	15%	15%	12%	18%	17%	30%	40%	39%
Percent acres/hectares treated									
Glyphosate herbicides	0%	148%	232%	264%	287%	291%	289%	244%	248%
All other herbicides	0%	−46%	−66%	−81%	−71%	−56%	−35%	−4%	5%
Average number of applications									
Glyphosate herbicides	0%	18%	27%	35%	53%	51%	37%	33%	32%
All other herbicides	0%	0%	0%	2%	3%	4%	5%	5%	9%
Average one-time rate of application									
Glyphosate herbicides	0%	10%	17%	20%	33%	43%	52%	43%	42%
All other herbicides	0%	−11%	−8%	−6%	−30%	−38%	−34%	−23%	−14%

the United States, herbicides bear most of the weed-management burden. Accordingly, farmers need accurate, timely information on how well their herbicide programs are working and whether costs are rising, falling, or stable.

Two unmistakable indicators of weed management problems are erratic efficacy and rising cost. The dominant cause for both is usually the emergence and spread of resistant weed phenotypes. Regrettably for US farmers, the response by industry to the emergence of glyphosate-resistant weeds has been to engineer additional herbicide-resistant genes into corn, soybean, and cotton varieties. This strategy is akin to pouring gasoline on a fire in the hope of putting it out. The consequences are evident in the recent, sharply upward trends in the number of herbicides required to bring a soybean crop to harvest, the pounds/kilograms applied and expenditures per acre, and rising environmental and public health problems stemming from herbicide use.

The upward trend in herbicide use cannot go on forever. But what comes next and what will begin to shift reliance from herbicides to prevention-based integrated weed management remains to be seen. Refinements in the ability to track changes in herbicide use and nonchemical control strategies will assist in accelerating positive change, as will a better understanding of the factors driving how farmers respond to changes in weed pressure and control problems.

References

1. Benbrook C, Groth E, Marquardt S. *Pesticide management at the crossroads.* Consumers Union; 1996.
2. Davis AS, Hill JD, Chase CA, Johanns AM, Liebman M. Increasing cropping system diversity balances productivity, profitability and environmental health. *PLoS One* 2012;**7**, e47149.
3. Liebman M, Baraibar B, Buckley Y, et al. Ecologically sustainable weed management: how do we get from proof-of-concept to adoption? *Ecol Appl* 2016;**26**(5):1352–69.
4. Dively GP, Kamel A. Insecticide residues in pollen and nectar of a cucurbit crop and their potential exposure to pollinators. *J Agric Food Chem* 2012;**60**:4449–56.
5. Lu C, Warchol KM, Callahan RA. Sub-lethal exposure to neonicotinoids impaired honey bees winterization before proceeding to colony collapse disorder. *Bull Insectol* 2014;**67**:125–30.
6. Acquavella J, Garabrant D, Marsh G, Sorahan T, Weed DL. Glyphosate epidemiology expert panel review: a weight of evidence systematic review of the relationship between glyphosate exposure and non-Hodgkin's lymphoma or multiple myeloma. *Crit Rev Toxicol* 2016;**46**(Suppl. 1):28–43.
7. Avila-Vazquez M, Maturano E, Etchegoyen A, Difilippo FS, Maclean B. Association between cancer and environmental exposure to glyphosate. *Int J Clinic Med* 2017; **08**(02):73–85.
8. Benbrook C. Impacts of genetically engineered crops on pesticide use in the U.S.—the first sixteen years. *Environ Sci Eur* 2012;**24**.
9. Benbrook C. Trends in glyphosate herbicide use in the United States and globally. *Environ Sci Eur* 2016;**28**(1):3.
10. U.S. Department of Agriculture. *Surveys: agricultural chemical use program*; n.d. https://www.nass.usda.gov/Surveys/Guide_to_NASS_Surveys/Chemical_Use/#data. [Accessed 21 September 2020].

11. California Department of Pesticide Regulation. *Pesticide use reporting (PUR)*; n.d. https://www.cdpr.ca.gov/docs/pur/purmain.htm. [Accessed 21 September 2020].
12. U.S. Environmental Protection Agency. *Pesticides industry sales and usage 2008–2012 market estimates*; n.d. https://www.epa.gov/pesticides/pesticides-industry-sales-and-usage-2008-2012-market-estimates. [Accessed 21 September 2020].
13. U.S. Geological Survey. *Estimated annual agricultural pesticide use*; n.d.; https://water.usgs.gov/nawqa/pnsp/usage/maps/county-level/ [Accessed 21 September 2020].
14. von Ehrenstein OS, Ling C, Cui X, et al. Prenatal and infant exposure to ambient pesticides and autism spectrum disorder in children: population based case-control study. *BMJ* 2019;**364**:l962.
15. Grube A, Donaldson D, Kiely T, Wu L. *Pesticides industry sales and usage: 2006 and 2007 market estimates.* Environmental Protection Agency (EPA); 2011.
16. Fisher M. *Many little hammers: fighting weed resistance with diversified management.* CSA News; September 2012.
17. Liebman M, Gallandt E. Many little hammers: ecological management of crop-weed interactions. In: Jackson LE, editor. *Ecology and agriculture.* Academic Press; 1997. p. 291–343.

Herbicide mode of action

András Székács
Agro-Environmental Research Centre, Institute of Environmental Sciences,
Hungarian University of Agriculture and Life Sciences, Budapest, Hungary

Chapter outline

Introduction

Pesticide mode of action is a fundamental aspect not only in the development but also in the exploration of the molecular mechanisms of the biological effect. Novel modes of action are essential to avoid the detrimental side effects of formerly introduced compounds and to prevent or mitigate pest resistance. The quest for novelty has been at the forefront since the beginning of the search for environmentally friendly agents. *The Pesticide Manual*,[1] the key reference book of pesticide active ingredients (AIs), not only lists the main physicochemical, biological, and environmental characteristics but specifies the chemical classification of each. In addition, excellent handbooks and monographs on pesticide chemistry and biochemistry,[2–4] ecotoxicity,[5] and use[6, 7] as

Herbicides. https://doi.org/10.1016/B978-0-12-823674-1.00008-0

well as public databases on pesticide properties[8] and registration status in the European Union (EU)[9] summarize the current knowledge of the given period on development, utilization, and consequences.

Herbicides have received particular emphasis within pesticide development[10–13] because of the biochemical difficulties in their development, their environmental impacts, and the corresponding weed resistance problems. To overcome the latter difficulty, herbicides with novel modes of action were introduced at the rate of one every 3 years until the 1990s,[14] but no new modes of action have appeared since the *p*-hydroxyphenylpyruvate dioxygenase inhibitors such as isoxaflutole in 1992 (see below). It was only in 2020 that Bayer announced the planned introduction of a new yet undisclosed herbicide AI with a novel mode of action. In comparison, the number of newly announced herbicide AIs (not necessarily with new modes of action) was 5.5 per year between 1980 and 2000, 4 per year by 2010, and 2 per year by 2018,[15] and this declining trend continues. Two factors for this are the increasing cost of registration and the emergence of genetically modified (GM) crops for herbicide tolerance (HT). The range of official test requirements in the toxicological assessment of new agents constantly broadens as knowledge of the side effects on nontarget organisms and the environment expands, requiring greater investment and compelling the pesticide industry to implement the intensive concentration of capital. This has shifted development toward computerized structure-activity relationship (SAR) assessment on the basis of automated high-throughput biological screening of large molecule libraries synthesized in combinatorial chemistry. Lead compounds for novel agents, so-called agrophores, are identified in quantitative SAR studies, but SAR is also effectively applied in health and environmental risk assessment. Due to the increase in research, development, and registration costs, the average cost of developing a new crop protection product rose from $152 million in 1995 to $286 million in 2010–14.[15] As a consequence, the interest of pesticide developers turned toward GM crops. Ironically, this further impeded the development of novel herbicides by boosting the potential uses of existing ones.

The global agricultural use of herbicides expanded from 730,000 tons of AIs in 1980 to 1,680,000 tons in 2016.[15] Leading herbicide AIs (in the order of their market share) within the top 10 pesticides were atrazine†, 2,4-D, trifluralin†, propachlor, dinoseb†, and chloramben† in 1968 and glyphosate, metolachlor, mesotrione, acetochlor†, and atrazine† in 2016. The wide application of HT GM crops substantially contributed to the global use of certain agents, particularly glyphosate, glufosinate, sulfonylureas, 2,4-D, and isoxaflutole, and to a lesser extent dicamba. However, the HT GM approach, although attempted, could not bring bromoxynil into use on the target crop (cotton).

Given the nearly two-thirds cut in the number of approved pesticides in the EU and the decline in new synthetic ones being introduced (a single novel herbicide AI registered every 4.5 years on average), along with the new strategic agricultural preferences, regulations are expected to become more stringent. Herbicides no longer authorized in the EU are marked with a "†," a few compounds not yet registered in the EU are marked with a "*," and compounds that have never appeared as candidates in EU regulatory documentation are marked with a "#" throughout this chapter.

Classification of target sites of action

The mode of action includes all processes that take place from the point of contact, through the stability of the herbicide on or in the plant, its translocation and metabolism within the plant, the biochemical effect of the AI at the target site, the physiological responses of the plant, to the final effect. Thus, to exert a physiological response, an agent must penetrate into the tissues, become translocated to the target protein, accumulate at the target protein location, bind to or react with the target protein, and finally produce cellular damage and necrosis. The biochemical site of action is of key importance for development, environmental and ecotoxicological assessment, and weed resistance. There is particular emphasis on target specificity and the corresponding reduction in side effects. As a rule of thumb, however, the more specific a substance is, the higher the likelihood of resistance, as the emergence of resistance through simple genetic changes (e.g., point mutation) is more probable against a single mode of action than against multiple biochemical mechanisms. Emerging weed resistance against herbicides with specific modes of action has gradually become of high concern, and an effective tool in resistance management is the combination or alternating use of herbicides of different biochemical mechanisms. Consequently, classification systems of herbicides according to their target sites were developed to delay the natural selection of resistant weeds. Schemes with agents acting at the same biochemical site assigned a group number were created.

A system approved by the Weed Science Society of America (WSSA)[16–18] designates mode of action groups by numbers. Another system from the international Herbicide Resistance Action Committee (HRAC),[19–22] founded by the agrochemical industry (Arysta LifeScience, BASF, Bayer CropScience Division, Corteva Agriscience, FMC, Makhteshim Agan/ADAMA, Syngenta Crop Protection, Sumitomo Chemical Company), used to assign letters to the groups. In March 2020, the HRAC system was changed from letters to numbers to harmonize with the WSSA codes to be more globally relevant and sustainable. After this transition, the previous letter-based HRAC classification is called "legacy HRAC" groups. Another letter-based system, the Australian Code System[23] (its categories are also called Aussie groups) was created in the early 1990s. The Australian Pesticides and Veterinary Medicines Authority[24] publicizes the official mode of action groups through CropLife Australia, a nonprofit organization funded by developers, manufacturers, formulators, and registrants of crop protection and agro-biotechnology products.[25] Currently, the HRAC system is the most prevalent worldwide while the WSSA system is used in the United States and Canada, and the Australian Code System is used in Australia. The systems substantially overlap, although there are differences.[26] In the HRAC and WSSA systems, 25 and 23 herbicide groups are defined, respectively.

According to their type of biochemical mechanism at the target site, herbicides include common cytotoxic agents, cellular metabolism disruptors, compounds acting by light activation of reactive oxygen species (including inhibition of photosynthesis), and cell division and growth modulators. Cellular targets of herbicide action include biochemical functions in the chloroplasts, microtubule organization, cell wall, mitochondria, cell nucleus, and hormonal transport (Table 3.1).

The discussion below of the herbicide type surveys AIs grouped according to their translocation in the plant, specificity of action, and actual biochemical mode of action. This listing follows a somewhat chronological order, as herbicide design shifted over time toward compounds with specific mechanisms of action, and as lessons learned from recognized side effects of previous agents were integrated into the development of new candidates.

Herbicides of broad cytotoxicity

Mineral oils

Long-chain hydrocarbons (mixtures of alkanes, alkenes, alicyclic, and aromatic compounds with nitrogen- and sulfur-containing components), mainly from refining of petroleum oils, are the oldest organic herbicides. They were used as contact agents on all crops, acting partly physically and partly biochemically. They work by blocking the airways of the leaves and damaging the lipid layer of the leaf epidermis. Also, becoming adsorbed into plasma membranes, they biochemically inhibit various membrane functions. Given their general cytotoxicity, they have also been used against insects as larvicides and ovicides. Although our current ecotoxicology approach renders them no longer applicable, used mineral oils were still recommended herbicides for railway tracks in the early 1980s.

Dinitrophenols (WSSA/HRAC group 24; legacy HRAC group M)

This group, among the oldest synthetic pesticides, is represented by dinitrophenol derivatives. The target is the inhibition of oxidative phosphorylation, a basic process in all living cells. Dinitrophenols decouple oxidation and phosphorylation during mitochondrial electron transport. As a result, cellular respiration is accelerated as the cell produces more and more NADH dinucleotides but is unable to restore its damaged ATP production. Due to the general mechanism of action, zoocidal and fungicidal agents are also found in the group. The main member of the group, dinoseb† (6-sec-butyl-2,4-dinitrophenol), was banned as a herbicide in the United States and the EU in 1986 and 1991, because of its toxicity on vertebrates and teratogenicity on mammals. A unique feature of its withdrawal in the United States is that it was the third case when the US Environmental Protection Agency (EPA) issued the strongest action under the federal pesticide law, an emergency order barring sale or use; the other two cases were 2,4,5-T† in 1977 and ethylene dibromide† in 1984.

Table 3.1 Modes of action of different classes of herbicides.

Main physiological function disrupted	Site of action	Mode of action	Group			Chemical family	Active ingredients (examples)	Reference
			WSSA/ HRAC	Legacy HRAC	Australian Code System			
Cellular metabolism	Inhibition of acetyl CoA carboxylase (ACCase)	Lipid synthesis inhibitors	1	A	A	Aryloxyphenoxypropionates, cyclohexanediones, phenylpyrazolins	Diclofop, fluazifop, quizalofop; sethoxydim, clethodim, cycloxydim	27
	Inhibition of acetolactate synthase (ALS) (acetohydroxyacid synthase, AHAS)	Amino acid (Leu, Ile, Val) synthesis inhibitors	2	B	B	Imidazolinones, pyrimidinylthiobenzoates, sulfonylamino carbonyltriazolinones, sulfonylureas, triazolopyrimidines	Imazaquin, imazapic, imazapyr, imazamox; chlorsulfuron	28
	Inhibition of enolpyruvyl shikimate phosphate synthase (EPSPS)	Aromatic amino acid (Phe, Trp, Tyr) synthesis inhibitors	9	G	M	Phosphonomethyl amino acid	Glyphosate	29
	Inhibition of very long chain fatty acid (VLCFA) synthesis	Shoot growth inhibitors	15	K3	K	Chloroacetamides, acetamides, oxyacetamides, tetrazolinones	*S*-Metolachlor, metolachlor, acetochlor, flufenacet, dimethenamid-P	30
	Inhibition of dihydropteroate synthase (DHPS)	Tetrahydrofolate synthesis inhibitors	18	I	R	Carbamates	Asulam	31
	Inhibition of cellulose synthase	Cell wall (cellulose) synthesis inhibitors	29	L	I, O, Z	Benzamides and nitriles	Dichlobenil	32

Continued

Table 3.1 Continued

Main physiological function disrupted	Site of action	Mode of action	Group			Chemical family	Active ingredients (examples)	Reference
			WSSA/ HRAC	Legacy HRAC	Australian Code System			
	Inhibition of fatty acid thioesterase (FAT)	Fatty acid synthesis inhibitors	30	Q	T	Benzyl ethers/cineoles	Cinmethylin	33
	Inhibition of serine threonine protein phosphatase	Cell cycle disruptors	31	R	Z	Endothalic acid	Endothal	34
	Inhibition of lipid synthesis (non-ACCase)	Shoot growth inhibitors	–	N	J	Phosphorodithioates and thiocarbamates	EPTC, triallate	35
	Inhibition of lipid synthesis (non-ACCase)	Lipid synthesis (e.g., fatty acid elongase) inhibitors	–	N	J	Benzofuranes, chlorocarbonic acids, phosphorodithioates, and thiocarbamates	EPTC, triallate	36
Light activation of reactive oxygen species	Inhibition of photosystem II—Site B	Photosynthesis inhibitors	5	C1	C, K	Triazines, phenylcarbamates, triazinones, pyridazinones, uracils	Atrazine, simazine, ametryn, prometon; metribuzin, hexazinone; terbacil, bromacil	37
	Inhibition of photosystem II—Site A, serine 264	Photosynthesis inhibitors	5	C2	C	Ureas, amides	Linuron, diuron, tebuthiuron, chlorotoluron, isoproturon	37
	Inhibition of photosystem II—Site C, histidine 215	Photosynthesis inhibitors	6	C3	C	Benzothiadiazinones, nitriles, phenylpyridazines	Bromoxynil, bentazon	37
	Inhibition of glutamine synthetase (GS)	Nitrogen metabolism inhibitors	10	H	N	Phosphorylated amino acid	Glufosinate, bialaphos	38

	Inhibition of phytoene desaturase (PDS)	Bleaching: Pigment (carotenoid) synthesis inhibitors	12	F1	F	Amides, anilides, furanones, phenoxybutanamides, pyridiazinones, pyridines	Diflufenican, flurtamone	39
	Inhibition of deoxy-D-xylulose phosphate synthase (DOXPS)	Pigment synthesis inhibitors/diterpene synthesis inhibitors	13	F4	F	Isoxazolidinone	Clomazone	40
	Inhibition of protoporphyrinogen oxidase (PPO)	Cell membrane disrupters (blocking heme biosynthesis for chlorophyll)	14	E	G	Diphenylether, aryl triazolinones, *N*-phenylphthalimides, oxadiazoles, oxazolidinediones, imines, phenylpyrazoles, pyrimidinediones, thiadiazoles	Acifluorfen, oxyfluorfen, lactofen, fomesafen, pyraflufen; sulfentrazone, carfentrazone; flumiclorac, flumioxazin; fluthiacet; saflufenacil	41
	Diversion of electron transfer in PS I	Cell membrane (thylakoid) disrupters	22	D	L	Bipyridiliums	Paraquat, diquat	42
	Inhibition of hydroxyph enylpyruvate dioxygenase (HPPD)	Pigment (carotenoid) synthesis inhibitors	27	F2	H	Isoxazoles, triketones, pyrazolones, callistemones	Isoxaflutole, pyrasulfotole; mesotrione, tembotrione, bicyclopyrone; topramezone	43
	Inhibition of solanesyl diphosphate synthase	Pigment (carotenoid) synthesis inhibitors	32	S	–	Diphenyl ethers	Aclonifen	44
	Inhibition of homogentisate solanesy ltransferase	Pigment (plastoquinone) synthesis inhibitors	33	T	–	Pyridazines	Cyclopyrimorate	45
	Inhibition of lycopene cyclase	Pigment (carotenoid) synthesis inhibitors	34	F3	Q	Triazoles, ureas, isoxazolidinones	Amitrole	46

Continued

Table 3.1 Continued

Main physiological function disrupted	Site of action	Mode of action	Group			Chemical family	Active ingredients (examples)	Reference
			WSSA/ HRAC	Legacy HRAC	Australian Code System			
Cell division and growth	Inhibition of microtubules (assembly)	Root growth inhibitors	3	K1	D	Benzamide, benzoic acid (DCPA), dinitroaniline, phosphoramidate, pyridine	Trifluralin, pendimethalin, etc.	47
	Auxin agonists (stimulation of transport inhibitor response protein 1, TIR1)	Plant growth inhibitors	4	O	I	Benzoic acids, phenoxycarboxylic acids, pyridinecarboxylic acids, quinolinecarboxylic acids, arylpicolinates, quinolines	Dicamba; 2,4-D, 2,4-DB (butyrac), 2,4-DP, MCPA, MCPP; picloram, clopyralid, fluroxypyr, triclopyr, aminopyralid, aminocy-clopyrachlor; quinclorac; halauxifen	48
	Inhibition of auxin transport	Long-range hormone signaling	19	P	P	Phthalamates, semicarbazone/aryl carboxylates	Naptalam; diflufenzopyr	49
	Inhibition of microtubules (polymerization, organization)	Mitosis inhibitors	23	K2	E	Carbamate, carbetamide	Propham, chlorpropham	50
	Uncoupling oxidative phosphorylation	Membrane disruptors	24	M	Z	Dinitrophenols	Dinoseb, DNOC	51
Unknown mode of action			0	Z	Z	Arylaminopropionic acids	Flamprop	
			0	Z	Z	Pyrazoliums	Difenzoquat	

Contact herbicides not translocated in plants

Disruptors of membrane processes

Bipyridyliums (WSSA/HRAC group 22; legacy HRAC group D)

The first two representatives, paraquat† and diquat† (Fig. 3.1A), were discovered in 1955, although paraquat as a chemical compound (methyl viologen) had been in use as a redox indicator since the 1930s.

These agents prevent the electron transfer step in photosynthesis photosystem I (PS I), thus the reduction of $NADP^+$ and ultimately the reductive conversion (assimilation) of carbon dioxide (the mechanism of photosynthesis and the attack points of herbicides in the process sequence are described in "Photosynthesis inhibitors" section). Instead of $NADP^+$, the herbicide is reduced and immediately oxidized with atmospheric oxygen. During the process, reactive superoxides are formed, which react with the unsaturated fatty acids of the membrane lipids. Under normal conditions, peroxides, superoxides, and hydrogen peroxide formed during the light phase of photosynthesis are degraded by an enzyme cascade. The first enzyme member of the cascade is superoxide dismutase, which catalyzes the association of two superoxide free radical anions (O^{2-}) into hydrogen peroxide. Hydrogen peroxide is reduced by ascorbic acid to water, which is reduced again by glutathione (reduced form), the latter being reduced by NADPH formed in the light phase. Thus, by inhibiting the reduction of $NADP^+$, bipyridyliums inhibit normal membrane function by stopping peroxide degradation.

As seen, an important element of the mechanism of action of bipyridyliums is that free radicals are formed and accumulated in the plant. The biodegradation of bipyridyliums in plants and animal organisms is very slow. In spite of its relatively high oral acute LD_{50} in mammals (20–150 mg/kg body weight), paraquat causes disturbances in nerve functions by inhibiting via free radical formation the enzyme DOPA decarboxylase, and this effect cannot be antidoted. For these reasons, and also because of its frequent use for suicide, paraquat formulations were requested to be colored with a blue additive, and were banned in 1991 in Hungary, ahead of most European countries. The marketing of paraquat was banned in the EU in 2007, and it was classified as "restricted use" in the United States. Interestingly, the EU ban was justified not for human health but for ecotoxicity reasons. There has been no revocation decision; the list status of the AI was simply terminated as a consequence of the court decision—a rare case in pesticide registration. Diquat in the form of dibromide, used mostly for desiccation, remained authorized in the EU until 2018.

These compounds are strongly bound to soil particles and therefore not accessible from the soil to plants or to microbial degradation. However, they are not bound to soil organic matter. Agents bound to soil particles lose their biological activity and accessibility, but they disappear from the soil only very slowly. Their decomposition is similarly slow in water. It is uncertain whether their degradation is the result of physical or biochemical processes. The half-life in water and soil is greater than 150 and 1000 days. Due to their high water solubility and slow degradation, they can

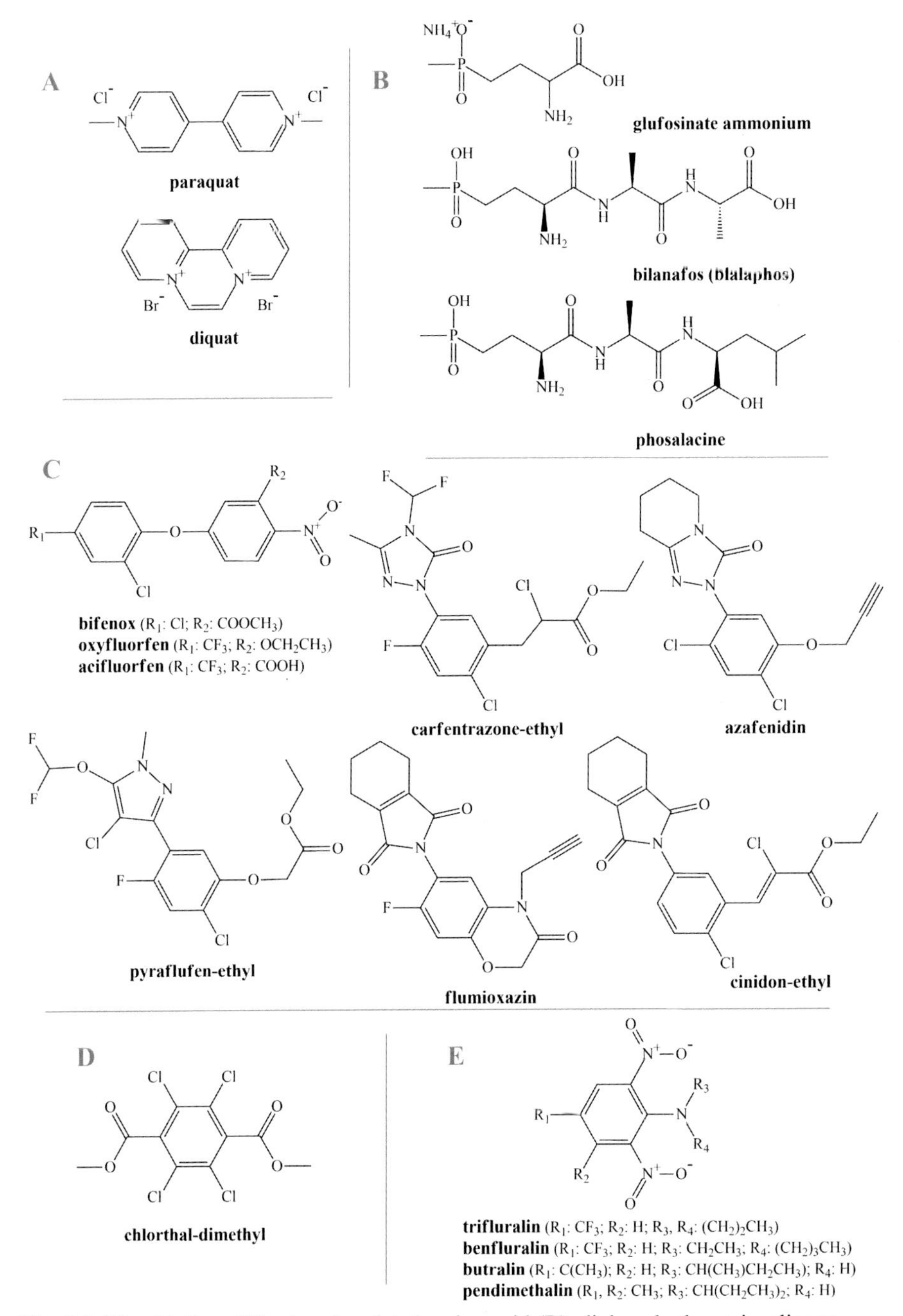

Fig. 3.1 Bipyridylium (A), phosphorylated amino acid (B), diphenyl ether, triazolinone, phenylpyrazole and *N*-phenylphthalimide (C), chlorinated terephthalic acid (D), and dinitroaniline and -pyridine (E) herbicide active ingredients.

accumulate in aquatic plants, which endangers animals that feed on them. Teratogenic effects of paraquat have been observed in mammals, but also in fish and frogs, and diquat has shown similar though milder effects.

Phosphorylated amino acids (WSSA/HRAC group 10; legacy HRAC group H)

The main members of these phosphinic acids are glufosinate[†] [(phosphinothricin), the methylphosphinoyl derivative of homoalanine (2-aminobutyric acid)], in addition to bilanafos[#] and phosalacine[#], similar tripeptide derivatives of L-Ala-L-Ala and Ala-Leu (Fig. 3.1B). The discovery of glufosinate, introduced in 1972 by Hoechst, was led by an antibiotic tripeptide isolated from the soil bacterium *Streptomyces viridochromogenes*: the tripeptide contained two alanine units and the novel amino acid moiety phosphinothricin. The new amino acid alone did not show significant antibiotic activity, but due to its structural similarity to glutamic acid, it inhibited the enzyme glutamine synthetase (GS) of bacterial or plant origin (interestingly, the tripeptide did not show the inhibitory effect on GS). The same tripeptide was also detected in an independent development project from an isolate of *Streptomyces hygroscopicus* from a Japanese soil sample, and was named bialaphos (later bilanaphos). It was described as a proherbicide of glufosinate, and introduced in 1973 as the first herbicide to be produced by fermentation. The former is marketed in the form of ammonium salt (glufosinate ammonium), the latter in the form of sodium salt. Bilanafos is a biological control agent produced by fermentation using the bacterium *S. hygroscopicus*. Phosalacine, announced in 1984, is a fermentation product of *Kitasatospora phosalacinea* (also known as *S. phosalacineus*), but to date no commercial herbicide product has been introduced with phosalacine as its AI.

Through blocking GS, which is necessary for the production of glutamine and for ammonia detoxification, these phosphinic acid derivatives inhibit ammonia assimilation in plants and are lethal to them. Ammonia is a substrate for many plant enzymes; of them all, GS has the highest affinity for it. In the first step of ammonia fixation, glutamic acid is formed from α-keto-glutaric acid by binding ammonia, and then GS forms glutamine by the addition of another ammonium molecule. Glutamine and α-keto-glutaric acid can further react with each other to form two molecules of glutamic acid in a process catalyzed by the enzyme glutamine-2-oxoglutarate-aminotransferase (also known as glutamic acid synthetase). The cycle of the two enzymes (GS and glutamine-2-oxoglutarate aminotransferase) ensures that ammonia assimilation is efficient and that the resulting glutamic acid product is available for the biosynthesis of amino acids, proline- and arginine-based nucleotides, and 5-aminolevulinic acid (a key intermediate in chlorophyll biosynthesis). In the presence of ammonium-binding scavengers, GS cannot function; therefore, ammonium ions accumulate, glutamine biosynthetic pathways are arrested, and photosynthesis is inhibited (due to the inhibition of chlorophyll formation). Several natural inhibitors of GS have been discovered, but none is as herbicidally active as phosphinothricin (glufosinate). Accordingly, glufosinate has a strong and broad-spectrum postemergent herbicidal effect against almost all weeds as a contact herbicide with a slight systemic effect in the plant (translocated within the leaf). Soil microbes decompose it rapidly; the half-life of glufosinate ammonium in soil is 7 days.

Glufosinate is not rapidly degraded metabolically within higher plant tissues. Interestingly, *S. viridochromogenes*, producing bilanafos, also produces an enzyme, phosphinothricin *N*-acetyltransferase (PAT), that can transacetylate glufosinate and bilanafos with an acetyl group from acetyl-CoA. A similar enzyme catalyzing the acetylation of glufosinate, but not of bilanafos, has been identified in *S. hygroscopicus*. The corresponding acetyl derivatives are inactive toward GS and hence are not phytotoxic. A particular application of these findings is that the genes (*pat* and *bar*) encoding these metabolitic enzymes have been inserted into transgenic plants, resulting in herbicide-resistant GM crops (e.g., glufosinate-tolerant cotton, maize, and canola, grown in North America), which broadened the agricultural use and therefore the environmental load of glufosinate. Its use for desiccation has also been blamed for enhanced environmental pressure. In the EU, banded application has been requested to protect mammals, but banded application of desiccants is pointless. In turn, glufosinate remained authorized until 2018, when the producer desisted from reregistration, and the other two compounds in the group have never received registered status.

Diphenyl ethers, triazolinones, phenylpyrazoles, and N-phenylphthalimides (WSSA/HRAC group 14; legacy HRAC group E)

The structure of diphenyl ethers is somewhat similar to those of aryloxyphenoxypropionates (see below), but differs from them in the mechanism of action and in application. The earliest members of the group still on the market today are bifenox, oxyfluorfen, and acifluorfen†, introduced in 1973, 1975, and 1980. Compounds with different structures but similar mechanisms of action are the triazolinones carfentrazone-ethyl and azafenidin, introduced in 1993 and 1997; the phenylpyrazole derivative pyraflufen-ethyl, introduced in 1993; and the *N*-phenylphthalimides flumioxazin and cinidon-ethyl†, introduced in 1991 and 1998 (Fig. 3.1C).

These compounds interfere with the metabolism of tetrapyrrole and inhibit the enzyme protoporphyrinogen oxidase (PPO or Protox), causing unprocessed protoporphyrinogen to accumulate in the photosynthetically active plant cells. The level of tetrapyrroles enhancing the photosensitivity of the plants also rises, which ultimately leads to photooxidation and leaf tissue death.

Their metabolism occurs by slow cleavage of the ether bond and subsequent conjugation of the nitrophenyl moiety to cysteine or glutathione. Diphenyl ethers decompose in soil photochemically and microbially, but relatively slowly. Because they are nonvolatile and strongly bind to soil colloids, they persist in soil and retain their phytotoxicity for 1–4 months. The half-lives in soil are 14–60 days for acifluorfen and 35 days for oxyfluorfen. Acifluorfen was withdrawn from the EU with essential use derogations and azafenidin was banned in 2002; cinidon-ethyl has not been approved since 2012.

Cell growth inhibitors

Herbicides that inhibit plant germination include agents with different mechanisms of action. Some of them (terephthalic acid derivatives, dinitroanilines, pyridine derivatives) exert phytotoxicity by inhibiting cell proliferation, affecting the most important steps of cell division, DNA replication, and RNA and protein synthesis, essential for cell growth. During mitotic cell division, they inhibit the formation of tubulin protein and thus microtubules, so that mitotic spindle strands that separate chromosome pairs to each daughter cell cannot be formed and mitosis is stopped. Microtubules are subcellular fibers composed of heterodimeric tubulin units that play a prominent role in a range of cellular physiological processes, such as maintaining cell structure, orienting and organizing cellulose (cellulose-hemicellulose-pectin) microfibrils, cell wall formation, intracellular cell motility, chromosome migration during cell division (mitosis), and cell differentiation. Other agents (thiolcarbamates) bind to the acetyl coenzyme A protein and inhibit its function, but not through inhibiting the acetyl CoA carboxylase (ACCase) enzyme, thereby blocking the biosynthesis of long chain fatty acids.

Chlorinated terephthalic acid derivatives (WSSA/HRAC group 3; legacy HRAC group K_1)

A representative member is the dimethyl ester of tetrachloroterephthalic acid, chlorthal-dimethyl[†] (also known as DCPA), introduced in 1960 (Fig. 3.1D). Its mitotic effect prevents the germination of plant seeds and has been used as a pre-emergent selective nonsystemic herbicide in vegetables (onions, tomatoes, peppers, cabbages, lettuce, legumes, potatoes, cucumbers, and horseradish), fruits (mulberry), cotton, and other crops to control mono- and dicotyledonous weeds. Its use is limited partly by its phytotoxicity (it was therefore unsuitable for the treatment of sugar beet, spinach, flax, and clover) and partly by its environmental stability (its half-life in soil is 100 days) due to its significant chlorine content. Its main degradation product is 2,3,5,6-tetrachloroterephthalic acid, which binds strongly to soil particles and has an even longer degradation time than chlorthal-dimethyl.

Dinitroanilines and -pyridines (WSSA/HRAC group 3; legacy HRAC group K_1)

The most important characteristic in the chemical structures of this family is that they contain two nitro groups in the ortho position to the aniline amino group and an alkyl group (methyl, trifluoromethyl, or short branched alkyl) in the para position [in addition, small substituents (methyl or amino groups) may also be present on the aromatic ring in the meta position], and that the nitrogen atom of the anilides is mono- or disubstituted with short alkyl groups (ethyl, propyl, isobutenyl, isopentenyl, etc.). Trifluralin[†], introduced in 1963, was the first and most significant representative; it already carried all the main criteria of the group. Benfluralin (benefin), introduced in 1963, also contains a trifluoromethyl group on the aromatic ring while butralin[†] and pendimethalin, introduced in 1971 and 1974, contain a simple alkyl substituent (Fig. 3.1E).

The phytotoxic effect of dinitroanilines is based on the inhibition of mitotic cell division by the inhibition of microtubule polymerization and by microtubule depolymerization. They bind to tubulin units, leading to a physically incorrect microtubule configuration and loss of tubulin fiber function; as a result, the mitotic spindle cannot be formed.

The compounds are decomposed by photodegradation and metabolism by soil bacteria. Microbial degradation is faster under anaerobic conditions. Soil persistence is stronger in cold, dry conditions. Depending on the circumstances, trifluralin can remain in the soil for several months. However, increased microbiological degradation may occur in areas previously exposed to the compounds, where the local microflora may specialize in more rapid degradation, which may significantly impair the effectiveness of herbicide treatments. The typical half-life in soil is 45 days (trifluralin). Certain plant species may become resistant through natural mutation. This can be achieved by point mutation by changing a single amino acid of tubulin, threonine at position 239, with isoleucine. The EU banned trifluralin in 2008 because of its toxic effects on aquatic life. In 2009, the EU also banned butralin, also used previously as a plant growth regulator to prevent sucker development in tobacco, because of its effects on human health. Both AIs remain approved in the United States.

Thiolcarbamates (WSSA/HRAC group 15; legacy HRAC group K_3)

These agents have a structural similarity to photosynthesis inhibitor carbamate derivatives (chlorpropham, see below), but display characteristic differences in structure and mode of action. They contain a thiol sulfur atom in their carbamate group, and substituents on the sulfur and nitrogen atoms can be alkyl (possibly benzyl), cycloalkyl, or heterocycloalkyl groups. The first member, with the simplest structure, was EPTC† or eptam (*S*-ethyl dipropylthiocarbamate), introduced in 1958. Due to its short alkyl groups, EPTC is the most volatile compound of the group. Butylate† (*N*,*N*-dibutyl *S*-ethyl derivative), introduced in 1962, is substituted on its nitrogen atom with normal alkyl groups; cycloate† contains a cyclohexyl group on its nitrogen atom; molinate†, introduced in 1964, has an azepane ring, an unusual, 7-membered heterocycle. Thiobencarb (benthiocarb)†, introduced in 1969, contains a benzyl (4-chlorobenzyl) moiety on its sulfur atom (Fig. 3.2A).

The effect of thiolcarbamates can be eliminated in some plants by antidotes. Examples are dichloroacetamides: dichlormid (*N*,*N*-diallyl-2,2-dichloroacetamide, R25788), the dimethyloxazolidine derivative furilazole#, the azepan derivative dahemid# (1-dichloroacetylazepane, TI-35), and two Hungarian inventions, the spirooxazolidine AD-67# and the dioxolane derivative MG-191#. As a result, certain crops (maize) become tolerant through the induction of metabolic enzymes (such as glutathione *S*-transferase) that can deactivate thiolcarbamates (Fig. 3.2B). It has to be noted that antidotes (safeners, or by a new terminology, protective agents) are legally covered by the plant protection product law (EC Regulation 1107/2009) in the EU, but

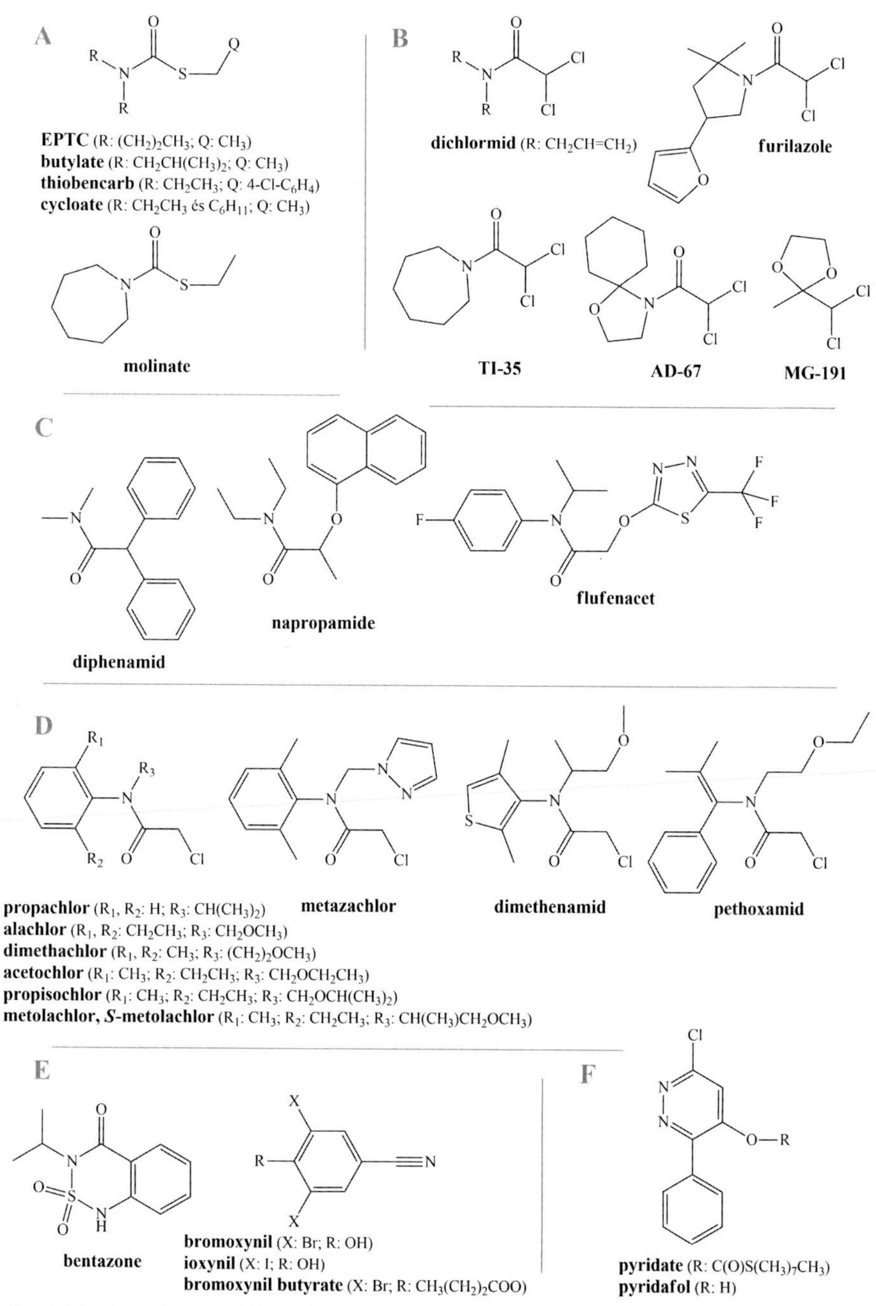

Fig. 3.2 Thiolcarbamate (A) and antidote (B), acetamide, oxyacetamide (C), and chloroacetamide (D), benzothiadiazinone and hydroxybenzonitrile (E), and phenylpyridazine (F) herbicide active ingredients.

remain under regulation at the level of member states. The list of safeners used in the EU is currently being compiled for an EU-level authorization system.

Thiolcarbamates are decomposed by certain soil microbes. These microorganisms adapt rapidly to members of the class. Therefore, as with dinitroanilines, increased microbiological degradation may occur in areas previously exposed. Such induced accelerated degradation capacity is characteristically specific to the compound type and not to the individual compound. Thiolcarbamates are typically degraded by hydrolytic decarboxylation, involving carbon dioxide formation, to the corresponding amine and thiol derivatives, which are then conjugated to cysteine, methionine, or glutathione. Soil microorganisms selected upon exposure to a given thiolcarbamate often can rapidly metabolize other thiolcarbamates as well. Another possible pathway for metabolism is oxidation of the sulfur atom to sulfoxide or sulfone. This process is partly "suicidal"; although the end product of the sulfone derivative is not phytotoxic, the toxicity of the sulfoxide intermediate is higher than that of the parent thiolcarbamate. Half-lives in soil are between 6 (EPTC) and 30 days (cycloate).

Acetamides, oxyacetamides, and chloroacetamides

A range of substituted amides began to be developed in the 1960s. The idea was that these relatively simple molecules would be easily degraded in the plant and soil. However, this has not always been the case. Diphenamid†, for example, introduced in 1975 as a pre-emergent agent, is extremely persistent in soil, lasting 3–12 months. The acid amide of the α-naphthyloxy group, napropamide, introduced in 1969, is slightly less persistent: 1–10 months, depending on soil quality. Both are inhibitors of plant cell division, yet are still classified as of unknown mode of action (WSSA/HRAC group 0; legacy HRAC group Z). Later developed acetamide derivatives are the so-called oxyacetamides, anilides of hydroxyacetic acid, in which the aromatic moiety is attached to the amide nitrogen atom. One of these is flufenacet, introduced in 1995, which contains a trifluoromethylthiadiazolyl group as a substituent on the hydroxyl group (Fig. 3.2C). It also inhibits cell division and growth, presumably through the inhibition of very long chain fatty acid biosynthesis (WSSA/HRAC group 15; legacy HRAC group K_3) or fatty acid metabolism.

Among the amide derivatives of aliphatic acids, a separate group is represented by the amide derivatives of chloroacetic acid. The first and simplest, propachlor†, introduced in 1964, is a chloroacetanilide (*N*-isopropyl-*N*-2-chloroacetanilide) that contains an unsubstituted aniline moiety on the aromatic ring. The others are all 2,6-dialkylanilides in which an alkoxyalkyl group is also found on the aniline nitrogen atom (Fig. 3.2D). These include alachlor†, metolachlor (now only the purified *S* enantiomer is allowed, called *S*-metolachlor), dimethachlor, metazachlor (which contains a heterocyclic pyrazolylmethyl group on the aniline nitrogen), acetochlor†, and propisochlor† (containing a branched alkyl group in its alkoxyalkyl moiety, developed in Hungary), introduced in 1966, 1974, 1977, 1982, 1985, and 1992. Instead of the anilide moiety, dimethenamid, introduced in 1991, contains a dimethylthienylamide group (and its pure *S*-isomer registered as a separate AI, dimethenamid-P, introduced in 2000), and pethoxamid, introduced in 2002, contains a phenylisobutenylamide moiety (Fig. 3.2D).

While the exact mode of action of diphenamid has not been identified, oxyacetamides and chloroacetamides inhibit the synthesis of very long chain fatty acids (VLCFAs) (WSSA/HRAC group 15; legacy HRAC group K_3). VLCFAs are saturated and unsaturated fatty acids longer than 18 carbons in length, important as components of various lipids (such as plant waxes and the typhine layer on pollen grains or plant surfaces that protect plants against external effects or water loss), triacylglycerols (for energy storage in seeds), sphingo- and phospholipids (in cell membranes), and lipids involved in phytohormonal regulation. They are synthesized from long chain fatty acid CoA thiolesters in the endoplasmic reticulum, and this biosynthesis is catalyzed by the concerted action of at least five enzymes (acetyl-CoA carboxylase, 3-ketoacyl-CoA synthase, 3-ketoacyl-CoA reductase, 3-hydroxyacyl-CoA dehydratase, 2,3-*trans*-enoyl-CoA reductase) in the membrane-bound, multi-enzyme acyl-CoA elongase complex. Chloroacetamides inhibit acyl-CoA elongase by carbamoylmethylating the thiol group of the cysteine moiety at the enzymatic site essential for fatty acid elongation. Their action can be countered by antidotes: the dioxolane derivative MG-191 or the dichloroacetamide furilazole increase the tolerance of certain crops (maize) to the AI acetochlor, and the use of the antidote fluxofenim[#] as a dressing agent makes some crops (*Sorghum*) tolerant to some chloroacetamides.

These agents are mainly degraded microbially. Prolonged persistence may occur under anaerobic and cool conditions. Their half-life in soil varies between 20 (dimethenamid) and 70 days (napropamide).

Acetochlor and propisochlor were banned in the EU in 2011. The decision was based on concerns over potential human exposure above the acceptable daily intake, and the potential for human exposure to the surface water metabolite t-norchloroacetochlor, the genotoxicity of which *cannot* be excluded. There is a high risk of groundwater contamination for several metabolites, a high risk for aquatic organisms, and a high long-term risk for herbivorous birds. The information available was not conclusive on the risk for groundwater contamination for metabolites t-norchloracetochlor and t-hydroxyacetochlor.

Benzothiadiazinones, hydroxybenzonitriles (WSSA/HRAC group 6; legacy HRAC group C_3)

This group includes contact herbicide benzothiadiazinones (bentazone, introduced in 1968) and nitriles (bromoxynil[†] and ioxynil[†], introduced in 1963) (Fig. 3.2E). Bromoxynil and ioxynil are formulated in the form of their octanoate, butyrate (butanoate), or potassium salts and octanoate, potassium, or sodium salts. They inhibit photosynthesis at site B of the second step of the light phase [photosystem II (PS II); see below] by disconnecting the electron transfer cascade, thus blocking the transfer of the light energy absorbed. They also decrease ATP levels in plants, suggesting that they inhibit both photosynthetic and oxidative phosphorylation. By inhibiting electron transfer processes, nascent oxygen or other strong oxidizing agents accumulate in the illuminated chloroplasts, leading to the peroxidation of membrane lipids, which results in the disruption of membrane processes, membrane rupture, and cell death. This oxidative membrane disruption may explain their rapid effect while the long-term effects are due to the induced decay in the ATP level, and thus the level of energy reserves; nucleic acid and protein synthesis decline as well.

The two nitriles are mainly degraded microbially. Bromoxynil is metabolized in several ways (by hydrolysis of the bromine substituents or the nitrile group), and it is decomposed by microorganisms to carbon dioxide. Its half-life in soil is 7–21 days. Due to their toxic effects in aquatic ecosystems, bromoxynil and ioxynil use has declined.

Tolerance to bromoxynil in GM crops was induced by deactivation by a transgenic nitrilase enzyme. A bromoxynil-tolerant cotton was the first GM crop approved in the United States by the EPA. However, bromoxynil could not be authorized for use on it because of the possibly carcinogenic metabolite 3,5-dibromo-4-hydroxybenzoic acid. The approval for ioxynil was discontinued in 2011 in the EU, and bromoxynil was also removed from the positive list in 2020, but the nonrenewal regulation has not yet been announced.

Phenylpyridazines (WSSA/HRAC group 6; legacy HRAC group C_3)

Pyridate, introduced in 1976, is a potent postemergent contact agent, mainly absorbed on the leaf surface and active against monocotyledonous and dicotyledonous weeds. It has some mobility in the plant, but when absorbed it is rapidly hydrolyzed to its nonmobile metabolite pyridafol (3-phenyl-4-hydroxy-6-chloropyridazine), which is responsible for the phytotoxic effect by inhibiting electron transfer due to blocking amino acid histidine 215 at the A site of PS II (see below) (Fig. 3.2F). Because of the rapid degradation, plant translocation is not enhanced by the addition of adjuvants. Pyridate is applied to cereals (maize, rice), cabbages (cabbage, kale, cauliflower, broccoli, Brussels sprouts), onions (onions, shallots, chives, leeks), and other crops (rapeseed, asparagus, fodder, medicinal plants). Its efficacy can be increased against certain weeds when used in combination with phenoxyacetic acids.

Acropetally translocated herbicides

Photosynthesis inhibitors

In green plants, two photosystems (PSs)—light-induced electron excitation mechanisms—exist in which chlorophyll is excited by sunlight and the excitation energy is transferred to the chemical reaction center (for the photolysis of water). PS I absorbs photons with a wavelength around 700 nm, has strong reducing properties, and produces NADPH; PS II absorbs at 670–680 nm and leads to the release of oxygen. The electron generated in PS II is transferred to PS I, where it results in the reduction of carbon dioxide to carbohydrate through the formation of NADPH (Fig. 3.3).

Inhibiting photosynthesis interferes with these two photosystems. Compounds that inhibit photosynthesis at site A of PS II (triazines, uracils, phenylcarbamates,

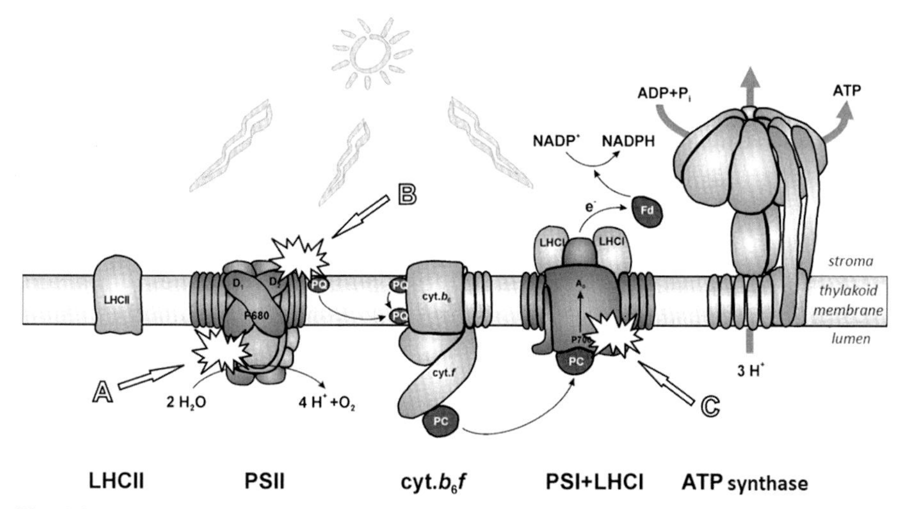

Fig. 3.3 Schematic of the cellular apparatus of the light reaction of photosynthesis in the thylakoid membrane.
Reproduced with permission from Darvas B, Székács A, editors. *Agricultural ecoloxicology*. Budapest, Hungary: L'Harmattan; 2010 [in Hungarian].

and pyridazinones) disrupt the electron transfer cascade, thus inhibiting the transfer of the absorbed light energy. Compounds that inhibit photosynthesis at site B of PS II act by a similar mechanism but at a different location (urea derivatives and other photosynthesis inhibitors). Compounds that inhibit photosynthesis at site C of PS I (bipyridylium, see above) inhibit electron transfer in the pre-$NADP^+$ reduction phase.

During the light reaction of photosynthesis, two photosynthetic reaction centers, PS I and II, are required to reduce $NADP^+$ to NADPH with the electron captured from water. In these centers, photosynthetic pigments, mainly chlorophylls [with a porphyrin ring structure complexing the central Mg^{2+} ion and containing a hydrophobic side chain (phytol) and their protein complexes] absorb the energy of light radiation in the red ($\lambda = 430$ nm) and blue ($\lambda = 660$ nm) spectral ranges. The light-collecting pigments, bound to a protein backbone, are organized into antenna complexes. Both PS I and PS II have an internal antenna complex in the photochemical reaction center or closely connected to it; also, each photochemical system has an external antenna complex. One of the complexes of PS I is LHC I (light-harvesting chlorophylls), which is permanently bound to the photochemical system. The other, LHC II, binds to the photochemical system only when the light demand of photosynthesis requires it; thus, LHC II also functions as a photosynthesis regulator. The internal antenna of PS II consists of two chlorophyll-α-binding protein units, CP43 and CP47, and its reaction center is a heterodimer of D1 and D2 proteins. From the chlorophyll molecules, the excited electron is transferred to a cascade of electron acceptors while the oxidized chlorophyll receives an electron from the decomposition of water through an electron donor. This is done by the redox components of the light-induced electron transport system bound to reaction center D1 D2. So, chlorophyll catalyzes the transfer of electrons between a donor and an acceptor using light energy (photosensitization). This process is similar in both photosystems but by different electron donors and acceptors that are activated by light at different wavelengths.

PS II (P680) absorbs the energy of light at wavelengths of 650–680 nm, thereby oxidizing water and reducing the electron acceptor plastoquinone (PQ). Its light-collecting chlorophyll-protein complex (LHC II) is located next to the PS II reaction center, which consists of several components [D1 and D2 heterodimeric proteins, photochemically active chlorophyll (P680), regulatory part]. In the PS II center, chlorophyll transfers the electron excited by light energy to the mobile PQ terpenoid benzoquinone molecule (reducing it to plastoquinol) and triggers the photosynthetic electron transport (PET) to PS I. Upon reducing PQ, chlorophyll recovers its electron deficiency with an electron from water decomposition through a manganese-containing complex. The reduction of PQ activates a protein kinase enzyme that phosphorylates the LHC II complex. PET between the two photochemical systems is mediated by a membrane-bound cytochrome *b6f* (cyt.*b6f*) complex. In the next step, the reduced PQ migrates to the cyt.*b6f* complex and transfers its absorbed electron to it, after which the electron is further transferred from this complex to a copper-containing metalloprotein, plastocyanin (PC). The mobile PC migrates to the next photochemical system, PS I. PS I (P700) absorbs the energy of light at wavelengths of 670–700 nm, oxidizing the PC protein and reducing the cysteine-iron sulfide complex ferredoxin (Fd). Similarly to PS II, PS I is composed of several components [heterodimeric proteins, photochemically active chlorophyll (P700), and the LHC I light-collecting chlorophyll-protein complex]. From ferredoxin, the electron is transferred partly to the enzyme ferredoxin-NADP oxidoreductase, which reduces the $NADP^+$ complex to NAPH. The final step in the light reaction is photosynthetic phosphorylation, in which the enzyme ATP synthase, also acting as a proton channel through the thylakoid membrane, phosphorylates the ADP nucleotide using H^+ ions from water degradation and inorganic phosphate. In the next steps of the photosynthesis process (dark reaction, not detailed in Fig. 3.3), the fixation and reduction of carbon dioxide take place, consuming the NADPH and ATP produced in the light reaction.

Phenylcarbamates and anilides (WSSA/HRAC groups 5, 23; legacy HRAC groups C_1, C_2, K_2)

Early representatives of photosynthesis-inhibiting herbicides are anilides, which are selective pre-emergent agents (although some also show postemergence effects). Their most important members are propham† and chlorpropham† (CIPC), introduced in 1945 and 1951 (Fig. 3.4A). Upon its introduction, the latter rapidly displaced the former. Close structural analogs are arylcarbamate derivatives represented by phenmedipham and its demethylated version desmedipham†, introduced in 1967 and 1969 (Fig. 3.4A).

The mechanism of action of this family is partly the inhibition of photosynthesis (phenmedipham and desmedipham, WSSA/HRAC group 5; legacy HRAC groups C_1, C_2) and partly the inhibition of cell division (propham and chlorpropham, WSSA/HRAC group 23; legacy HRAC group K_2). Growth inhibition of plant tissues is also detectable. Thus, chlorpropham gained importance as an inhibitor of potato

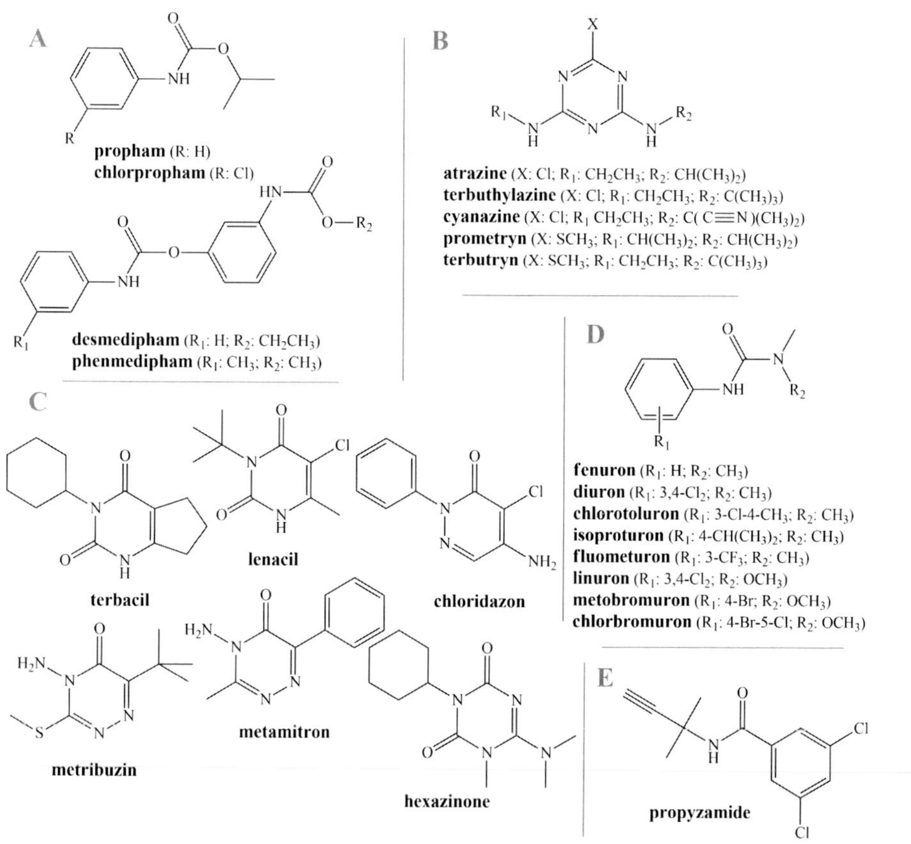

Fig. 3.4 Phenylcarbamate and -anilide (A), triazine (B), pyridazinone and triazinone (C), urea (D), and benzamide (E) herbicide active ingredients.

germination. Desmedipham and phenmedipham are unavoidable participants in sugar beet weed control. Phenylcarbamates are primarily degraded microbially in the soil. In sandy soils, their half-life is 2–5 weeks, which can be enhanced by extenders. One extender is *O*-(4-chlorophenyl)-*N*-methylcarbamate (resembling *N*-methylcarbamate insecticides in its chemical structure), which substantially prolongs the half-lives of propham and chlorpropham in soil. Propham was banned in the EU in 2003, chlorpropham was removed from the positive list in 2019, and the authorization of desmedipham was not renewed in 2019.

Triazine herbicides (WSSA/HRAC group 5; legacy HRAC group C_1)

Symmetrical triazine derivatives include a large number of compounds with high structural similarity. On the rigid and planar 1,3,5-triazine ring, they contain two substituted amino groups and an electrophilic substituent (chlorine, methoxy, or methylthio). Many agents have been developed in this group, the most important chlorine-substituted derivatives are atrazine†, terbuthylazine, and cyanazine†, introduced in 1957, 1966, and 1967. The most important methylthio derivatives are promethrin† and terbutryn†, introduced in 1962 and 1965 (Fig. 3.4B).

Along with the substitution pattern, the selectivity properties of the compounds change. Chloro-substituted derivatives are generally used on maize because maize can degrade them; methylthio derivatives have other selectivities (e.g., rice, cotton, beets) while methoxy derivatives are less selective. Due to their mechanism of action, triazines are phytotoxic to all photosynthesizing plants, yet some plants are tolerant to some triazines, providing application selectivity. Triazines bind, competing with plastoquinone, to protein D1, a stable secondary electron acceptor of PS II in the thylakoid membrane at the quinone B binding site. By stopping the flow of electrons and causing the accumulation of free radicals, they impede the PS II process (Fig. 3.3), preventing the transfer of the excited electron either through the plastoquinone cascade or to the PS I center. Another atrazine binding site, protein D2, has also been identified, but the affinity to this site was much lower than that to protein D1.

Members of the group are degraded mainly microbially, but at low pH values hydrolysis becomes the major route. One of the strongest limiting factors of application is their environmental persistence: the half-life of atrazine in subsurface water can be as high as 3–12 months, and these agents remain in treated soils for months and even beyond vegetation periods. Ironically, this was considered an advantage (lasting efficacy) in the 1970s. Cyanazine is the least persistent in soil (half-life of 14 days) while the half-life of other members is 60 days (atrazine, prometryn). Persistence is even more severe under dry conditions, lower temperatures, and higher pH, and in soils with low organic matter and clay content. Under anaerobic conditions, decomposition stops. As the compounds reach the anaerobic soil zone, their metabolism stops, and the persistent residue level leaches into groundwater; atrazine was the most significant groundwater contaminant in the 1990s. As a result, the EU withdrew prometryn, cyanazine, and terbutryn with "essential use" derogations in 2002, and atrazine in 2004.

A GM canola tolerant to atrazine was devised by the introduction of an insensitive target, a mutated D1 protein. It is encoded by a chloroplastic gene originated from naturally emerged triazine-resistant weeds.

Pyridazinones and triazinones (WSSA/HRAC group 5; legacy HRAC group C_1)

Like symmetrical triazines, pyridazinone and triazinone derivatives also have *N*-heterocyclic elements. The former contain two oxo groups in a uracil-type (pyrimidine-2,4(1*H*,3*H*)-dione) cyclic carbonylurea structure. The two most important members are terbacil† and lenacil, introduced in 1962 and 1964 (Fig. 3.4C). Regarding the relationship between their chemical structure and biological effect, the presence of a carbonyl group (carboxamide) is required to elicit the effect; nitrogen atom 1 of the 1,3-pyridazinone ring cannot be substituted. Among derivatives, alkyl substituted on nitrogen atom 3 of the ring branched alkyl groups enhances the biological effect. Chloridazone†, introduced in 1962, is not a uracil but a 1,2-pyridazinone

(Fig. 3.4C). Asymmetric triazine derivatives are metribuzin and metamitron, introduced in 1968 and 1975, respectively (Fig. 3.4C). Hexazinone†, introduced in 1964, is a somewhat similar symmetric triazinedione (Fig. 3.4C).

The mechanism is the same as that of the symmetrical triazines (PS II inhibitors), but some also interfere with carotenoid biosynthesis (similarly to aminotriazoles). Uracil derivatives are similarly degraded by living organisms: microsomal monooxygenase enzymes hydroxylate the molecules at various sites. The primary route of degradation in soil is microbial. Their half-life in soil is 1–3 months. Hexazinone was withdrawn from the EU with essential use derogations in 2002. Terbacil was not included in the positive list in the EU and had to be withdrawn by member states by 2003. Chloridazone, used primarily on the sugar beet, was phased out in 2018.

Ureas (WSSA/HRAC group 5; legacy HRAC group C_2)

Typical members are aromatic urea derivatives that contain a phenyl ring (possibly substituted with halogens and alkyl or haloalkyl groups) on one of the urea nitrogen atoms and two short alkyl or alkoxy groups on the other. One of these alkyl groups has to be a methyl, and members of the family are usually categorized by the other short alkyl group, that is, divided into the groups of dimethylureas fenuron†, diuron†, chlorotoluron, isoproturon†, and fluometuron, introduced between 1951 and 1971, and methoxyureas linuron†, metobromuron, and chlorbromuron†, introduced between 1962 and 1967 (Fig. 3.4D).

These agents inhibit photosynthesis at the B site of the second step of the light reaction (PS II) (Fig. 3.3) by disrupting the PET process at the PQ cascade. It should be noted that these compounds at higher concentrations also inhibit oxidative phosphorylation in plants, thus their phytotoxic effect is exerted on the nonphotosynthetic parts of the plant. Phenylureas are mostly selective pre-emergent agents, well absorbed into the soil and from there to the roots. They are resistant to chemical and environmental influences and therefore degrade slowly.

The chemical stability and strong soil adsorption capacity of ureas are significant (an exception is fenuron, which has relatively high water solubility and is only weakly adsorbed on soil particles; as a result, it readily leaches from soil). The primary route of decomposition is microbial, mainly dealkylation, hydroxylation on the aromatic ring, and subsequent hydrolysis. Their half-life in soil varies between 1 and 3 months. The European Food Safety Authority (EFSA) recommended in 2015 that isoproturon be classified as toxic to reproduction with potential endocrine-mediated effects on fertility (mild "gender-bending" antiestrogenic and antiandrogenic properties); it was prohibited by moratorium in the EU in 2016. Fenuron and chlorbromuron were withdrawn from the EU with essential use derogations in 2002. Approval for linuron was not renewed in 2017 because it was classified as toxic for reproduction and a suspected carcinogen to humans. Diuron has been assessed as very toxic to aquatic life with long-lasting effects; it is also suspected of causing cancer, and was banned in 2018–19.

Cell growth inhibitors

Benzamide derivatives (WSSA/HRAC group 3; legacy HRAC group K_1)

An uncommon benzamide containing a dimethylpropargyl group is propyzamide (also known as pronamide) (Fig. 3.4F), introduced in 1969. It is a selective, systemic, pre- and postemergent agent for controlling monocotyledonous and dicotyledonous weeds, entering plants by being absorbed from the soil by the root. The mechanism is the same as that of dinitroanilines: it inhibits cell division by inhibiting the assembly of mitotic spindle-forming tubulin fibers into microtubules.

Propyzamide has been a leading agent in lettuce for decades but has been criticized for its toxic effects on endangered amphibians. Degradation in soil occurs by aerobic metabolism and in water by photolysis. Its half-life in soil is 30 days, but it is moderately persistent under anaerobic conditions.

Basipetally translocated herbicides

Lipid synthesis inhibitors

The use of herbicides with specific hormonal effects (such as phenoxyacetic acids, see below) has led to ecological problems. These agents mainly control dicotyledons while other weeds (mainly perennial monocotyledons, especially those capable of vegetative reproduction), undisturbed by their natural competitors, can spread in the cultivation area. Monocotyledonous weeds were commonly controlled by preemergent treatment with quaternary ammonium compounds (paraquat) or glyphosate. However, the use of these total herbicides posed ecotoxicity problems due to their propensity to contaminate surface and subsurface water, which was resolved partly by the discovery of novel postemergence agents. These are aryloxyphenoxypropionates (e.g., diclofop-methyl) and cyclohexanediones (e.g., sethoxydim, alloxydim), which act by inhibiting lipid biosynthesis. The main mechanism of action is blocking the activity of the enzyme ACCase, thereby interfering with fatty acid biosynthesis. In green plant parts, fatty acid biosynthesis takes place in the chloroplast membranes. Not only the structure but also the composition of the thylakoid membranes differs significantly from those of other boundaries in plant and animal cells. Their main phospholipid is phosphatidylglycerol, and they are rich in galactosyldiacyl triglycerides. ACCase, composed of seven subunits and an acid transport protein, binds to this membrane.

Aryloxyphenoxypropionates (WSSA/HRAC group 1; legacy HRAC group A)

The main members are diclofop-methyl, introduced in 1975 (and its isolated *R* stereoisomer diclofop-P-methyl in 2005); fluazifop-butyl, introduced in 1980 (and its *R* stereoisomer fluazifop-P-butyl in 1981); quizalofop, introduced in 1983 (and its *R* stereoisomer quizalofop-P, ethyl and tefuryl esters quizalofop-P-ethyl in 1989, and quizalofop-P-tefuryl in 1989); haloxyfop, introduced in 1985 (and its *R* stereoisomer haloxyfop-P and methyl ester haloxyfop-P-methyl in 1992); propaquizafop, introduced in 1987 (as its *R* stereoisomer); and fenoxaprop, introduced in 1988 (and its *R* isomer fenoxaprop-P in 1989 and ethyl ester fenoxaprop-P-ethyl in 1990) (Fig. 3.5A). Most aryloxyphenoxypropionates are commercialized as esters (to achieve better foliar uptake), although the corresponding carboxylic acids are believed to be the true AIs.

The activity of the biotin-dependent carboxylase enzyme ACCase, catalyzing the transcarboxylation of malonyl-CoA from acetyl-CoA in the chloroplast as the first step of fatty acid biosynthesis in plants, is particularly sensitive to the effects of aryloxyphenoxypropionates. The composition of fatty acids produced in the thylakoid membrane—from acetate and malonate parent compounds—changes significantly upon exposure to fluazifop even at low concentrations, resulting in corrupted membrane functions. The herbicidal effect is principally due to the *R* stereoisomers. What molecular differences (presumably in the structure of the enzyme or transporter protein) may cause the outstanding selectivity of these compounds (primarily for monocotyledons) has not yet been elucidated, as the exact structure of the chloroplastic ACCase in monocotyledons has not been revealed by X-ray crystallography; therefore, the structure of a yeast ACCase has been used as a surrogate. The AIs have been proven (by the examples of diclofop and haloxyfop) to cause significant conformational changes in the active site of the carboxylate transferase domain of ACCase upon binding. The compounds are readily absorbed in plants, but translocation is only partial. They degrade rapidly microbially. Anaerobic conditions can slow degradation. Degradation of quizalofop is slow (half-life of ~60 days). They are less mobile in soil.

Cyclohexanediones (WSSA/HRAC group 1; legacy HRAC group A)

These are cyclohexane-1,3-dione derivatives: alloxydim†, sethoxydim†, cycloxydim, and clethodim, introduced in 1976, 1983, 1985, and 1987 (Fig. 3.5B). The main mechanism of the phytotoxic effect is interference with the biosynthesis of fatty acids in chloroplasts by inhibiting ACCase. They degrade microbially and are generally not persistent in soil. An HT GM crop tolerant to sethoxydim was achieved using a transgene encoding a mutated ACCase.

Benzofurans (WSSA/HRAC group 15; legacy HRAC group K_3)

The dimethyldihydrobenzofuran derivative ethofumesate, developed in the late 1960s and introduced in 1969, has an ethoxy group on the furan ring and a methylsulfonyl group on the fused benzene ring (Fig. 3.5C). Similarly to oxyacetamides and chloroacetamides (see below), it inhibits the biosynthesis of VLCFAs, interacting with the acyl-CoA elongase enzyme (essential for fatty acid elongation) by carbamoylmethylating it on the thiol group of the cysteine moiety at the enzymatic active site.

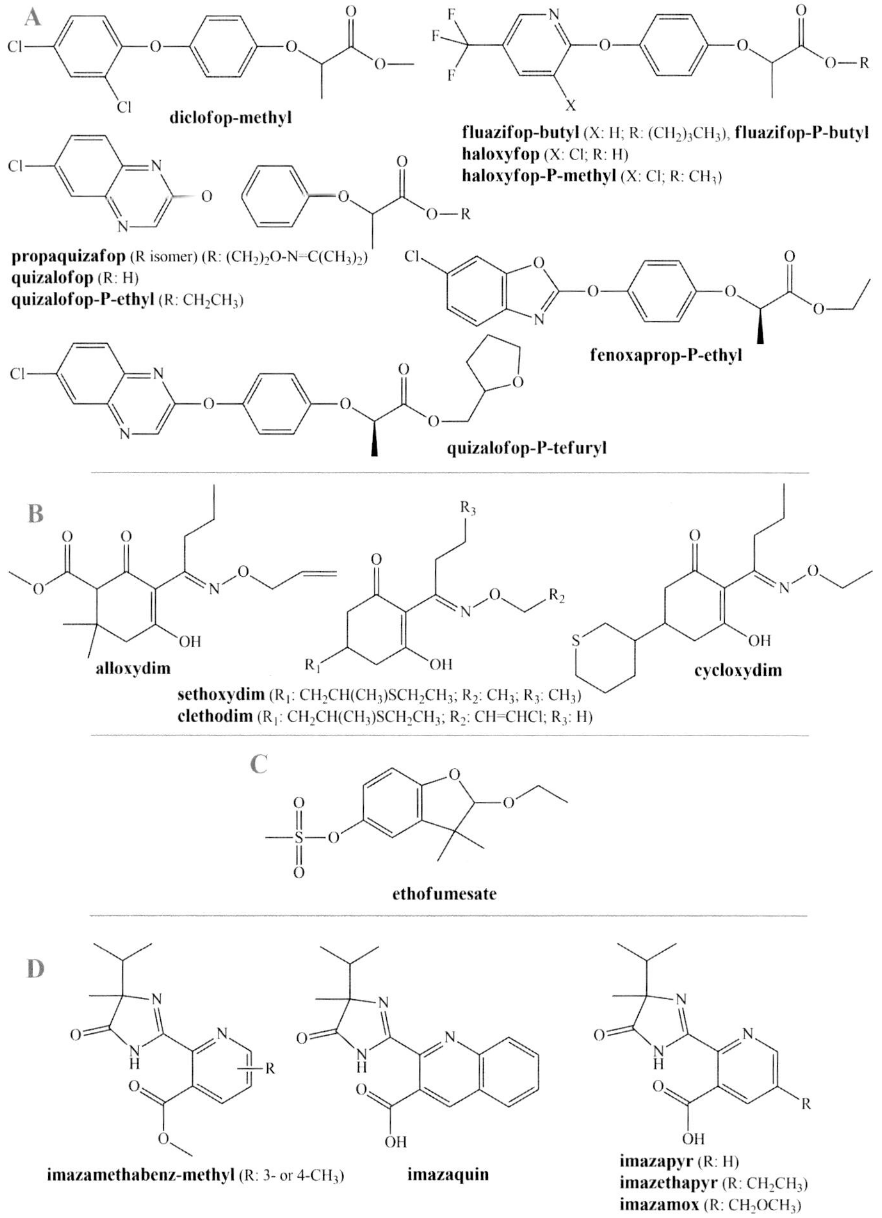

A

diclofop-methyl

fluazifop-butyl (X: H; R: $(CH_2)_3CH_3$), **fluazifop-P-butyl**
haloxyfop (X: Cl; R: H)
haloxyfop-P-methyl (X: Cl; R: CH_3)

propaquizafop (R isomer) (R: $(CH_2)_2O\text{-}N{=}C(CH_3)_2$)
quizalofop (R: H)
quizalofop-P-ethyl (R: CH_2CH_3)

fenoxaprop-P-ethyl

quizalofop-P-tefuryl

B

alloxydim

sethoxydim (R_1: $CH_2CH(CH_3)SCH_2CH_3$; R_2: CH_3; R_3: CH_3)
clethodim (R_1: $CH_2CH(CH_3)SCH_2CH_3$; R_2: $CH{=}CHCl$; R_3: H)

cycloxydim

C

ethofumesate

D

imazamethabenz-methyl (R: 3- or 4-CH_3)

imazaquin

imazapyr (R: H)
imazethapyr (R: CH_2CH_3)
imazamox (R: CH_2OCH_3)

Fig. 3.5 Aryloxyphenoxypropionate (A), cyclohexanedione (B), benzofuran (C), and imidazolinone (D) herbicide active ingredients.

Ethofumesate is metabolized in plants and animals by oxidative hydrolysis of its 2-ethoxy group to the corresponding 2-oxo-lactone as the major metabolite. Although it binds strongly to soil particles, it is not persistent and, due to its rapid degradation, does not leach into groundwater.

Inhibitors of branched-chain amino acid biosynthesis

These compounds impede the synthesis of branched-chain amino acids (leucine, isoleucine, valine). They block the first biosynthetic step by inhibiting the enzyme acetolactate synthetase (ALS), catalyzing it. Biosynthesis starts from α-ketoacids: pyruvate for valine and leucine, and α-ketobutyrate (which is formed from threonine) for isoleucine.

Imidazolinones (WSSA/HRAC group 2; legacy HRAC group B)

These have a typical heterocyclic ring system, derivatives of 1,3-imidazolin-5-one. It seems to be a structural requirement for biological activity to have an isopropyl and a methyl substituent in the 4 position of the imidazoline ring, and to have a substituted toluic acid, picolinecarboxylic acid, or quinolinecarboxylic acid moiety in the 2 position, which is ortho to the imidazole group relative to the carboxyl group of the aromatic ring. The first compound in the family was imazamethabenz-methyl[†] (imazethabenz), introduced in 1982; two subsequent members are imazaquin[†] and imazapyr[†] (the former a quinolinecarboxylic acid, the latter a picolinecarboxylic acid), both introduced in 1983 (Fig. 3.5D). Later members are all alkylated derivatives of imazapyr at the 5-position of the pyridine ring, with an ethyl group for imazethapyr[†] and a methoxymethyl group for imazamox, introduced in 1984 and 1995 (Fig. 3.5D).

These compounds inhibit the ALS enzyme. They degrade primarily microbially, with minimal anaerobic degradation. They are strongly bound to soil organic matter. The half-life in soil is 1–3 months.

Imazapyr was withdrawn from the EU with essential use derogations in 2002, imazethapyr was withdrawn in 2004, imazamethabenz-methyl was banned in 2005, and imazaquin was discontinued in 2011. Thus, the only surviving member of the group is imazamox. It is of great importance in sunflower weed control as resistant non-GM sunflower varieties have been bred and can be kept free of weeds with imazamox without major damage to the crop. Large companies sell seed and the herbicide together. France does not allow this technology. HT GM crops against imidazolinones have also been developed using the insensitive target approach, a transgene encoding a mutant ALS enzyme.

Sulfonylureas (WSSA/HRAC group 2; legacy HRAC group B)

These are structurally closely similar compounds, urea derivatives having a heterocyclic moiety on one of the nitrogen atoms and an arylsulfonyl derivative moiety on the other. The best known and earliest marketed are chlorsulfuron[†], bensulfuron-methyl, and metsulfuron-methyl, which already show the most significant structural elements (Fig. 3.6A). The heterocyclic substituent group can be a 4-methoxy-6-methyl-1,3,5-triazin-2-yl (e.g., chlorsulfuron) or a 4,6-dimethoxy-1,3-pyrimidin-2-yl (e.g., bensulfuron methyl), and the benzenesulfonyl moiety may contain an electrophilic substituent in the ortho position, such as a chlorine atom (chlorsulfuron), a methoxycarbonyl (metsulfuron-methyl, bensulfuron-methyl,

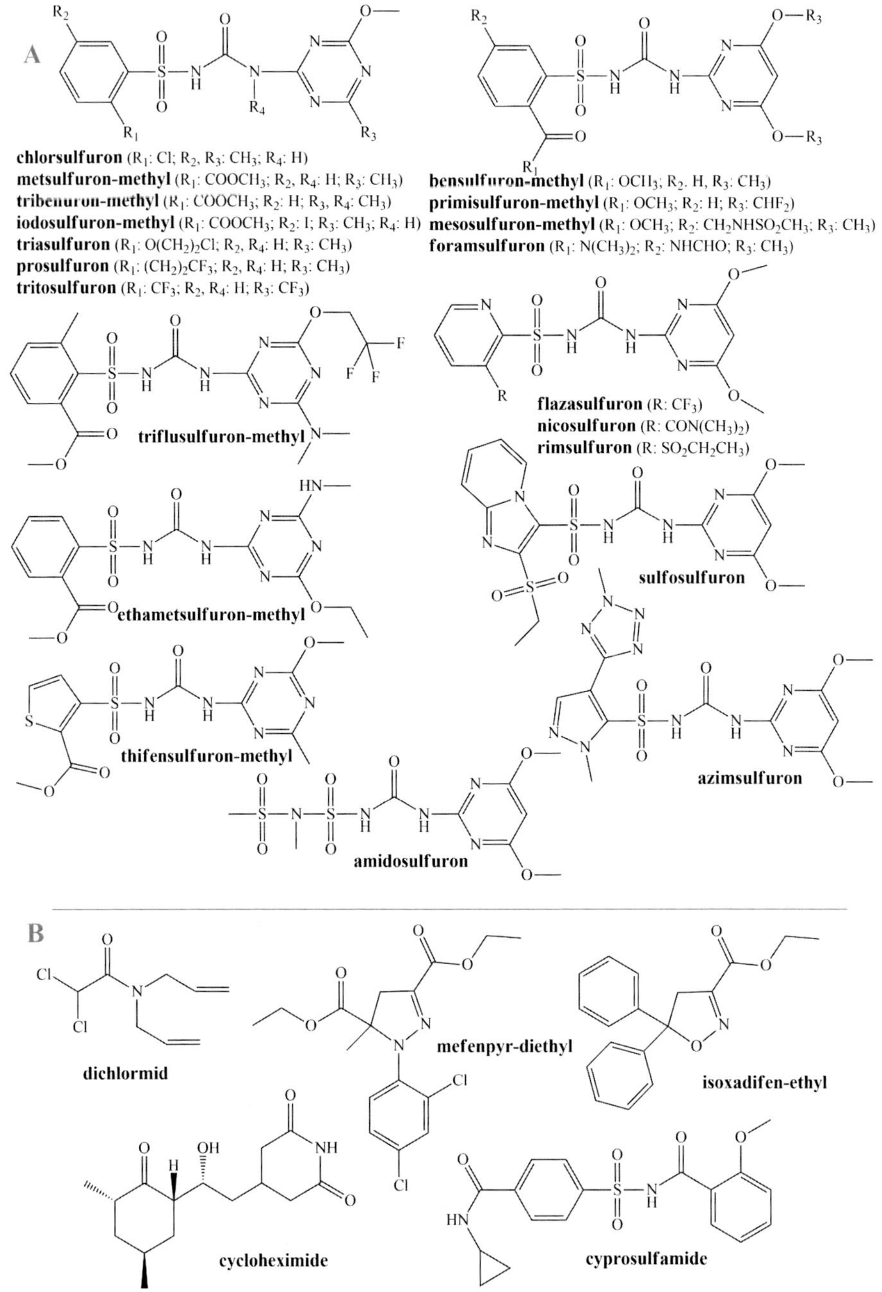

A

chlorsulfuron (R_1: Cl; R_2, R_3: CH_3; R_4: H)
metsulfuron-methyl (R_1: $COOCH_3$; R_2, R_4: H; R_3: CH_3)
tribenuron-methyl (R_1: $COOCH_3$; R_2: H; R_3, R_4: CH_3)
iodosulfuron-methyl (R_1: $COOCH_3$; R_2: I; R_3: CH_3; R_4: H)
triasulfuron (R_1: $O(CH_2)_2Cl$; R_2, R_4: H; R_3: CH_3)
prosulfuron (R_1: $(CH_2)_2CF_3$; R_2, R_4: H; R_3: CH_3)
tritosulfuron (R_1: CF_3; R_2, R_4: H; R_3: CF_3)

bensulfuron-methyl (R_1: OCH_3; R_2: H, R_3: CH_3)
primisulfuron-methyl (R_1: OCH_3; R_2: H; R_3: CHF_2)
mesosulfuron-methyl (R_1: OCH_3; R_2: $CH_2NHSO_2CH_3$; R_3: CH_3)
foramsulfuron (R_1: $N(CH_3)_2$; R_2: NHCHO; R_3: CH_3)

triflusulfuron-methyl

flazasulfuron (R: CF_3)
nicosulfuron (R: $CON(CH_3)_2$)
rimsulfuron (R: $SO_2CH_2CH_3$)

ethametsulfuron-methyl

sulfosulfuron

thifensulfuron-methyl

azimsulfuron

amidosulfuron

B

dichlormid

mefenpyr-diethyl

isoxadifen-ethyl

cycloheximide

cyprosulfamide

Fig. 3.6 Sulfonylurea (A) and antidote (B) herbicide active ingredients.

tribenuron-methyl, iodosulfuron-methyl, primisulfuron-methyl†, triflusulfuron-methyl, ethametsulfuron-methyl†), a chloroethoxy (triasulfuron†), a trifluoropropyl (prosulfuron), or the unusual dimethylaminocarbonyl and formylamino groups (foramsulfuron). The sulfonyl moiety may also be a pyridinesulfonyl group containing a trifluoromethyl (flazasulfuron), a dimethylaminocarbonyl (nicosulfuron), or an ethoxysulfinyl group (rimsulfuron) in the ortho position, or it may be another heterocyclic sulfonic acid such as a thiophenesulfonyl (thifensulfuron-methyl) or an imidazopyridinesulfonyl group (sulfosulfuron). Optionally, an aromatic sulfonyl moiety may not even be required, but an *N*-methyl-*N*-methanesulfonamidyl group (amidosulfuron) may be included in the structure. An atypical representative within the family of compounds is tribenuron methyl, which contains a methyl substituent on the nitrogen atom of the urea toward the heterocyclic part (triazine).

Sulfonylureas, although structurally reminiscent of heterocyclic ureas, do not block photosynthesis, but inhibit ALS. They are called microherbicides because they are effective in unusually low doses—10–100 g/ha.

The structure is such that relatively small modifications may cause significant differences in stability and solubility, resulting in a wide range of soil persistence durations. Members degrade primarily by hydrolysis and microbial pathways. At low soil pH, they bind more strongly to soil particles and organic matter, and degradation is slowed. At high pH, when acid hydrolysis stops, persistence even between vegetation periods may develop. Hydrolysis is fastest at pH <6.8 and accelerates with increasing soil temperature. If the pH changes within an area, persistence may appear at high pH spots. Their half-life is usually short, 2–12 days for triflusulfuron, tribenuron, prosulfuron, and thifensulfuron-methyl, but may be longer, 21–40 days for nicosulfuron, primisulfuron-methyl, and chlorsulfuron.

Several members of this family of compounds have been withdrawn in the EU registration. Primisulfuron-methyl was discontinued in 2004, chlorsulfuron in 2009, triasulfuron in 2016, and ethametsulfuron-methyl was to be withdrawn in 2020.

Phytotoxic effects on certain plants can be reduced by antidotes (Fig. 3.6B). For example, dichlormid# (*N*,*N*-diallyl-2,2-dichloroacetamide) protects maize from the effects of chlorsulfuron by inducing glutathione to enhance the rate of degradation of the compound. The same effect is obtained with the pyrazoline derivative mefenpyr-diethyl and the isoxazolecarboxylic acid isoxadifen-ethyl, but not with the cycloheximide protein synthesis inhibitor or the recently developed sulfonamide safener cyprosulfamide. An approach in agricultural biotechnology is to genetically modify crops (particularly maize) with a gene encoding an ALS enzyme not susceptible to the inhibitory action of a given sulfonylurea.

Triazolopyrimidines (WSSA/HRAC group 2; legacy HRAC group B)

This group was developed in a bioisosteric agrophore design approach from sulfonylureas and then optimized in structure-activity relationship studies. They are sulfonylanilides, mostly of 1,3,4-triazolopyrimidine-2-sulfonic acid with 2,6-disubstituted anilines; 1,3,4-triazolopyrimidin-2-amine with 2,6-disubstituted benzenesulfonic acid; or 3-pyridinesulfonic acid. Four main members authorized in the EU are the triazolopyrimidinesulfonyl derivatives flumetsulam[†] and florasulam, introduced in 1994 and 1999, and the triazolopyrimidinamine derivatives penoxsulam and pyroxsulam, introduced in 2004 and 2008 (Fig. 3.7A) although the authorization of flumetsulam was recently revoked.

The mechanisms and symptoms are the same as those of imidazolines and sulfonylureas. They inhibit the biosynthesis of branched-chain amino acids in plants by inhibiting ALS or the hydroxyacetic acid synthase (AHAS) enzyme, their primary sites of action being chloroplasts and meristems. The first representative, flumetsulam, was developed for weed control in maize and soybeans, and later florasulam was developed to control dicotyledonous weeds in maize. The selectivity of florasulam in maize is due to differences in metabolism; the hydrolysis of the compound is rapid in maize. Flumetsulam shows a similar, metabolism-based selectivity in soy. The transfer of the triazolopyrimidine moiety to the sulfonamide nitrogen in the two lastly developed compounds significantly modified their selectivity: they are effective against mono- and dicotyledonous weeds, penoxsulam in rice and pyroxsulam in wheat; and the chemical modification relative to flumetsulam and florasulam had a beneficial effect on mammalian toxicity as well.

Degradation is primarily microbial. Phytotoxicity and rate of degradation increase at higher soil pH, as the compounds have lower binding capacity to soil particles under alkaline conditions, rendering them more readily available for both absorption into plants and degradation by soil microorganisms. Any factor that increases microbial activity in the soil accelerates degradation. At low soil pH, however, decomposition is slowed to such an extent that persistence between vegetation periods may develop, which may hinder subsequent planting in the case of sensitive crops. The half-life of flumetsulam is 1–3 months (shorter at high pH values).

Sulfonylaminocarbonyltriazolinones (WSSA/HRAC group 2; legacy HRAC group B)

The development of this group goes a long way back in time and in organic synthetic chemistry. In the 1970s, bicyclic derivatives containing azepam and triazolinone rings on a common N-pillar atom were synthesized in research on chemical applications of ε-caprolactam, an intermediate in nylon production. The new N-heterocycles were first attempted in the development of fungicides, and then were subjected to various N-acylation, sulfonation, and arylation reactions. This yielded, inter alia, sulfonylaminocarbonyltriazolinone, which was not effective as a fungicide, but inhibited ALS. Ironically, the starting azepam ring with which application development had initially been launched was eventually removed during herbicidal activity optimization, and systematic agrophore development led to flucarbazone-sodium[†] and propoxycarbazone-sodium, both introduced in 1999, followed by thiencarbazone-methyl introduced 10 years later (Fig. 3.7B).

Like sulfonylureas, these compounds inhibit ALS or AHAS enzymes. When absorbed through leaves or roots, they translocate into the plant, but thiencarbazone-methyl also exerts a contact and lasting effect. It is also used in combination with isoxaflutole (see Isoxazoles). Flucarbazone-sodium is mainly used in Canada, propoxycarbazone-sodium is widely used in the United States and Europe,

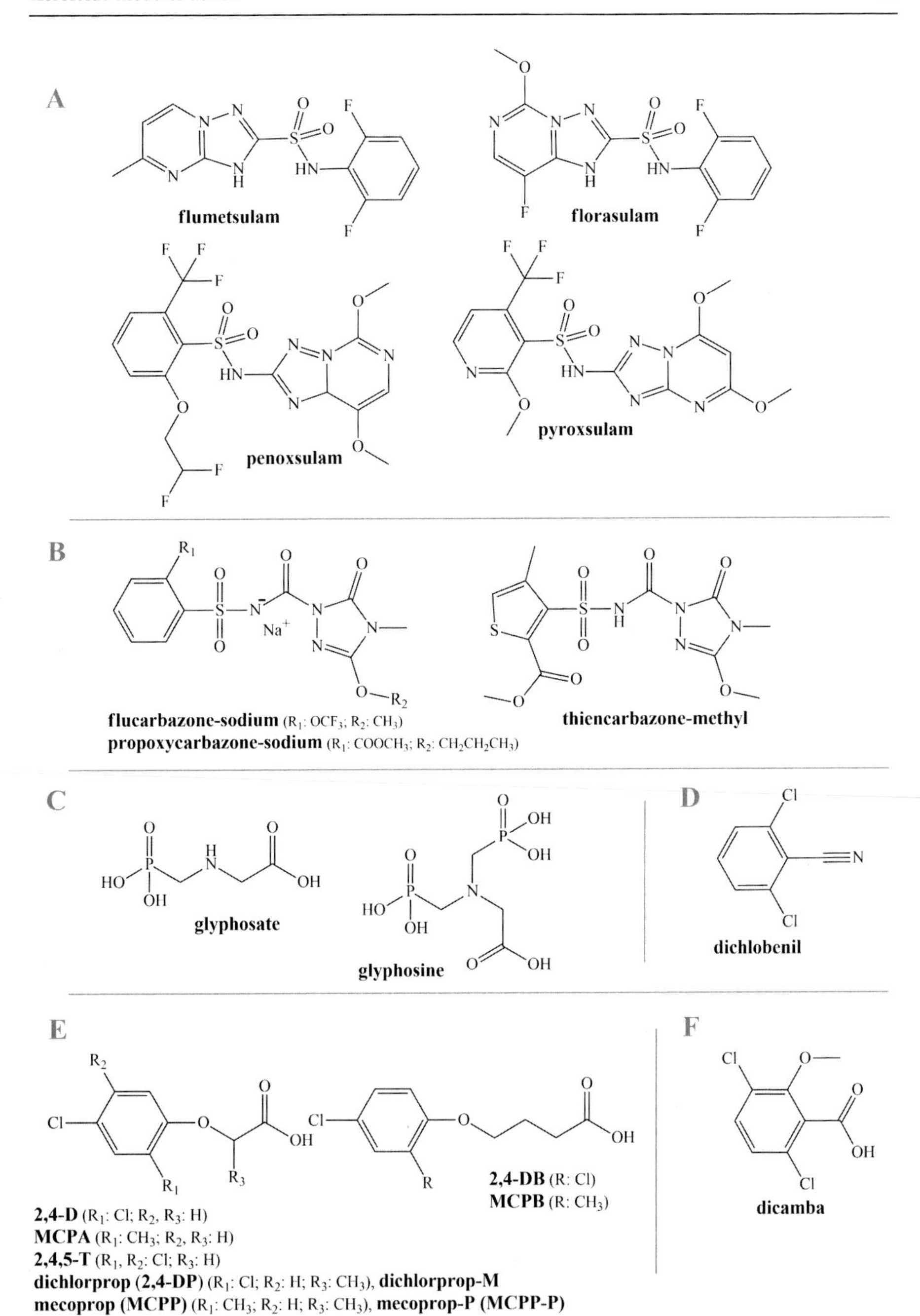

Fig. 3.7 Triazolopyrimidine (A), sulfonylaminocarbonyltriazolinone (B), amino acid (C), aromatic nitrile (D), phenoxyacetic and -butyric acid (E), and benzoic acid (F) herbicide active ingredients.

and thiencarbazone-methyl is used worldwide. There is considerable interest in sulfonylaminocarbonyltriazolinones, primarily based on the biological profile. While the best-known ALS inhibitors in the early 1990s, such as sulfonylureas, were useful against dicotyledons, this family is effective against monocotyledonous weeds as well and can be used in cereals due to their selectivity or antidotability. Thiencarbazone-methyl particularly may be phototoxic to the crop, but that can be suppressed with another sulfonylaminocarbonyl, cyprosulfamide (Fig. 3.6B), as an antidote in maize. Other safeners such as the 2-isoxazoline derivative isoxadifen-ethyl and the 2-pyrazoline derivative mefenpyr-diethyl (Fig. 3.6B), with no herbicidal activity of their own, can antidote sulfonylaminocarbonyltriazolinones in maize and in wheat.

Inhibitors of aromatic amino acid biosynthesis

Amino acid derivatives (glycines) (WSSA/HRAC group 9; legacy HRAC group G)

Practically the only member of the group is glyphosate, introduced in 1971 (Fig. 3.7C). It is one of the most translocated herbicides in plants and is the most widely used agrochemical in the world. It is a phosphonated derivative of the simplest amino acid, glycine, marketed in various formulations as highly water-soluble ammonium, isopropylammonium, sodium, and trimethylsulfonium (trimesium) salts. An interesting aspect is that the developer also patented its bisphosphonomethylated derivative glyphosine[†] as a plant growth regulator because it enhances carbohydrate production in the sugar beet.

Glyphosate inhibits the biosynthesis of the essential aromatic amino acids phenylalanine, tryptophan, and tyrosine, and thereby inhibits nucleic acid and protein synthesis. The central intermediate in the biosynthesis of these aromatic amino acids is 5-enolpyruvyl-shikimic acid-3-phosphate (EPSP), the formation of which is catalyzed by the chloroplast enzyme EPSP synthase. Glyphosate interferes with the function of this enzyme as a transition state analog inhibitor of the enzyme-catalyzed process. Due to its general biochemical mechanism of action, glyphosate is a total and systemic foliar and soil herbicide.

The low apparent residue level used to be mentioned as an environmental advantage but actually reflected an analytical difficulty, as the parent compound and its metabolites, with their water solubility, were difficult to handle in sample preparation and traditional instrumental analysis (gas chromatography). Residue levels appeared to be low and/or infrequent because they were undetected. With advanced analytical techniques, glyphosate and its main metabolite aminomethylphosphonic acid (AMPA) are now found to be ubiquitous environmental and frequent food and feed contaminants. Glyphosate decomposes slowly in plants to its main degradation product, AMPA. It binds to soil particles as it forms a strong complex with most metal ions (aluminum, iron, zinc, manganese), and complexation increases its environmental stability. It decomposes slowly in sterile soils, the degree of microbial degradation being strongly dependent on the soil type and the nature of the microorganisms. Due to strong soil adsorption, long-term degradation processes do not play a role in its bioavailability. Because of its water solubility, it appears rapidly in groundwater and surface waters. Its half-life in soil is 47 days.

Immediately upon glyphosate's introduction, its sales rose rapidly, and it became one of the top three herbicides by market volume. Market position was further strengthened after the 1990s by the introduction of glyphosate-tolerant GM ("Roundup Ready") crop varieties (soybean, maize, cotton, and many others) by the insertion of transgenes encoding either EPSP synthase enzymes not susceptible to glyphosate or metabolic enzymes glyphosate oxidoreductase and glyphosate acetyltransferase to enable its rapid degradation. Glyphosate is the leading pesticide worldwide. Its consumption exceeds 800,000 tons per year, and it is used not only in agriculture but also by industry and households. Such immense application volume resulted in the broad occurrence of the compound and its residues in environmental matrices and food, making extended exposures possible. Thus, glyphosate became an ubiquitous surface water contaminant, and its peak levels correlated with the geographical distribution of glyphosate-tolerant GM crop cultivation. The cytotoxicity of glyphosate has been demonstrated on various cell lines, the inhibition of aromatases has been shown at very low concentrations, and the estrogenic potential of glyphosate (and its formulated products) has been indicated in estrogen receptor activation tests, implying hormonal disrupting effects. Formulated glyphosate products—especially with polyethoxylated tallowamine (POEA)—have been shown to cause stronger cytotoxic and endocrine disrupting effects than the AI alone, an indication that glyphosate is less toxic than its common coformulant.

As a result of the above concerns, reregistration of glyphosate has been problematic in the EU. Its periodic reregistration due in 2012 was postponed, along with 38 other pesticide AIs, to 2015, and then resulted in a partial, 5-year reregistration period only. The major cause of the assessment debate has been conflicting views by different regulatory agencies. While the EFSA concluded in 2015 that glyphosate is "unlikely to pose a carcinogenic hazard to humans," the International Agency for Research on Cancer (IARC) classified it as "probably carcinogenic to humans" in the same year. The subsequent assessment by the European Chemicals Agency (ECHA) in 2017 corroborated the opinion of EFSA. A possible cause of the dissent is that the genotoxicity of the formulated products was in some cases erroneously attributed entirely to glyphosate, not considering the fact that the coformulant POEA exerted several orders higher DNA-damaging and genotoxic effects in various in vitro bioassays. This anomaly came to an end in 2016 when POEA was banned from use in glyphosate-based herbicides in EU member states. There still exists, however, uncertainty about product names: Roundup products that used to be formulated with POEA are still commercialized under the same name but do not now contain POEA. Moreover, a Roundup variety not containing glyphosate as its AI is also marketed, which can further confuse consumers. Another source of diverging opinions among regulatory agencies is that the EFSA and ECHA based their opinions on the data submitted by the applicant for the reregistration while IARC analyzed data in peer-reviewed scientific publications. The debate on glyphosate among regulatory agencies mostly focused on carcinogenicity, but a variety of other effects (e.g., endocrine disruption, modification of cell adhesion through interaction with integrin receptors, possible induction of non-alcoholic fatty liver disease) have also been found. Currently, an assessment group of EU member states (France, Hungary, the Netherlands, and Sweden) evaluates the renewal application for the next authorization period after 2022. The Austrian government passed a decision for a total ban on the use of glyphosate in agriculture and horticulture, but this notion was rebuffed by the European Commission, as such a national ban on an approved AI is incompatible with applicable EU law.

Inhibitors of cellulose biosynthesis

Aromatic nitriles (WSSA/HRAC group 29; legacy HRAC group L)

A member of substantial importance has been dichlobenil† (2,6-dichlorobenzonitrile) (Fig. 3.7D), introduced in 1960. Its chemical structure is partly reminiscent of the photosynthesis inhibitors bromoxynil and ioxynil (see above), yet it is a specific inhibitor of cellulose biosynthesis (cell wall construction).

Dichlobenil is strongly bound to the lignin and humus content of soil and is therefore immobile and persistent where organic matter content is high. As a potential groundwater pollutant, its authorization has been terminated. Also, it was included in the Rotterdam Convention on the export and import of hazardous substances, which can only be exported to countries of the Convention with the prior informed consent of the importing party.

Modifiers of hormonal activity

The most important growth hormone in plants is auxin (indolylacetic acid), which promotes cell elongation. Synthetic auxin derivatives produce the same effect. As a result, cell division is modified, phosphate and nucleic acid degradation is activated, and rapid RNA production is initiated, which results in the rate of growth exceeding the rate of food intake, causing disproportionate shoot growth, loose plant structure, and collapsed water balance.

Phenoxyacetic and -butyric acids (WSSA/HRAC group 4; legacy HRAC group O)

The first phenoxyacetic acid was 2,4-D, developed in the early 1940s. It was followed by several other derivatives, the common structural element being a chloro-substituted phenoxy group usually attached to the β-carbon of an aliphatic carboxylic acid (acetic acid, propionic acid, butyric acid) (Fig. 3.7E). These include 2,4,5-T† and 2,4-DB, as well as MCPA and MCPB. 2,4-DB (2,4-dichlorophenoxybutyric acid) and MCPB are butyric acid analogs of 2,4-D and MCPA; they are metabolically degraded in the plant by β-oxidation to the corresponding acetic acid derivatives and therefore are precursors (proherbicides) of 2,4-D and MCPA. In addition to phenoxyacetic acids, 2-phenoxypropionic acids such as dichlorprop† (2,4-DP) and mecoprop† (MCPP) are included in the group. Dichlorprop is marketed as the free acid (or its potassium and ammonium salts) as well as its methyl and other (isooctyl and butothyl) esters; mecoprop is marketed only as free acid. Phenoxypropionic acids contain a chirality center, so they are marketed as racemates or as their optically pure *R* isomers (mecoprop-P, dichlorprop-P). Legumes (peas, lentils) are unable to β-oxidize phenoxybutyric acids, resulting in crop selectivity.

In contrast to natural auxin, 2,4-D and MCPA are persistent, as their hormonal effect is essentially concentration-independent above their concentration limit, whereas natural auxin produces strong deforming effects (similar to those induced by 2,4-D) only at very high concentrations. This may be due to the fact that natural auxin exerts its hormonal effect by interacting with the plasma membrane of the cells. From the membrane, a secondary messenger enters the nucleus from the cytoplasm and induces corresponding modifications of protein biosynthesis. Persistent auxins also elicit the above membrane effect; however, the induced transcription abnormally persists. Based on the results of structure-activity relationship studies, the auxin receptor on the cell membrane is thought to contain three recognition regions, interacting with the ionic carboxyl group, with the aromatic ring, and with the carbon atom in the α-position relative to the ring. This also explains why only one of the stereoisomers (*R*) of compounds with an optical chirality center on the α-carbon atom has significant phytotoxicity. In addition, certain compounds show selectivity due to differences in their uptake in monocotyledonous and dicotyledonous plants.

Phenoxyacetic acids decompose microbially, accelerated at higher temperature, humidity, pH, and organic matter. They are in principle mobile in soil; when applied to the soil, MCPA or 2,4-D is rapidly leached or microbially degraded. Degradation by soil microbes is more rapid under conditions that promote microbial growth (temperature, humidity, and air content). Accelerated degradation may occur by selected microflora after repeated applications. The half-lives of these synthetic auxins in soil range between 5 and 21 days.

The most severe toxicological difficulty associated with this family of compounds is not attributable to the AIs themselves, but to chlorinated dibenzodioxin derivatives, their potential manufacturing contaminants. The most toxic of these is 2,3,7,8-tetrachlorodibenzodioxin (TCDD), a strong teratogen and one of the strongest synthetic toxins known today ($LD_{50} = 0.0006$ mg/kg). It is formed during the synthesis of 2,4,5-T. By modifying the manufacturing technology, contamination can be reduced; however, toxic dioxin derivatives may occur not only as industrial contaminants from manufacturing but also by photochemical formation after application. Consequently, the EPA issued an emergency order in 1986 barring the sale or use of 2,4,5-T, the strongest restriction possible by the federal pesticide law (similar to their action on dinoseb[†] in 1977). The compound had been banned in 45 counties and withdrawn in 16 countries by 1995, and it was included in the Rotterdam Convention on the export and import of hazardous substances. Exposure to chlorodibenzodioxin contaminants has also been attributed to 2,4-D, but could be eliminated through strict production specifications, and the compound has been reregistered in the EU. Dichlorprop and mecoprop are only approved in the form of their purified isomers dichlorprop-P and mecoprop-P. The registration status of 2,4-DB has been renewed until 2032.

HR GM crops have been developed against 2,4-D by using transgenic aryloxyalkanoate dioxygenases to decompose it. Coincidentally, these metabolitic enzymes also decompose certain ACCase-inhibiting herbicides and therefore confer tolerance to aryloxyphenoxy proprionate herbicides as well.

Benzoic acids (WSSA/HRAC group 4; legacy HRAC group O)

Chlorinated benzoic acid derivative herbicides, mainly dicamba (3,6-dichloro-2-methoxybenzoic acid), are similar to phenoxyacetic acids in terms of chemical structure and hormonal effects in plants (Fig. 3.7F). These compounds also exert an auxin effect.

The compounds are dissipated from soil primarily through evaporation and microbial degradation. Dicamba is highly mobile, especially in sandy soils, but the risk of leaching is low because of rapid degradation. Persistence may occur under dry conditions (low rainfall, desiccated soil). Its half-life is 14 days.

Tolerance to dicamba has been achieved in GM crops using the deactivation approach by dicamba monooxygenase. A dicamba-tolerant GM cotton has been commercialized, but it has been criticized for causing substantial environmental phytotoxicity through leaching.

Pyridine- and quinolinecarboxylic acids (WSSA/HRAC group 4; legacy HRAC group O)

Members of this family are structurally very similar to benzoic acids. Pyridine derivatives include pyridinecarboxylic acids and pyridyloxyacetic acids, the former mainly picloram (5-amino-3,4,6-trichloropicolinic acid), clopyralid (3,6-dichloropicolinic acid), and aminopyralid (5-amino-3,6-dichloropicolinic acid in the form of its potassium salt), introduced in 1963, 1977, and 2005, while the latter are triclopyr (3,4,6-trichloro-2-pyridyloxyacetic acid) and fluroxypyr (4-amino-3,5-dichloro-6-fluoro-2-pyridyloxyacetic acid), introduced in 1975 and 1983 (Fig. 3.8A). Quinolinecarboxylic acid, reminiscent of the structure of clopyralid, is quinmerac (7-chloro-3-methyl-8-quinolinecarboxylic acid), introduced in 1993. The latest representatives are two pyridinecarboxylic acid esters, halauxifen-methyl and florpyrauxifen-benzyl, introduced in 2014 and 2016.

These compounds show a strong auxin effect in susceptible plants, stimulating plant cell elongation even at a concentration as low as 1 μM, which is reflected in the disturbance of plant growth processes. They also disturb the synthesis of enzymes and other proteins, nucleic acid synthesis, and metabolism, which can be detected in changes in gene expression. Cell- and tissue-level effects include changes in cell wall elasticity; induced growth of photosynthetically nonproductive, nonassimilating tissue stocks (stems, roots, and, in part, fruit) that increases their proportion in the total plant biomass; and asymmetric changes in growth, for example, more in the upper part (leaves, petioles) than in the lower part, leading to drooping of leaves (epinasty). As a result of circulatory (phloem) disorders, photosynthesis products are not transported properly, which can cause tissue death in weeks or even a few days, leading to possible

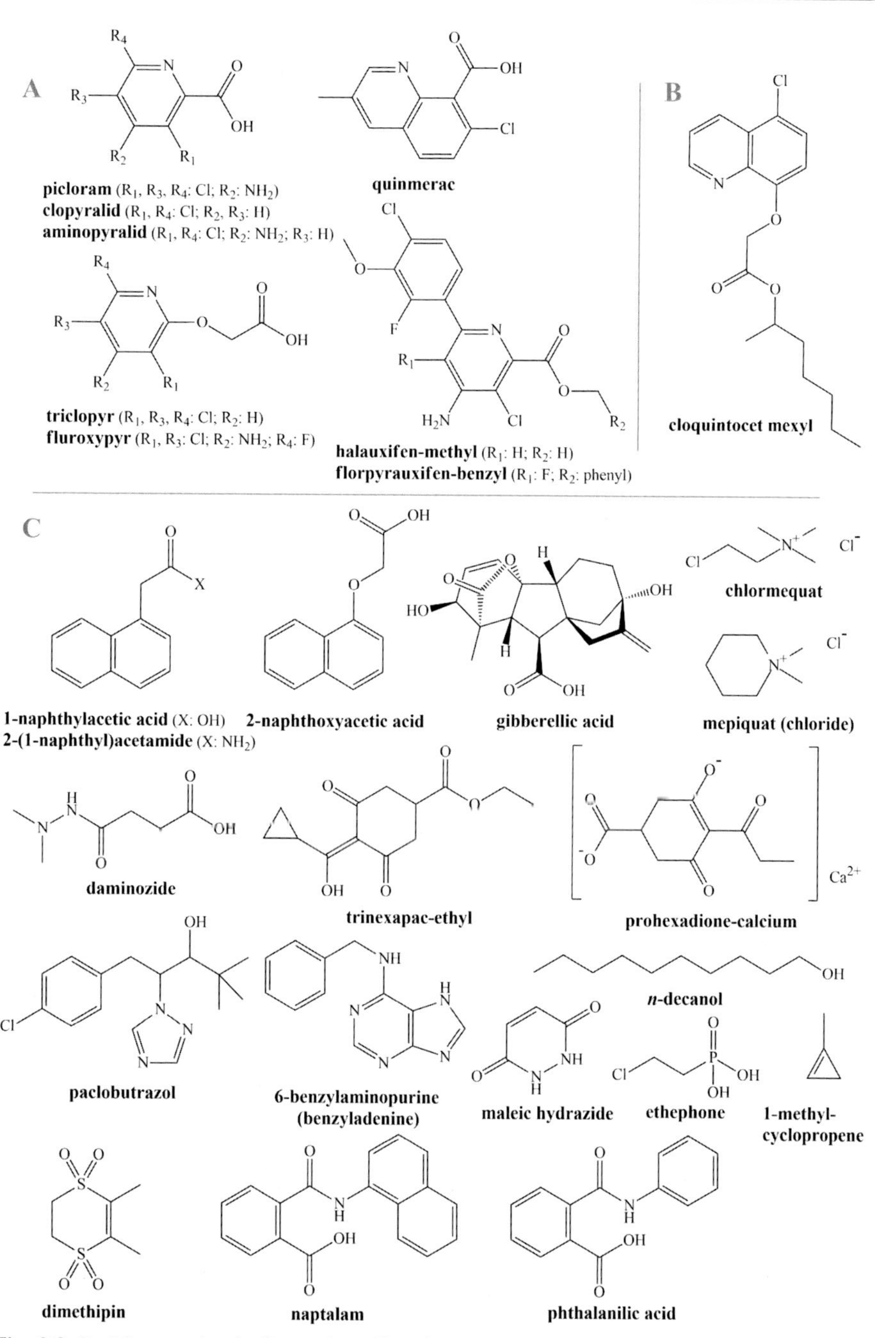

Fig. 3.8 Pyridine- and quinolinecarboxylic acid (A), antidote (B), and other plant regulator (C) herbicide active ingredients.

necrotization of the entire plant. A special feature of halauxifen-methyl is that it is also effective against mature ragweed (*Ambrosia artemisiifolia*) in sunflower fields. This is an extremely important property, as the common ragweed has become aggressive in Central Europe, particularly in Hungary. It is a unique feature also because no other herbicides are suitable for postemergent treatment of fully developed ragweed in crops. The basis of the individual effect is presumably that the sunflower, unlike ragweed and other dicotyledonous weeds, rapidly degrades it. Halauxifen-methyl has been approved by the EU, for example in combination with the antidote cloquintocet mexyl (Fig. 3.8B). However, because it can pollute water, especially in the case of permeable soils and near-surface groundwater levels, it can be applied only at a certain distance (buffer zone) from lakes and streams.

These compounds decompose partly by photodegradation and partly by microbial pathways, but much more slowly than phenoxyacetic acids and benzoic acids. Their half-lives in soil are 30–90 days. The degradation product of triclopyr is persistent in soil for up to a year.

Other plant growth regulators

1-Naphthylacetic acid (1-NAA), 2-(1-naphthyl)acetamide and 2-naphthoxyacetic acid† (Fig. 3.8C) are synthetic auxin analogs. 1-Naphthylacetic acid is an important AI as a plant rooting agent. The plant hormone gibberellic acid itself can be used as a growth regulator; its biosynthesis is disrupted by several registered plant growth regulators, including the quaternary ammonium salts chlormequat (chlorocholine chloride, CCC) and mepiquat, introduced in 1966 and 1980, in the form of chloride salts. Both are applied in stem hardening in cereals by inhibiting stem elongation, and although toxicological concerns have been raised on several occasions about chlormequat chloride, it has been on the market since the 1960s. A simple succinic acid derivative, daminozide, introduced in 1962, interferes with gibberellic acid biosynthesis. It was associated with carcinogenicity by the EPA in the 1980s, and the manufacturer voluntarily withdrew it. Currently, it is registered in the EU in horticulture. Cyclohexanedione derivatives are trinexapac-ethyl (cimetacarb ethyl) and prohexadione-calcium, introduced in 1992 and 1994. Paclobutrazole, reminiscent in the structure of triazole fungicides and introduced in 1986, is a most potent gibberellic acid biosynthesis inhibitor. A natural growth regulator is 6-benzylaminopurine (benzyladenine). Others are normal decanol (*n*-decanol), antiproliferative maleic hydrazide (1950), the ethylene precursor ethephon (2-chloroethylphosphonic acid), and 1-methylcyclopropene. Maleic hydrazide was reregistered in the EU with restriction applying to its allowed hydrazine content in order to meet tightening restrictions on metabolites. It remained in use to prevent sucker development in tobacco, and it may also replace the recently banned chlorpropham (see above) as an inhibitor of potato germination.

Other compounds with effects linked to plant growth regulators, but discontinued in the EU, may also be mentioned (Fig. 3.8C). Dimethipin†, a cyclic disulfone, inhibits protein biosynthesis. Naptalam†, a naphthyl monoamide of phthalic acid, inhibits auxin transport (WSSA/HRAC group 19; legacy HRAC group P), thus disturbing the formation of the normal (downward) curvature of roots and the growth of shoots toward light. Its half-life in soil is 14 days. Another phthalic acid derivative, phthalanilic acid† (*N*-phenylphthalamic acid), developed in 1982 in Hungary, is a pollination and fruit bonding enhancer that was applied on various cultures (sunflower, soy, alfalfa, rice, tomatoes, peppers, beans, peas, lentils, apples, cherries, grapes, clover) to prevent early fruit rejection.

Pigment inhibitors

"Bleaching" herbicides restrain the formation of plant dyes (pigments), thus the biosynthetic cycles of terpenes and carotenoids, thereby inhibiting the formation of chlorophyll and, ultimately, photosynthesis. In the reaction centers of photosynthesizing (oxygen-producing) organisms and in the antennas of prokaryotes, in addition to chlorophylls, certain carotenes are also present as auxiliary pigments. These absorb the energy of sunlight in the blue range ($\lambda = 400$–500 nm) and have a dual role: they provide more efficient light absorption, on the one hand, and they protect chlorophyll molecules from light damage in the excited state, on the other. This is because chlorophyll, excited by light energy, easily reacts with atmospheric oxygen to form reactive oxygen species (e.g., hydroxyl radicals, •OH). These are captured by carotenoids as antioxidants. In their absence, reactive oxygen damages chlorophyll and cell membranes, leading to the whitening of green (chlorophyll-rich) plant tissues and to subsequent cell and tissue death. The inhibition of photosynthesis by pigment inhibitors is accomplished by inactivating certain enzymes, such as lycopene cyclase (LCC), phytoene desaturase (PDS), and (*p*-hydroxyphenyl)pyruvate dioxygenase (HPPD), and by repressing folic acid and carotenoid biosynthesis.

Triazoles (WSSA/HRAC group 34; legacy HRAC group F_3)

Amitrole† (3-amino-1,2,4-triazole) (Fig. 3.9A), introduced in 1953, is a weak base that can take a zwitterionic form, so it has high water solubility. It acts primarily on plant pigments, inhibiting carotenoid biosynthesis by hampering lycopene cyclization. It leads to lycopene accumulation in plant tissues in vivo at a temperature-dependent rate, which can be a result of the effective direct inhibition of LCC but can also occur by indirect inhibition of an early step in carotenoid biosynthesis. The action of amitrole still allows the reduced production of photoprotective pigments, which permits the preservation of small chlorophyll quantities and some inner chloroplast membranes, although with substantially eroded membrane functions. Whichever inhibition takes pace, amitrole causes strong bleaching. It also interferes with normal levels of auxin, which disturbs growth.

Its water solubility makes amitrole moderately mobile in the environment, with some potential for leaching to groundwater. Chemical, microbial, and leaching processes contribute to its dissipation from soil, where the half-life is 14 days. Amitrole is goitrogenic (toxic to the thyroid gland) and was shown to be carcinogenic in rodents in 1959 and classified as a probable human carcinogen. The EPA banned amitrole in 1971 for crop use, but reregistered it for nonagricultural uses. It was [illegible] as not

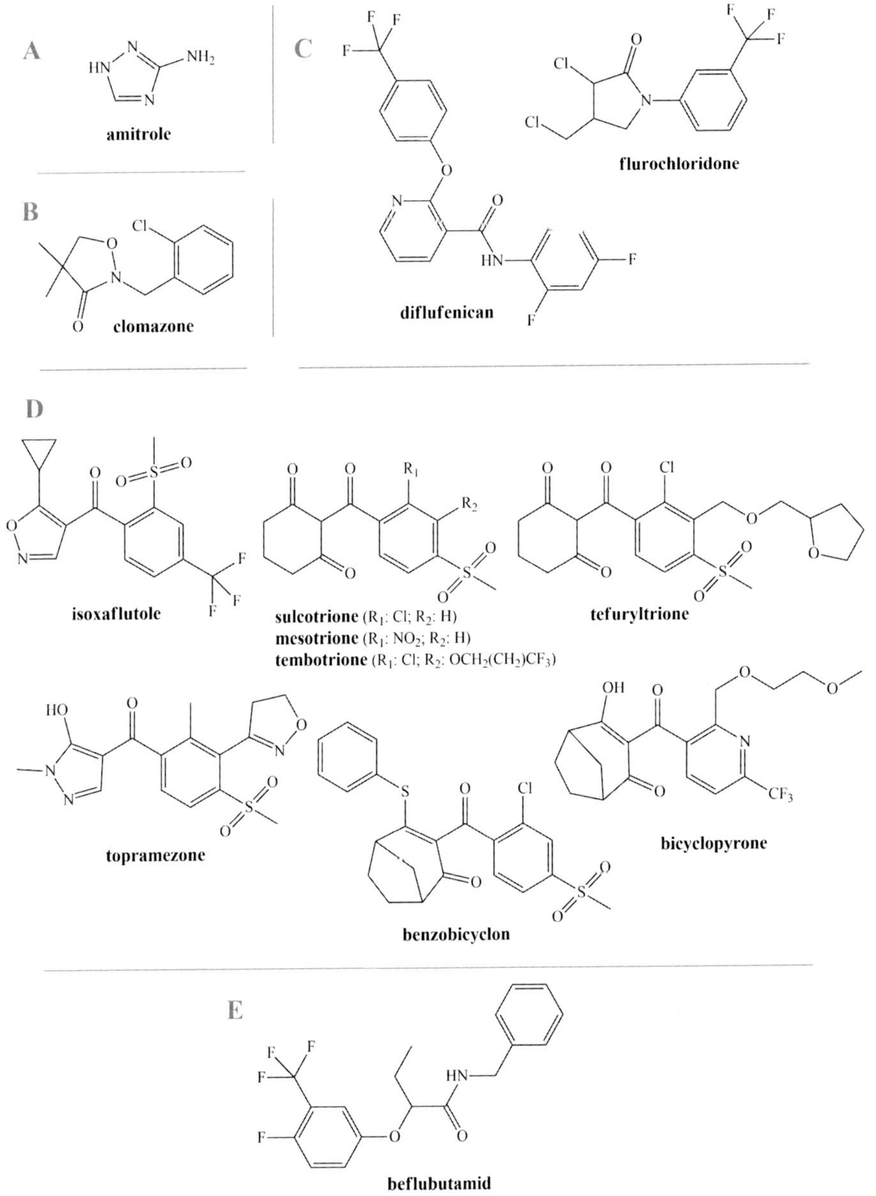

A amitrole

B clomazone

C diflufenican flurochloridone

D isoxaflutole

sulcotrione (R_1: Cl; R_2: H)
mesotrione (R_1: NO_2; R_2: H)
tembotrione (R_1: Cl; R_2: $OCH_2(CH_2)CF_3$)

tefuryltrione

topramezone benzobicyclon bicyclopyrone

E beflubutamid

Fig. 3.9 Triazole (A), isoxazolidinone (B), pyridinecarboxamide and pyrrolidinone (C), isoxazole and triketone (D), and aromatic oxycarbonamide (E) herbicide active ingredients.

likely to be carcinogenic to humans at doses that do not alter rat thyroid hormone homeostasis. Though approved by the EU, it was assessed as toxic to aquatic life, with long-lasting effects, and was finally withdrawn in 2016 due to its endocrine effects (possible teratogenicity, toxic to the thyroid and reproductive organs through prolonged or repeated exposure). Because of the EU restriction, amitrole was suggested in 2019 to be included on the list of hazardous substances distributable under prior informed consent within the Rotterdam Convention, but no further action was taken due to a lack of sufficient endorsement.

Isoxazolidinones (WSSA/HRAC group 13; legacy HRAC group F_4)

The isoxalidinone derivative clomazone (Fig. 3.9B), introduced in 1986, inhibits carotenoid biosynthesis. Its effect is similar to that of triazoles, but both the active form and the inhibition site are different from those of amitrole. It is a proherbicide that requires metabolic conversion to 5-ketoclomazone, the actual AI. That AI inhibits 1-deoxy-D-xyulose-5-phosphate (DOXP) synthase, a key enzyme in plastid isoprenoid synthesis that is an early step in the biosynthesis of carotenoid pigments. Thus, it is applied broad-spectrum and used on rice, peas, pumpkins, soybeans, sweet potatoes, winter squash, cotton, tobacco, and fallow wheat fields to control annual grasses and broadleaf weeds The enzymatic cleavage of clomazone, particularly in cotton, can be suppressed by the safener dietholate#. In the environment, clomazone is primarily degraded microbially and to a lesser extent by photodegradation. Its half-life in soil is 24 days.

Pyridinecarboxamides and pyrrolidinones (WSSA/HRAC group 12; legacy HRAC group F_1)

This type of compound includes diflufenican, which is no longer authorized (Fig. 3.9C). It was introduced in 1995, and inhibits carotene biosynthesis by blocking the enzyme phytoene desaturase. It is a selective contact herbicide with the advantage of a low dose (125–350 g/ha) and so is called a microherbicide. Also of significance is the cyclic anilide pyrrolidinone derivative flurochloridone, introduced in 2002, which differs from pyridinecarboxamides in its efficacy by being better absorbed by the plants.

Diflufenican decomposes in soil by the cleavage of its amide bond while flurochloridone undergoes full degradation. Their soil half-lives are 100–210 and 9–70 days, depending on the soil type and water content.

Isoxazoles and triketones (WSSA/HRAC group 27; legacy HRAC group F_2)

These relatively new carbonyl compounds, developed in the mid-1990s, contain an isoxazole or cyclohexanedione ring (Fig. 3.9D). The latter are called triketones, as they contain three adjacent carbonyl groups. The isoxazole derivatives are isoxaflutole and topramezone, introduced in 1992 and 2006. Isoxaflutole is a ketone containing an aromatic moiety substituted with a haloalkyl and a methanesulfonyl group as well as a rather unusual cyclopropylisoxazole ring. Topramezone contains given structural elements of isoxaflutole and triketones (but not the triketone moiety itself), a 2-hydroxypyrazole ring instead of the isoxazole ring of isoxaflutole, a methanesulfonyl group in the para position of the benzene ring, and a dihydroisoxazole ring in the meta position. Triketones are significantly different from this in their chemical structure. Sulcotrione (or by its other name, chloromesulon), introduced in 1991, is a ketone containing an aromatic moiety substituted with a chlorine atom, a methanesulfonyl group, and a 2,6-dioxocyclohexane ring. Its derivatives are mesotrione containing a nitro group in place of the chlorine atom, tembotrione containing a 2-(1,1,1-trifluoromethoxy) substituent on the carbon atom next to the chlorine atom, and tefuryltrione† containing a furfurylmethoxymethyl group in the same meta position, introduced in 1991, 1997, and 2009. The family also includes two cycloaliphatic members containing a fused bicyclo[3.2.1]oct-3-en-2-one ring, benzobicyclon* and bicyclopyrone*, introduced in 2001 and 2014, respectively. Benzobicyclon contains an enthioether, bicyclopyrone an enol moiety instead of one of the carbonyls, and the latter contains a substituted pyridine ring instead of the benzene ring in the former.

Both isoxaflutole and triketones, or rather their metabolites, inhibit the enzyme *p*-hydroxyphenylpyruvate dioxygenase (HPPD), which plays a key role in plastoquinone biosynthesis. Without this enzyme, the electron transferring substance consisting of isoprene units and a hydroquinone moiety, similar to the structure of the Q-coenzyme, cannot be formed. Therefore, carotenoid biosynthesis is indirectly damaged with the electron transfer between PS II and PS I and ultimately the process of photosynthesis being blocked. These compounds or their metabolites are time-dependent (tight-binding) inhibitors of HPPD. Isoxaflutole is a proherbicide, as its primary metabolite (3-cyclopropyl-2-[2-(methylsulfonyl)-4-(trifluoromethyl) benzoyl]-3-oxopropanenitrile), in which the diketonitrile structure occurs by the cleavage of the isoxazole ring under alkaline conditions, is the herbicidally active entity. A similar activation occurs for topramezone.

Sulcotrione, mesotrione, and tembotrione show a synergistic effect with triazine herbicides (atrazine†, terbuthylazine) and, like sulfonylureas, are often coformulated with the antidote isoxadifen-ethyl. However, the biodegradation of sulcotrione and mesotrione by cytochrome P450 enzymes is so rapid in some plants (e.g., maize) that those compounds can be applied selectively without the use of an antidote. They are particularly popular for weeds that have become resistant to sulfonylureas; however, because of intensive use, residues are beginning to emerge in drinking water.

Isoxaflutole is degraded microbially in soil with a half-life of 28 days. Resistance development has not yet been observed. Triketones decompose rapidly to the corresponding benzoic acid via cleavage of the 2,6-dioxocyclohexane ring, but other degradation products have also been identified. One of the main photolytic metabolites of sulcotrione, along with cyclization products, is 2-chloro-4-mesylbenzoic acid.

The use of isoxaflutole has been boosted by the introduction of GM crop varieties designed to tolerate it by the use of mutated HPPD of microbial (*Pseudomonas*) or vegetal (oat) origin, the latter resulting in tolerance also to mesotrione. Tefuryltrione, mainly used in rice, is not authorized in the EU, and benzobicyclon* and bicyclopyrone* are not yet authorized in the EU either.

Aromatic oxycarbonamides (WSSA/HRAC group 12; legacy HRAC group F_1)

The exact agrophore of this family is not known, as their single agent, beflubutamid* (Fig. 3.9E), introduced in 1999, is a distant derivative of diaryl heterocyclic compounds (e.g., fluridone†) that inhibit the enzyme phytoene desaturase (PDS). During carotenoid biosynthesis, the plant produces the tetraterpene phytoene, of eight isoprene units, from isopentenyl diphosphate via intermediates geranyl pyrophosphate (C_{10}) and prefitogen pyrophosphate (C_{40}). Phytoene is subsequently reduced by PDS to phytofluene and then to ξ(zeta)-carotene. The inhibition of PDS prevents the formation of this key intermediate of carotenoids (lycopene and other carotenes such as lutein and zeaxanthin), so chlorophyll remains unprotected in the chloroplast against autooxidation. According to this mechanism, beflubutamide also acts by inhibiting carotenoid biosynthesis; when used alone or in combination with other agents (isoproturon†, ioxynil†, MCPP), it is effective against pre- and early post-emergent dicotyledonous and single monocotyledonous weeds in cereals (wheat, barley). Its half-life in soil is 5.4 days, but it may persist in surface water. The issue of its association with developmental and reproductive disorders in humans has been raised.

Conclusions

The history of herbicides, as of all pesticides, is a chronology of the introduction of active ingredients, a few of them acting by novel biochemical mechanisms. In parallel, increased concerns and corresponding regulatory changes occur as previously unknown side effects emerge, which eventually results in the withdrawal or banning of the related AIs. The modes of action of herbicides are aspects of essential importance not only in evaluating their physiological effects, but also in assessing and mitigating weed resistance that can occur after a period of use. The trend in herbicide development has been a continuous shift from substances of broad phytotoxicity toward agents with specific modes of action. Currently authorized and applied herbicidal compounds may disrupt cellular metabolism processes, inhibit photosynthesis and interfere with the light activation of reactive oxygen species, or modulate cell division and growth. Cellular targets of herbicide action include a large number (at least 28) of biochemical functions, yet the development of compounds with novel modes of action has slowed substantially during the last decades, partly due to increasing rigor and corresponding costs in herbicide registration triggered by concerns about health and environmental impacts. The forthcoming introduction of a compound with a new mode of action was heralded this year, nearly three decades after the last registration of an agent with a novel biochemical mechanism at that time.

The main intention of the more restrictive registration requirements is to assure that newly approved AIs do not exert detrimental side effects identified in previous substances. In turn, herbicide design turned toward biorational approaches that include the agrophore concept, molecular modeling, and combinatorial chemistry, often resulting in chemically more complex and herbicidally more efficient molecules that allow considerably lower technical dosages as an environmental advantage. New modes of action can provide more targeted phytotoxicity; however, they cannot guarantee the avoidance of unintended ecotoxicity. Therefore, the vicious circle of registration and withdrawal continues, although at improved levels.

The obsolescence of current herbicides led not only to the registration of new ones but also to reaching back for old, already outdone AIs. For example, the loss of triazines and chloroacetamides brought phenoxyacetic acids back into use. And a conceivable future withdrawal of glyphosate is likely to result in the revival of currently overshadowed AIs. Such reachback is also seen in response to weed resistance, and among crops genetically modified for herbicide tolerance.

Acknowledgments

The author expresses his sincere appreciation to Prof. Tamás Kőmíves at the Plant Protection Institute of the Centre for Agricultural Research in Hungary, to Prof. Béla Darvas at the Hungarian Society of Ecotoxicology, to Dr. Gábor Tőkés at the Directorate for Plant Protection, Soil Conservation and Agri-Environment of the National Food Chain Safety Office in Hungary, and to Dr. Gerald T. Pollard in collaboration with RTI International in the United States for their valuable comments.

References

1. Turner JA, editor. *The pesticide manual: a world compendium*. 18th ed. Hampshire, UK: British Crop Production Council; 2018.
2. Matolcsy G, Nádasy M, Andriska V. In: Antus S-N, Székács A, editors. *Pesticide chemistry*. Amsterdam, The Netherlands: Elsevier Science; 1989.
3. Hassal KE. *The biochemistry and uses of pesticides: structure, metabolism, mode of action and uses in crop protection*. Weinheim, Germany: VCH; 1990.
4. Jeschke P, Witschel M, Kramer W, Schirmer U, editors. *Modern crop protection compounds*. Vols. 1–3. Weinheim, Germany: Wiley-VCH; 2019.
5. Römbke J, Moltmann JF. *Applied ecotoxicology*. Boca Raton, FL, USA: CRC Press; 1995.
6. Ware GW, Whitacre DM. *The pesticide book*. 6th ed. Willoughby, OH, USA: Meister Publishing Co; 2004.
7. Matthews G. *Pesticides. Health, safety and the environment*. 2nd ed. Wiley-Blackwell: Chichester, UK; 2016.
8. University of Hertfordshire. *PPDB: pesticide properties database*; 2020 http://sitem.herts.ac.uk/aeru/ppdb/en/atoz.htm. [Accessed 29 August 2020].
9. European Commission. *EU pesticides database*; 2020 https://ec.europa.eu/food/plant/pesticides/eu-pesticides-database/public. [Accessed 29 August 2020].
10. Fedtke C. *Biochemistry and physiology of herbicide action*. Berlin, Germany: Springer Verlag; 1982.
11. Kirkwood RC, editor. *Target sites for herbicide action*. New York, USA: Springer Science-Business Media, LLC; 1991.
12. Böger P, Wakabayashi K, Hirai K, editors. *Herbicide classes in development. Mode of action, targets, genetic engineering, chemistry*. Berlin, Germany: Springer Verlag; 2002.
13. Duke SO, editor. *Weed physiology. Vol. II. Herbicide physiology*. Boca Raton, FL, USA: CRC Press; 2018.

14. Duke SO. Why have no new modes of action appeared in recent years? *Pest Manag Sci* 2012;**68**(4):505–12.
15. Phillips McDougall. *Evolution of the crop protection industry since 1960.* London, UK: Phillips McDougall Agrobusiness Intelligence; 2018.
16. Retzinger Jr EJ, Mallory-Smith CA. Classification of herbicides by site of action for weed resistance management strategies. *Weed Technol* 1997;**11**(2):384–93.
17. Mallory-Smith CA, Retzinger Jr EJ. Revised classification of herbicides by site of action for weed resistance management strategies. *Weed Technol* 2003;**17**(3):605–19.
18. WSSA (Weed Science Society of America). *Herbicide Site of Action (SOA) classification list*; 2020 http://wssa.net/wp-content/uploads/WSSA-Herbicide-SOA-2020-5-12.xlsx. [Accessed 29 August 2020].
19. Schmidt RR. HRAC classification of herbicides according to mode of action. In: *Proceedings of the Brighton crop protection conference: weeds.* vol. 3. Farnham, UK: British Crop Protection Council; 1997. p. 1133–40.
20. Menne H, Köcher H. HRAC classification of herbicides and resistance development. In: Krämer W, Schirmer U, editors. *Modern crop protection compounds.* 3rd ed. Weinheim, Germany: Wiley-VCH Verlag GmbH & Co. KGaA; 2007.
21. Beffa R, Menne H, Köcher H. Herbicide resistance action committee (HRAC): herbicide classification, resistance evolution, survey, and resistance mitigation activities. In: Jeschke P, Witschel M, Krämer W, Schirmer U, editors. *Modern crop protection compounds.* 3rd ed. vol. 1. Weinheim, Germany: Wiley-VCH Verlag GmbH & Co. KGaA; 2019. p. 5–32.
22. HRAC (Herbicide Resistance Action Committee). *HRAC mode of action classification*; 2020 https://hracglobal.com/files/HRAC_Revised_MOA_Classification_Herbicides_Poster.pdf. [Accessed 29 August 2020].
23. GRDC (Australian Government Grains Research and Development Corporation). *Herbicide resistance. Mode of action groups*; 2020 https://grdc.com.au/resources-and-publications/all-publications/bookshop/2010/12/herbicide-resistance-modes-of-action-groups. [Accessed 29 August 2020].
24. APVMA (Australian Pesticides and Veterinary Medicines Authority). *'Mode of action' indicator*; 2020 https://apvma.gov.au/node/934#Mode_of_action_indicator. [Accessed 29 August 2020].
25. CropLife Australia. *Herbicide mode of action table*; 2020 https://www.croplife.org.au/resources/programs/resistance-management/herbicide-moa-table-4-draft-2. [Accessed 29 August 2020].
26. Forouzesh A, Zand E, Soufizadeh S, Samadi Foroushani S. Classification of herbicides according to chemical family for weed resistance management strategies – an update. *Weed Res* 2015;**55**(4):334–58.
27. Stoltenberg DE, Gronwald JW, Wyse DL, Burton JD. Effect of sethoxydim and haloxyfop on acetyl-CoA carboxylase activity in *Festuca* species. *Weed Sci* 1989;**37**(4):512–6.
28. LaRossa RA, Schloss JV. The sulfonylurea herbicide sulfometuron methyl is an extremely potent and selective inhibitor of acetolactate synthase in *Salmonella typhimurium*. *J Biol Chem* 1984;**259**(14):8753–7.
29. Amrhein N, Deus B, Gehrke P, Steinrücken HC. The site of the inhibition of the shikimate pathway by glyphosate: II. Interference of glyphosate with chorismate formation *in vivo* and *in vitro*. *Plant Physiol* 1980;**66**(5):830–4.
30. Husted RF, Olin JF, Upchurch RP. A new selective preemergence herbicide: CP-50144. In: *Proceedings of the northern central weed control conference, vol. 21*; 1966. p. 44.
31. Fedtke C. *Biochemistry and physiology of herbicide action.* Berlin, Germany: Springer-Verlag; 1982. p. 137–8.

32. Heim DR, Skomp JR, Tschabold EE, Larrinua I. Isoxaben inhibits the synthesis of acid-insoluble cell wall materials in *Arabidopsis thaliana*. *Plant Physiol* 1990;**93**(2):695–700.
33. Campe R, Hollenbach E, Kämmerer L, Hendriks J, Höffken HW, Kraus H, et al. A new herbicidal site of action: Cinmethylin binds to acyl-ACP thioesterase and inhibits plant fatty acid biosynthesis. *Pestic Biochem Physiol* 2018;**148**:116–25.
34. Erdődi F, Tóth B, Hirano K, Hirano M, Hartshorne DJ, Gergely P. Endothall thioanhydride inhibits protein phosphatases-1 and -2A *in vivo*. *Am J Physiol* 1995;**269**(5 Pt 1):C1176–84.
35. Wilkinson RE, Smith AE. Reversal of EPTC induced fatty acid synthesis inhibition. *Weed Sci* 1975;**23**(2):90–2.
36. Gronwald JW. Lipid biosynthesis inhibitors. *Weed Sci* 1991;**39**(3):435–49.
37. Devine M, Duke SO, Fedtke C. *Physiology of herbicide action*. New Jersey: Prentice Hall; 1993.
38. Tachibana K, Watanabe T, Sekizawa Y, Takematsu T. Accumulation of ammonia in plants treated with bialaphos. *J Pestic Sci* 1986;**11**(1):33–7.
39. Sandmann G, Böger P. Inhibition of carotenoid biosynthesis by herbicides. In: Boger P, Sandmann G, editors. *Target sites for herbicide action*. Boca Raton, FL, USA: CRC Press; 1989. p. 25–44.
40. Mueller C, Schwender J, Zeidler J, Lichtenthaler HK. Properties and inhibition of the first two enzymes of the non-mevalonate pathway of isoprenoid biosynthesis. *Biochem Soc Trans* 2000;**28**(6):792–3.
41. Duke SO, John L, José MB, Timothy DS, Larry PLJR, Hroshi M. Protoporphyrinogen oxidase-inhibiting herbicides. *Weed Sci* 1991;**39**(3):465–73.
42. Dodge AD. The role of light and oxygen in the action of photosynthetic inhibitor herbicides. In: Moreland DE, St. John JB, Hess FD, editors. *Biochemical responses induced by herbicides. ACS Symposium Series*, vol. 181. Washington, DC, USA: American Chemical Society; 1982. p. 57–77.
43. Viviani E, Little JP, Pallett KE. The mode of action of isoxaflutole II. Characterization of the inhibition of carrot 4-hydroxyphenylpyruvate dioxygenase by the diketonitrile derivative of isoxaflutole. *Pestic Biochem Physiol* 1998;**62**(2):125–34.
44. Kahlau S, Schröder F, Freigang J, Laber B, Lange G, Passon D, et al. Aclonifen targets solanesyl diphosphate synthase, representing a novel mode of action for herbicides. *Pest Manag Sci* 2020. https://doi.org/10.1002/ps.5781. online first.
45. Shino M, Hamada T, Shigematsu Y, Hirase K, Banba S. Action mechanism of bleaching herbicide cyclopyrimorate, a novel homogentisate solanesyltransferase inhibitor. *J Pestic Sci* 2018;**43**(4):233–9.
46. Rudiger W, Benz J. Influence of aminotriazol on the biosynthesis of chlorophyll and phytol. *Z Naturforsch C* 1979;**34**(11):1055–7.
47. Vaughn KC, Lehnen JR. Mitotic disrupter herbicides. *Weed Sci* 1991;**39**(3):450–7.
48. Gleason C, Foley RC, Singh KB. Mutant analysis in *Arabidopsis* provides insight into the molecular mode of action of the auxinic herbicide dicamba. *PLoS ONE* 2011;**6**(3), e17245.
49. Keitt Jr GW, Baker A. Auxin activity of substituted benzoic acids and their effect on polar auxin transport. *Plant Physiol* 1966;**41**(10):1561–9.
50. Morejohn LC, Fosket DE. Tubulins from plants, animals and protists. In: Sharp JW, editor. *Cell and molecular biology of the cytoskeleton*. New York, USA: Plenum Press; 1986. p. 257–329.
51. Simon EW. Mechanism of dinitrophenol toxicity. *Biol Rev* 1953;**28**(4):453–79.

Coformulants in commercial herbicides

Robin Mesnage
King's College London, London, United Kingdom

Chapter outline

Introduction

Herbicides are rarely single ingredients in their commercial forms. They are sold as complex mixtures of compounds, each having a specific role in herbicidal activity. The cost-effectiveness of a spray application not only depends on the toxicity of the active substance, but it also depends on the solubility, volatilization, adherence, penetration, rainfastness, foaming, metabolism, and drift of the spray mixture. This can be affected by many factors such as climatic conditions and the type of soil.[1] Even if an active substance is highly toxic for plant metabolism, it will act efficiently only if it can penetrate the plant epidermal surface, which is covered by a waxy layer; the cuticle; or enter the plant via the soil-root interface. In the case of foliar applications, active substances also need to have a certain degree of retention on the leaf surface and persist long enough to penetrate weed-leaf tissue regardless of climatic conditions.

No single compound or formulation will reliably assure consistent attainment of all these functions across the diversity of cropping systems, soil types, weed communities, application equipment, and weather patterns that exist. Formulation chemistry and innovation are the mechanisms through which the industry incrementally enhances the "fit" between a given formulation and a specific farmer's weed management challenges.

Herbicides. https://doi.org/10.1016/B978-0-12-823674-1.00010-9

Ingredients other than active substances are added to improve the herbicidal performance of the final preparation. They are called adjuvants, inerts, coformulants, formulants, and other ingredients. Some of these terms should not be used interchangeably (Table 4.1). For instance, the difference between coformulants and adjuvants is that coformulants are added when the herbicide product is formulated in a chemical plant by the manufacturer. Adjuvants are added afterward, when the farmer or applicator pours the formulated herbicide into a spray tank and then adds the recommended amount of water as well as any other pesticides or fertilizers that will be applied along with the formulated herbicide.

Adjuvants serve multiple purposes. Some help keep the rest of the contents of the spray-tank mix properly suspended and mixed while others adjust the pH of the final

Table 4.1 Legal definitions used to regulate the commercialization of herbicides in the European Union. Terms are from European Commission Regulation No. 1107/2009 on the placing of plant protection products on the market.[2]

Term	Definition
Pesticide	Any substance or preparation intended for repelling, destroying, or controlling any pest or regulating plant growth. "Technical pesticide" refers to technical materials and technical concentrates. "Formulated pesticide" refers to any formulation containing a pesticide. The term "pesticide" can refer to an active substance.
Herbicide	A type of pesticide classified as a plant protection product that is used to destroy undesired plants or parts of plants.
Active substance	Also known as active ingredient. A chemical element and its compounds as they occur naturally or by manufacture, including any impurities. It has specific effects on plant metabolism resulting in plant death.
Coformulant	A nonactive ingredient intended to be used in a plant protection product or adjuvant but is not itself an active substance, safener, or synergist. Common examples are surfactants, emulsifiers, and dyes.
Safener	A substance or preparation added to a plant protection product to eliminate or reduce the phytotoxic effects of the product on certain plants.
Synergist	A substance or preparation that, while showing no or only weak activity, can enhance the activity to the active substances in a plant protection product.
Adjuvant	A substance or preparation consisting of coformulants that is placed on the market to be mixed by the user with a plant protection product to enhance its effectiveness.
Impurity	Any component other than the pure active substance and/or variant that is present in the technical material, including components originating from the manufacturing process and from degradation during storage.
Surfactant	A compound used in herbicides or adjuvants that reduces surface tension, thereby increasing the emulsifying, spreading, dispersibility, and/or wetting properties of liquids or solids. They can be included in formulated products or adjuvants.
Metabolite	A degradation product of an active substance, safener, or synergist, formed in either living organisms or the environment.

mixture or block adverse chemical reactions across the active agents in the tank mix. Others promote better penetration, reduce spray drift, and enhance rainfastness. Some are intended to reduce applicator exposure or otherwise promote safety. An early modern research effort to improve formulation was done at the end of the 19th century, when Millardet created the Bordeaux mixture, the first fungicide, by mixing copper sulfate and milk of lime.[3]

The inclusion of coformulants or their addition to tank mixes as adjuvants has become a very important aspect of the development of modern pesticides. It is a principle method manufacturers use to differentiate their products, target specific market segments, and compete for market share. The market for agricultural adjuvants was valued at $3.3 billion in 2019, and is projected to increase by 5.8% to reach $4.7 billion by 2026.[4] This chapter is an overview of the variety of coformulants, many of which are also used as adjuvants. Their toxicological profiles for nontarget organisms are also discussed.

Classification and function of coformulants

A large variety of coformulants are included in formulated pesticides. During the 18th century, resins and molasses were used as adjuvants with inorganic pesticides (sulfur, copper, and arsenates) to improve the adhesion and rainfastness of the mixture.[5] The list of coformulant categories is continuously expanding as pesticide technology evolves. For instance, recent studies are focused on the deployment of the bio-encapsulation process from nanotechnologies.[6]

There is no consensus on the classification of adjuvants.[7] They may be grouped by function (activator or utility), chemistry (such as alkoxylated surfactants), or source (vegetable, animal, or petroleum oils). Most are classified as activators or utility modifiers. Activators directly enhance biological activity by influencing the wetting, spreading, sticking, and humecting property of sprays, ultimately improving penetration. Utility modifiers improve the application process but have no direct effect on pesticidal activity. They can be pH buffering agents, compatibility agents to allow mixing with other pesticides, drift retardants, antifoams, preservatives, and dyes (Table 4.2).

The most common coformulants are surfactants (Fig. 4.1). Soap made from animal fat was used before 1900 to improve the performance of arsenate pesticides.[15] This was not without problems because soap can be highly phytotoxic, which impairs the delivery of the active substance. Further studies comparing different types of potassium soaps revealed that foliar damage is correlated with the length of alkyl chains.[16]

Petroleum fractions were used as weed killers in the mid-20th century, before they were used as adjuvants. First were waste materials such as acid sludge from tar refining and old engine oil.[17] These were replaced by refined petroleum products such as kerosene, diesel, and Edeleanu extracts. Oils with a high fraction of sulfonatable molecules were the most effective. Although oils were considered excellent herbicides, their limited availability and high cost was an obstacle to wide application. A technical

Table 4.2 Types and functions of compounds used as coformulants or adjuvants. This nonexhaustive list groups them by function. Different scientific societies have established their own definitions and classifications.[7]

Compound	Use and examples
Wetters and spreaders	Work by reducing surface tension between the spray droplet and the target surface, allowing the spray droplet to have a lower contact angle with the substrate. This increases the surface of exposure. As wetting increases, the droplet begins to spread. These coformulants are mostly surfactants.
Penetration agents	Crucial to allow active substances to diffuse through the waxy cuticle of plants and reach more hydrophilic structures. They are generally surfactants that act by reducing the surface tension. They may also dissolve wax to increase the movement of the active substance. Nonionic surfactants are the most widely used with systemic herbicides. They generally have long aliphatic chains; shorter-chain surfactants tend to be phytotoxic.
Stickers	Viscous mixtures allowing pesticide droplets to adhere for longer exposure. Molasses, syrups, resins, flour pasta, and petroleum oils have been classically used. Some were used in the 18th–19th century to improve the adhesion and rainfastness of inorganic pesticides (sulfur, copper, and arsenates).[5] Modern stickers can be based on latex, terpenes, or pyrrolidone polymers. Terpene autopolymerizes with exposure to air and UV light when deposited on a leaf surface.[8]
UV absorbents	Natural sunlight can degrade some herbicide ingredients. Compounds such as substituted benzophenones can absorb the UV light. The half-life of cyclohexanedione herbicides due to photodegradation can be as low as 6 min.[9] The application of UV absorbers as adjuvants of deltamethrin, used for tsetse fly control on cotton fabric, reduced photolytic degradation from 90% to 7%.[10]
Drift retardants	Increase spray droplet size to make them less susceptible to being carried and spread by the wind. They have been showed to reduce the drift of the herbicide propanil.[11] Dicamba drifting from application on dicamba-tolerant crops is a major source of crop damage; the additions of the retardant Intact, ammonium sulfate, and pH modifiers have been tried with various degrees of success.[12]
Dyes	Added to track sprayed areas. Basic Violet 3, Rhodamine B, and Hi-Light Blue, among others, have been used in various applications by the US Forest Service. This practice has been proposed to allow a reduction of pesticides used.[13]
Antifoam agents	Reduce the formation of foam in spray tanks. They are usually silicones (e.g., dimethylpolysiloxane) and fatty acids.
pH buffers	Change the pH of the spray solution. They can also be called compatibility agents, and can also be used as drift retardants. pH modifiers have been recommended to reduce dicamba drifting from applications on dicamba-tolerant crops.[12]

Table 4.2 Continued

Compound	Use and examples
Compatibility agents	They allow mixing and the simultaneous application of two or more active substances, and ensure compatibility with hard water. They can be emulsifiers. For instance, when a 2,4-D amine herbicide is mixed with another product that contains nonamine cations, such as a fertilizer, a precipitate can form in the spray tank. A quaternary surfactant can be added to increase the compatibility (US Patent 9012365).
Water conditioners	They minimize the ability of ions from the pesticide spray, originating from the formulated product itself or from hard water, to form precipitates, which reduces the efficacy of the active substance. Ions can also come from the plant itself. High levels of cations in the water reduce the efficacy of glyphosate, which has chelative properties. The addition of ammonium sulfate prevents the precipitation of glyphosate.[14]
Thickening agents	They modify the viscosity of spray solutions. They also reduce drift in aerial applications. They are claimed to provide greater stability, enabling storage for a longer period. Polymers such as polyacrylics, polyacrylamides, hydrophobically modified cellulose derivatives, modified starches, and copolymers of cellulose derivatives are among the most widely used thickeners (US Patent 20150011599A1).
Odor masking agents	They are used to control malodorous ingredients in agricultural pesticides, notably 2,4-dichlorophenoxyacetic acid. Examples are compounds having "green notes" such as galbanum oil, phenyl ethyl methyl ether, phenyl acetaldehyde, phenyl acetaldehyde dimethyl acetal, cis-3-hexenyl acetate, and cinnamic alcohol.
Preservatives	Hexamethylenete is a stabilizer for carriers in solid pesticide formulations. Potassium benzoate is used as a preservative in pesticide formulations applied pre- and postharvest.

bulletin from the US Department of Agriculture (USDA) reports details of experiments to control the common cattail from 1947 to 1959.[18] Stove and diesel oils produced a rapid kill of top growth weed parts. However, vigorous regrowth appeared rapidly, as these compounds were mostly damaging foliage, not the roots. To circumvent this issue, reduce cost, and improve efficiency, mixing oils with substituted phenols was proposed.[17] These mixtures were designed as "fortified oils," which allowed the use of lower doses. Although the classification as inert or active did not exist, the active substance in modern agriculture could be seen as the adjuvant back then. Conversely, petroleum distillates now classified as inerts were then primarily active substances to kill weeds. Oils are still used to kill pests. In the 1950s, the situation changed when more-toxic synthetic chemicals started to be produced, such as 2,4-D and the triazines. The enhancement of herbicide activity by surfactants became an intense topic of interest with the development of formulations of 2,4-D.[19] A 1945 Monsanto patent on emulsifiable oils provides the first notions of what would become modern

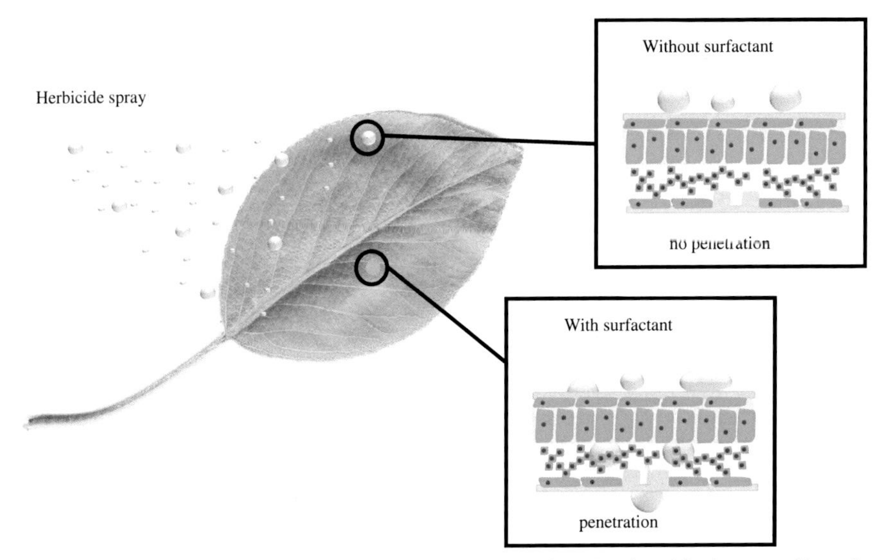

Fig. 4.1 The effects of a surfactant on wetting and penetration. Depicted is the spreading of a plain water droplet on a leaf's surface in comparison to a water droplet containing a surfactant.

formulated herbicides (US patent 2,447,475). It reports new oil compositions that "may contain various active ingredients dissolved therein, such as oil-soluble insecticides, plant hormones, fungicides, herbicides and the like." A mixture of diesel oil with 2,4-D became one of the most frequently used weed killers. The USDA report on cattail control says that in 1951, the use of 4 pounds of 2,4-D alone dissolved in 400 gal of water was rather ineffective, and that efficacy was increased considerably by adding 0.5 gal of nonionic emulsifier or 5 gal of diesel oil. The use of petroleum distillates as adjuvants was also reported in 1964 when atrazine was mixed with the petroleum phytobland oil SunSpray for weed control in corn.[20]

The most commonly used surfactants in agriculture are nonionic. They rapidly emerged as effective adjuvants because of their wetting and emulsifying properties as well as their lower toxicity compared to ionic surfactants. Most of those in agricultural products and cosmetics nowadays are made from the alkoxylation (e.g., ethoxylation for the polymerization of ethylene oxide) of a range of precursors. In the early days, sucrose was also used instead of ethylene oxide.[21] In the 1950s, compared with ethylene oxide at 15–50 US cents per pound, sugar cost 3–5 cents and was the lowest-cost hydrophilic substance for the preparation of nonionic surfactants. Animal fat extract (tallow) was one of the cheapest hydrophobic substances at about 7 cents, which is why tallow-based surfactants rapidly became the most popular. Ethoxylated amines, produced from tallow, are also called polyoxyethyleneamines (POEAs).

Although coformulants are used in a large range of products containing different active substances, information about the final composition of commercialized products is scarce because the nature of coformulants is often considered confidential business information. Because of the many debates on the toxicity of glyphosate coformulants, more data have been made available for public scrutiny. More than

750 pesticides have glyphosate as an active ingredient for use in agriculture, amenities (such as on golf courses, parks, and sports grounds), and home applications.[22] Glyphosate was first commercialized as the formulated product Roundup, also known as MON 2139, which contained a surfactant made of a mixture of POEAs that has an average of 15 ethylene oxide units, MON 0818. The toxicity of MON 0818 was rapidly identified as a worker safety issue because it contained an ethoxylated tallowamine that was far more toxic than glyphosate. The California Department of Pesticide Regulation determined that the top two causes of occupational pesticide incidents were eye (50%) and skin (35%) injuries largely attributed to MON 0818. Glyphosate-based herbicides ranked third among agents involved in pesticide-related accidents. Monsanto mitigated the problem by introducing compounds known to lessen eye irritation, such as phosphate-esters in POEA-based products (US Patent 5683958A, 1990). Tallowamine ethoxylates tend to solidify at cold temperatures to form a stiff gel. This was mitigated by adding polyethylene glycol to first-generation Roundup formulations. Polyethylene glycol was rendered unnecessary by replacing ethoxylated alkylamines with ethoxylated etheramines, which also increased glyphosate loading to cover a larger agricultural area with the same volume of pesticide (US Patent 5750468A, 1995). Roundup surfactants have been further diversified to include propoxylated quaternary ammonium compounds, which are much less toxic than ethoxylated compounds.[23,24]

Glyphosate-formulated products do not always need to be mixed with adjuvants; they already contain a range of surfactants suitable for different uses. However, some surfactants can be added to improve the efficiency for some targets.[25] For instance, rhododendron control benefits from the addition of nonionic surfactants such as Mixture B (a mix of ethoxylated alcohols and isobutanol) to improve glyphosate penetration into the thick waxy leaf surface.[26] A variety of surfactant mixtures are sold in the United Kingdom to be mixed with glyphosate (Fig. 4.2). They include alkoxylated products (sorbitan esters, tallow amines, alcohols, and soybean triglycerides), trisiloxane organosilicone, and the styrene-butadiene copolymer. Few formulations use ethoxylated surfactants in Europe because of their known health risk among applicators. Rapeseed oil emulsions and pinene oligomers are sold as sticking and wetting agents. Ammonium sulfate is frequently used with glyphosate as a compatibility agent.

POEA surfactants: Structure-toxicity relationship

Most POEA surfactants are made from the reaction of tallow that is first reacted with ammonia.[27] The resulting product, a tallowamine, is oxidized by the successive addition of ethylene oxide. Although a type of POEA surfactant became infamous for causing the acute toxic effects of Roundup, the class contains many compounds with a common structural feature[27] represented by the general formula R-N-$(CH_2CH_2O)_xH$, where R-N is an amine and $(CH_2CH_2O)_x$ is a chain of ethylene oxide molecules (i.e., a polyoxyethylene). The chains are of different lengths depending on the size of the polyoxyethylene moiety and the type of fatty acid used; polyethoxylated

Fig. 4.2 Wordcloud of adjuvant component groups registered for use by the United Kingdom Health and Safety Executive.
Downloaded April 23, 2020, from https://secure.pesticides.gov.uk/adjuvants/.

coco amine from coconut oil or polyethoxylated soya amine from soya. Tallow amines are mostly C16 and C18 primary alkyl amines. The structures of the molecules of POEA mixtures can be the same regardless of the origin of the amine (vegetal or animal fat). If propylene oxide is added, the surfactant is propoxylated.

Surfactants have a common toxicity mechanism: they disturb cell membrane integrity.[28,29] This is directly linked to their primary function as surface active agents, which they do by reducing the surface tension between two liquids. Surface tension is the cohesive force at the surface of a liquid, allowing the assembly of the molecules in the liquid. For instance, surface tension allows molecules of water to take a spherical shape, forming drops. A surfactant molecule is both lipid- and water-soluble. The ethoxylated portion of a POEA surfactant is water-soluble. By contrast, the fatty amine part of the POEA is lipid-soluble. This bifunctionality explains how surfactants mix and stabilize water-soluble and lipid-soluble molecules. The resulting product is an emulsion.

A double layer of lipids forms the basic structure of a cell membrane. They are predominantly phospholipids, the structure of which allows them to form bilayers. The physicochemical properties of phospholipids and some surfactants are somewhat comparable in that phospholipids have a lipid-soluble extremity (fatty acid tail) and a water-soluble extremity (phosphate head). This similarity means that surfactants can insert themselves into cellular membranes and alter their structures, leading to membrane collapse and cell death. The alteration of cell membrane structure explains why surfactants irritate skin and are toxic to human cells.

Although most surfactants share a common toxicity mechanism, the severity of their toxic effects can vary widely. The acute toxicity of POEA surfactants follows general rules. The toxicity of ethoxylated surfactants decreases when chemical groups are attached to the alkyl chains (e.g., methyl groups) because it impairs cell membrane insertion. Longer water-soluble parts (ethylene oxide chain) have lower toxicity because they interact less with cell membranes. This is the case for human cells and for whole organisms such as aquatic species.[30]

In fish and aquatic invertebrates, surfactants emulsify the mucus on gill surfaces. The gill membrane surface may also be penetrated by the hydrophobic portions of surfactant molecules. As a consequence, respiration is impaired and the animal dies of suffocation. A longer ethoxylated chain makes the surfactant less fat-soluble, which hinders emulsification and/or penetration into the gill membrane.[31]

These general toxicity properties are highly dose-dependent because the physicochemical interactions between surfactant molecules change dramatically when they reach the critical micelle concentration (CMC). The CMC is the concentration at which surfactant molecules form micelle structures—in other words, detergent vesicles. The CMC of POE-15 tallowamine is correlated with toxic effects to human cells.[32] Although the compound is not toxic at concentrations below 1 mg/L, it rapidly becomes toxic when the concentration likely to disturb cell membranes, the CMC, is reached. These mechanisms explain human cell toxicity in in vitro assays, and thus predict irritant properties and acute toxic effects rapidly, but they may not fully reflect toxicity to whole organisms.

The genotoxicity of POEA surfactants has been debated extensively in recent decades to understand whether higher rates of cancer in some populations occupationally exposed to formulated pesticides can be attributed to cancer-initiating or promoting effects of the active substance or of the coformulants. This was the case recently for glyphosate, which was held responsible for the development of non-Hodgkin lymphoma in some pesticide workers.[22] The results of animal carcinogenicity studies are equivocal. Because it is not clear that some tumors are related to glyphosate exposure, it has been hypothesized that the carcinogenic effects of glyphosate-formulated products might be due to the coformulants. Some studies have suggested that the ethoxylated tallowamine surfactant contained in the adjuvant MON 0818 has genotoxic effects. Although a study in fish blood cells by Guilherme and colleagues suggested that MON 0818 could contribute to the genotoxicity of Roundup, little evidence exists regarding the relevance of these effects for occupational exposure in humans. Most of the studies pointing to the increased genotoxicity of surfactants used high concentrations that could be over the CMC, thus causing general cytotoxicity. A recent project from the US National Toxicology Program, unpublished as of this writing, compared the effects of glyphosate on measures of genotoxicity, oxidative stress, and cell viability with the effects of glyphosate-based formulations in several models. Preliminary results were presented at the 2019 Society of Toxicology meeting. Thirteen formulations were tested in human HepaRG (hepatocyte) and HaCaT (keratinocyte) cell lines using assays that detect DNA double strand damage and H_2O_2 production, reflecting oxidative stress. Glyphosate and its formulations did not induce DNA damage and marginally increased oxidative stress only after a significant loss of cell viability.[33] Another presentation from this study reported that some formulated products showed genotoxic activity in the TK6 human B lymphoblastoid cell line.[34] Other studies providing evidence of genomic damage induced by glyphosate on human lymphocytes will not be reviewed here, for the sake of clarity.[35] Collectively, more than 300 assays were performed to study the genotoxicity of glyphosate.[35] Around half of these assays (51%) reported positive results. Discrepancies in the results of these assays could be attributed to their sensitivity, the nature of the compounds studied (formulated products or glyphosate), or to the choice of the tested concentrations. Among notable examples of studies that compared the toxicity of formulated products to their active substances, Bolognesi et al. showed an induction of sister chromatid exchanges in human lymphocytes starting at 1 g/L for glyphosate and 0.1 g/L for Roundup.[36] This suggested that Roundup was at least 10 times as genotoxic as glyphosate in this study. This difference was confirmed by other studies of Roundup UltraMax (Austria)[37] and Roundup.[38,39] However, some other studies found no difference when they compared the genotoxic effects of glyphosate to those of Roundup[40] or Roundup transorb.[41] The topic is controversial, and further experiments will be needed to understand the dose relationship of the potential surfactant-related genotoxic effects induced by glyphosate-based herbicides. The possibility that the greater genotoxicity of some formulated products can be due to their composition in surfactants is also cautiously mentioned by the European Food Safety Agency (EFSA). In 2015, the EFSA was asked by the European Commission to review the toxicity of the supposedly inert ingredient POE-tallowamine. They concluded that "POE-tallowamine is clearly more toxic than glyphosate when tested in glyphosate-based formulations" and that "there is evidence that products which contain cytotoxic tallowamine (or other) surfactants might produce, mostly in very high concentrations, DNA damage."[42]

Impurities in surfactants

Surfactant mixtures made from the ethoxylation of fatty amines can be contaminated by impurities such as 1,4-dioxane and ethylene oxide. The most frequent impurities are polyethylene glycols, which are formed when ethylene oxides polymerize instead of reacting with a fatty amine.[43] The presence and levels of impurities depend on the manufacturing procedure used, including the type of catalyst, the reaction temperature and timing, and the quality or purity of the amines. For instance, the level of dioxane can vary from undetectable to high (~3%) for different catalysts. Consequently, it is very difficult to establish a typical profile of impurities.

Dioxanes are common byproducts of the ethoxylation process. They are cyclic molecules created by the dimerization (combination) of two ethylene oxide molecules. Although 1,2-dioxane and 1,3-dioxane exist, they are minor, and the term dioxane generally refers to 1,4-dioxane, also known as 1,4-dioxyacyclohexane, diethylene oxide, dioxyethylene ether, and *p*-dioxane. The amount of dioxane is very variable depending on the quality of raw materials. Little information is available on the current contamination of POEA-based adjuvants with dioxane contaminants. Documents from the USDA reported that contamination by 1,4-dioxane was relatively high several decades ago, at the upper limit of 300 ppm in some Roundup formulations. However, manufacturing processes were improved and the level was reduced to 23 ppm.[44]

Literature on ethylene oxide surfactant contamination is scarce. Most reports, including supplier's material safety data sheets, only indicate an "upper bound concentration" or "typical maximum," usually 1–10 ppm, without indicating a specific value.[31] Actual levels depend on the chemical procedure and the grade of the ingredients used in the surfactant preparation. An unpublished study available on the website of the Taiwan Agricultural Chemicals and Toxic Substances Research Institute showed that 44 surfactants tested were not contaminated by ethylene oxide. Only one sample was positive and contained 7.52 ppm.

Ethylene oxide is frequently found in the environment, released from automobile exhaust and cigarette smoke while also being used to sterilize medical equipment. The excess cancer risk for a lifetime exposure to average ambient airborne concentrations is estimated to be 6–8 cases per million persons exposed.[45] It is even produced by the human body from the transformation of ethylene. It has been argued that the dose of ethylene oxide received from the use surfactants in cosmetic products containing 1 ppm is negligible in comparison to the quantity produced by the body.[46]

Studies have raised concern about the contamination of herbicides by heavy metals, which could contribute to chronic kidney disease.[47] Urinary heavy metal concentrations were higher in children and adolescents most exposed to glyphosate in rural regions of Slovenia.[48] Two studies have investigated heavy metal contamination.[49,50] Among the 36 pesticide formulations tested, including 11 glyphosate-based agents, arsenic was found at a median level of 52 μg/L.

Heavy metal contamination in herbicides is unlikely to be a significant risk for human health, especially from dietary exposure. When glyphosate is sprayed on a crop, its residues are generally found at levels several orders of magnitude lower than the spray dose. For instance, levels in supervised residue trials on conventional wheat grains cultivated according to good agricultural practices (harvested 7 days after an application of ~1 kg glyphosate/ha) reached a median of 1 mg/kg (max 18 mg/kg).[51] Because arsenic levels in a bottle of concentrated glyphosate are more than a million times lower than that of glyphosate (commonly 360 g/L), the degree of foodstuff contamination is negligible. By comparison, arsenic in soil and irrigation water at the mg/kg level is the major contributor to wheat contamination because arsenic is naturally occurring in the environment.[52]

Enhancement of dermal penetration by surfactants

The reduction of surface tension explains the phenomenon of allowing the active substance (the water-soluble part) to penetrate plant cell membranes (the lipid-soluble part). In the presence of a surfactant, the forces holding together water molecules are decreased. This allows the water to spread on the surface of the plant leaves, increasing the penetrability of the active ingredient in the droplets. The skin is made of three layers. The outer one, the epidermis, is a barrier providing protection against environmental pathogens. The upper stratum of the epidermis is constituted of dead keratinocytes with low water content, making the surface of the skin very lipophilic. The action of surfactants is comparable on human and leaf epidermis. Surfactants are added to cosmetics and drugs to increase the penetration of active substances. POEA surfactants with different structural characteristics have different absorption abilities. A study of ibuprofen showed that dermal penetration was most effective with POEA surfactants with an ethoxylated chain (POE) length of 2–5 EO units and an alkyl chain of 16–18 carbon groups.[53] Adjuvants also increase the penetration of herbicides into human skin, as shown in a study comparing atrazine, alachlor, and trifluralin to their commercial formulations Aatrex, Lasso, and Treflan.[54]

The interaction of surfactants upon contact with skin has been reviewed. Five key features that influence the dermal penetration of active substances were proposed[55,56]:

- Binding to the surface proteins of the skin.
- Denaturing skin surface proteins.
- Solubilizing or disorganizing the intercellular lipids of the skin.
- Penetrating the epidermal lipid barrier.
- Interacting with the living cell.

Hydrophilic active substances such as glyphosate have very low dermal penetration when the epidermis is intact. In the first study to evaluate the dermal penetration of Roundup MON 2139 (the Roundup original containing 15% POE-15 tallowamine), 0.063% of the formulation concentration and 0.152% of a spray dilution were absorbed by fresh human abdominal skin in vitro.[57] Regulatory studies were used to set a dermal penetration factor of 3% in the European Union (EU) monograph for the 2002 reauthorization of glyphosate.[58] The dermal absorption factor was changed in the most recent peer review of the pesticide risk assessment of glyphosate in Europe because MON 2139 is no longer used there. The representative formulation chosen is MON 52276, with dermal absorption set at 1% for concentrate and in-use spray dilutions. The toxicity of glyphosate and MON 52276 was recently studied.[59] Although MON 52276 is used as a reference formulation, others are sold in Europe. They contain different surfactants (such as etheramine ethoxylated) that may have a different effect on glyphosate dermal penetration.

Confusion on the composition of formulated herbicides

Although the different compounds in the composition of formulated pesticides have clear regulatory definitions, the difference between an active and an inert ingredient is not always clearly reflected in their toxicological properties. Ingredients labeled as inerts can be toxic. A classic example is the toxicity of surfactants from the adjuvant preparation MON 0818, which was included in the first generation of glyphosate products, and which contributed substantial acute toxic effects to the final preparations while glyphosate did not. "Inert" gave a false impression of safety, and therefore was changed to "other ingredients" by the US Environmental Protection Agency (EPA). "Coformulant" is more used in European regulations.

According to the International Union of Pure and Applied Chemistry (IUPAC), an active ingredient is "the part of a pesticide formulation from which the biological effect is obtained." For a herbicide, this definition holds true at recommended agricultural dilutions at which the active substance causes toxic effects in plants while exposure to the inert ingredients has no herbicidal effect. However, inert ingredients can have herbicidal properties at higher concentrations, although this generally arises through a phytotoxicity mechanism, which is not desirable for the efficiency of weed removal. The phytotoxicity of surfactants was reported in 1888 in early experiments on the addition of soapy solutions to pesticide sprays.[15] Foliar damage due to the presence of a phytotoxic surfactant can impair the translocation of active substances to deeper tissues, leading to only partial damage to the unwanted plant.

In addition, the fact that a herbicide is active in plants does not mean a priori that it is the most toxic element of the mixture for humans or other organisms in ecosystems. There is an unexpressed assumption that the active substance against plant metabolism is the most toxic compound of the formulations on nontarget species. For instance, quaternary ammonium compounds (QACs) have a variety of uses: cationic surfactants, disinfectants, biocides, herbicides (growth regulators), and fabric softeners. Some are surfactants that promote the action of pesticides by increasing penetration. This is the case for Dodigen. However, QACs might have herbicidal properties at high concentrations above agricultural dilutions. They can be very toxic depending on the dose and the organism, which is why some are considered active substances in biocides. They act on mitochondrial bioenergetics by inhibiting mitochondrial respiration initiated at complex I at concentrations below the CMC in mammalian epithelial cell cultures.[60] QACs are not herbicidal alone at recommended agricultural dilutions, which is why they are not considered active substances. However, when a worker is accidentally exposed to a concentrated product, QACs can be present at a dose toxic to humans. This is the case of most surfactants, which is why coformulants are known to be responsible for the acute toxic effects of some herbicide formulations, rather than the active ingredient.

Confusion is frequent in academic articles. Authors say that they tested "glyphosate," which suggests glyphosate alone, when in fact they used a commercial formulation containing additional ingredients. Very few authors provide information on where and when the product was manufactured and procured, introducing a source of uncertainty. These shortcomings can impair the evaluation of risks associated with herbicide exposure. Guidance on the minimum quantity of information to allow reproducibility across herbicide toxicity studies can be found in a recent paper by Mesnage et al.[23]

Some active substances such as glyphosate are sold in different salt forms such as glyphosate-isopropylammonium, glyphosate-potassium, and glyphosate-ammonium.[61,62] All can be contaminated by impurities. Collectively, formulated products should be considered mixtures of compounds with different toxicity profiles. Toxicity is always a matter of dose, and this cannot be reflected by the binary classification of active or inactive.

Regulation of coformulants

Before the EPA was created by Richard Nixon's administration in 1970, the registration of pesticides was managed by the Food and Drug Administration (FDA). In 1961, the FDA published notice 26 FR 10640 stating that the USDA had determined that each component of a registered product, including its coformulants, was considered a pesticide chemical and subject to the same requirements for tolerance under section 408 of the Federal Food, Drug, and Cosmetic Act. This notice stipulates that sufficient information be available on some of these ingredients to conclude that their use will result in no hazard to health. In addition, notice is given that the Commissioner of Food and Drugs, on his own initiative, proposes to add certain inert ingredients used in pesticides to the list of chemicals exempted from the requirement that tolerances be determined. In 1969, the FDA established more detailed policies on inert ingredients, stipulating that data requirements would depend on the toxicity of the chemical. The current regulation of inert ingredients has its origins in the EPA inert strategy from 1987, which grouped inert ingredients in different toxicity categories, from List 1, the most toxic (known to cause cancer or other chronic health effects), to List 4, minimal risk (inerts), which are generally recognized as safe (GRAS).

The inerts list is regularly updated to reflect new tolerance exemptions. For instance, the residues of phosphate ester ethoxylated tallowamine introduced to reduce the toxicity of ethoxylated tallowamines were exempt from tolerance in regulation 75 FR 22234 of the Federal Register of April 28, 2010. Interestingly, the toxicological assessment of phosphate ester ethoxylated tallowamine established a chronic reference dose (cRfD) of 1 mg/kg/day based on mortalities and clinical signs in a reproduction and developmental study in rats. Although this profile suggested that the surfactant was as least as toxic as glyphosate if not more so (cRfD of 1.75 mg/kg/day), it was exempt from tolerance while the tolerance for residues of glyphosate-based herbicides is established for glyphosate only.

Pesticide toxicity tests follow a common set of standards adopted by the Organisation for Economic Co-operation and Development (OECD) countries to facilitate registration and international trade. OECD guidelines define the protocols of tests that must be performed on active ingredients and formulated products. In the United States, this is managed by the EPA Office of Chemical Safety and Pollution Prevention, which publishes regulatory guidance documents. These documents are a blend of EPA data requirements and OECD guidelines.

Companies applying for marketing authorization are responsible for testing any pesticide they want to register. The results are considered confidential business information. Companies seeking regulatory approval analyze their test results and submit their interpretations to regulatory authorities.

Active ingredients are tested in a battery of cell assays and rodent bioassays to establish health-based guidance values such as the chronic reference dose. A wide range of tests are used such as long-term toxicity and carcinogenicity and in vivo and in vitro genotoxicity. Coformulants are generally exempt. Because their risks are considered to be negligible, only a few tests are performed on the commercialized product (the final combination of inerts and active ingredients). The only toxicity testing of formulated products required by OECD guidelines is for short-term toxicity to determine dermal and ocular irritation properties. Chronic and subchronic effects are not evaluated.[63]

Tests to evaluate the toxicity of pesticides are very expensive and involve the use of many laboratory animals. Approximately 1 million animals are used every year in Europe. A single carcinogenicity test for a pesticide active ingredient can require 200–400 rats. Performing the complete battery on every pesticide formulation is not feasible. Alternatives need to be found. There is no magic bullet, but some solutions could be rapidly implemented at low cost. This includes disclosure on the chemical composition of pesticides.

The lack of transparency limits the scientific evaluation of pesticides.[64] Because the exact composition of formulated products is often confidential business information, most scientific studies do not report information sufficient for other researchers to reproduce results.

Another problem is the lack of postmarket monitoring of compounds used as coformulants or adjuvants. Although contamination by active substances is regularly monitored in foodstuffs and water,[65] very little information is available on the fate of coformulants in the environment and the potential human exposure. In a study of groundwater contamination by surfactants in agricultural areas, the total concentration of six alcohol ethoxylates was found to be 710 ng/L.[66] Only one study was done to understand the persistence of POEA surfactants in the environment after agricultural application. US Geological Survey scientists found that POEA persisted in surface soil from planting season to planting season but did not substantially migrate into deeper soil. The study was done on the soils of six states: Georgia, Hawaii, Iowa, Mississippi, North Carolina, and South Carolina.[67] They found that POEA was lost more slowly than glyphosate.[68]

The lack of monitoring of toxic substances entering the food chain can have far-reaching consequences. A broad range of toxic substances can contaminate food, water, or cage materials, increasing the so-called spontaneous background of pathologies that can affect control animals in scientific studies. We encountered this issue some years ago. The control animals of a chronic toxicity study in rats to evaluate Roundup use on a Roundup-tolerant genetically modified crop were developing many tumors. The literature gave inconsistent rates of tumors for control animals across experiments. For instance, the incidence of pituitary adenomas in control groups of Sprague-Dawley rats from Charles River ranged from 26% to 93%.[69] These incidences were not stable over time, and we hypothesized that they could be due to the presence of toxic levels of environmental contaminants in the rodent diets. We described the contamination of 13 laboratory rodent diets by pesticides, heavy metals, PCDD/Fs, and PCBs.[70] Methods were not available to measure coformulants such as surfactants, and further studies will have to be undertaken to tackle this issue. The use of QAC disinfectants in rodent husbandry can affect the sensitivity of the animals and cause reproducibility problems.[71] It is not clear if the use of formulated products as preharvest desiccants will carry over coformulants into the food supply. Interestingly, POE-tallowamine surfactants were prioritized for biomonitoring in the European Human Biomonitoring Initiative.[72]

Confidential classification can sometimes serve an important and positive function within the industry by providing a mechanism through which manufacturers can gain and hold market share via innovative formulation technology. Tweaking formulations to enhance the efficacy of a given herbicide for certain uses (e.g., treating a pond, tree bark, or weed), in specific environments (e.g., low-pH soils), applied via certain equipment (airplane vs. backpack sprayer), and sold in other agrochemical market segments can confer competitive advantage and foster innovation that improves product efficacy and advances the agricultural sector. There are also compelling reasons to put in place new pesticide product labeling requirements that include a mandatory, accurate description of the full chemical composition. A compromise should be found with policymakers by coupling the requirement for full disclosure with policy reforms to reward innovation.

Another solution would be to prioritize the testing and evaluation of the different pesticide mixtures. Although they have different commercial names, many formulated products have similar compositions. It should be possible to cluster glyphosate-based herbicides into categories based on compositional similarities and to submit representative formulations to long-term toxicity testing.

Regulatory guidance values established to evaluate health risks often combine the results of animal toxicity tests with safety factors. Currently, the toxicity threshold taken from an animal study is divided by a factor of 100 to consider the diversity of responses in animal species and the human population to set a "safe" dose (also called acceptable daily intake). It would not be unreasonable to add another safety factor to take into account the uncertainty caused by the addition of coformulants.

Recent studies have made clear that there are gaps in the regulatory assessment of pesticides. Important remedial steps have already been undertaken. The EU has the most protective pesticide regulations in the world; 1107/2009 is underpinned by the precautionary principle. In July, the European Commission presented a black list of "unacceptable coformulants," which was discussed with EU government representatives to modify Annex III of 1107/2009.[73] The list includes POE-tallowamine, nonylphenol, formaldehyde, *n*-butyl phthalate, and some benzene derivatives. This is a big step, showing that EU authorities are starting to embrace the problem of coformulants.

Toxicity in humans

Because petroleum distillates are frequently used as pesticides, several epidemiological studies have investigated the association between their use and the occurrence of various chronic diseases. Among 22,134 men interviewed in the Agricultural Health Study (AHS, conducted by four US government agencies beginning in 1993), the use of 78 pesticides was reported; of these, petroleum distillates had the highest odds of being associated with nonallergic and allergic wheeze.[74] In another part of the AHS, residential proximity to agriculture-related applications of pesticides containing paraffin-based petroleum oil and the adjuvant polyoxyethylene sorbitol was associated with elevated risk of hypospadias in the Central Valley of California.[75] Serum antinuclear autoantibodies were elevated in 668 male farmers of an AHS subcohort who reported the use of petroleum oil distillates as a herbicide.[76] This association was corroborated by studies in laboratory animals showing that a byproduct of petroleum distillation can induce lupus-like autoantibodies.[77] Another subcohort showed an increased prostate cancer risk with high compared to no petroleum oil or distillate exposure among men carrying a variant allele in xenobiotic metabolic enzyme genes.[78] These results may not account for the inclusion of petroleum distillates in coformulants. It is thus plausible that some associations between the use of some pesticide active substances and increased risks of disease development are due to coformulants, or to their promotional effect on the toxicity of the active substances. Petroleum distillates (known collectively as naphtha) are among the most commonly used coformulants. We found differential cytotoxicity between formulations of pesticides and their active substances and attributed the effect to the inclusion of naphtha.[79]

The confidentiality of compounds used as coformulants means that health risks to farm workers are not well characterized. Active substances are the compounds predominantly involved in systemic toxic effects observed after accidental ingestion. Most of the pesticides implicated in patients admitted to the Ain Shams University Poisoning Treatment Center in Cairo in 2006 were organophosphates.[80] In the US National Institute of Occupational Safety and Health SENSOR-Pesticides program, pyrethroid insecticides were most often implicated in acute occupational pesticide-related illness (28.7% of all cases) during the period 2005–09 ($N = 9906$).[81] It is also widely recognized that herbicides containing paraquat are among the pesticides most frequently found responsible for self-poisoning.[82] However, surfactants can also cause systemic toxicity after accidental exposure. Among 107 patients with glyphosate poisoning studied at Soonchunhyang University Cheonan Hospital in South Korea, surfactant volume was found to be an essential element to predict toxic outcomes.[83] Although the surfactants are known to contribute to the acute toxicity of glyphosate-formulated products, they are thought not to potentiate the toxicity of glyphosate in accidental ingestion.[84] Overall, penetrant agents such as heavy petroleum oils and ethoxylated surfactants, which are eye and skin irritants, account for a large proportion of the formulated product acute toxicity reports.

Some surfactants have congener-specific toxic properties. For instance, nonylphenol polyethoxylates are degraded in the environment to breakdown products that are endocrine disruptors.[85] An in vitro study in human cells found a disturbance of the aromatase enzyme, which controls the androgen-estrogen sex hormone balance.[24] Nonylphenol polyethoxylate surfactants (e.g., Agral 90) are used with pyrethroid insecticides (Ambush 500EC) and glyphosate herbicides (Touchdown 480), among other formulated products. A variety of nonionic surfactants were found to promote triglyceride accumulation and/or preadipocyte proliferation in 3T3-L1 cells at environmentally relevant concentrations.[86] The authors suggested that they promoted triglyceride accumulation through a PPARγ-independent mechanism linked to the activation of thyroid hormone receptor beta.

Although quaternary surfactants have favorable toxicity profiles by comparison to ethoxylated surfactants, they also have specific toxicological properties. The introduction of QAC disinfectants in rodent husbandry was linked to an increased rate of reproductive and developmental effects in mice, first reported by Patricia Hunt at Washington State University in Pullman.[87] ADBAC (*n*-alkyl dimethyl benzyl ammonium chloride) and DDAC (didecyl dimethyl ammonium chloride) caused a massive decline in fertility. Decreased fertility was also observed in the Hrubec laboratory at Virginia Polytechnic Institute. This group replicated the effect in a controlled experiment.[88] If animals from different facilities are exposed to different toxic compounds, thereby changing their sensitivity differentially, it can have far-reaching negative consequences on the reproducibility of results. An example is a study on allergic lung inflammation in mice induced by a model allergen.[71] Similar reproducibility problems were hypothesized to be caused by the uncontrolled contamination of rodent diets by pesticide residues.[70]

Inhalation is the main accidental exposure route for aromatic hydrocarbons such as toluene and xylene. Petroleum distillates are highly volatile and are responsible for the strong odors of some formulated pesticides. The most common features of hydrocarbon pneumonitis caused by accidental inhalation of these compounds are rapid respiration, cyanosis, tachycardia, and low-grade fever.[89] The chlorinated solvents in some formulated products add significantly to the negative health consequences of accidental exposure, especially ingestion.[81] Nonetheless, the addition of other coformulants such as alcohols, glycols, and ethers is often considered not to significantly change the acute mammalian toxicity of the products. Similarly, substances used as stickers and spreaders such as flour, gelatin, oils, gums, resins, clays, polyoxyethylene glycols, and terpenes are also unlikely to substantially change the toxic effects of formulated products.

Conclusions

The use of coformulants in herbicides, and the addition of adjuvants to tank mixes, is a cornerstone of weed management. Little progress has been made on active substances. In 2012, no major new mode of action had been introduced to the marketplace for about 20 years.[90] Behind the scenes, continuous progress in the science of adjuvant technology is increasing the safety and efficiency of herbicides. However, the success of adjuvant technology is hindered by controversies about the neglect of their potential human health effects. When surfactants and oils started to be used at the beginning of the 20th century, they were recognized as an important source of toxicity. Early regulations considered that each component of a formulated herbicide should be subjected to the same safety requirements. This is in contrast to the modern regulation of adjuvants, in which they are considered nontoxic, inert substances and information on them is held confidentially for business reasons. Although this confidentiality serves an important purpose for manufacturers to gain and retain competitive advantages, the recent debate about the toxicity of glyphosate-formulated products shows that it can also cause confusion in regulatory and academic circles. The debate on glyphosate carcinogenicity would certainly be very different if pesticide-product labeling requirements included a description of the full chemical composition, and if all ingredients were included in the postmarketing surveillance program. Improving transparency would also be key to building public trust in agricultural technologies.

Close-up: The first long-term toxicity study of a glyphosate-formulated product

One of my earliest research studies when I joined Professor Gilles-Eric Séralini's group at the University of Caen in France was to compare the toxicity of glyphosate-formulated herbicides on cultures of human cells. We observed that formulated products were more toxic than glyphosate alone, and that some products were more toxic than others.[32] We tested seven glyphosate formulations. Some were as toxic as glyphosate, and others were approximately 300 times more toxic. We did not stop at the cytotoxicity profiles. We looked beyond the glyphosate label by analyzing the composition of the coformulants by mass spectrometry. This revealed that the toxic effects were linked to the presence of ethoxylated tallowamine surfactants.[32] Although this was not the first study showing the toxicity of these surfactants, it revealed that they were more toxic than other surfactants in other glyphosate-based herbicides sold for the same purposes. So ethoxylated tallowamines were not essential and could be substituted. This observation contributed to the banning of ethoxylated tallowamines from glyphosate-based herbicides in Europe.

Because formulated glyphosate products were not tested for their long-term effects, concerns were raised about the potential health consequences of the consumption of their residues. This debate is closely connected to the debate on the health effects of genetically modified plants, most of which are engineered to tolerate glyphosate agricultural sprays, thus enhancing dietary exposure to their residues. When I started my PhD in 2010, the controversy on GM crops was raging. No long-term studies were available to understand whether the residues from glyphosate spray on GM plants could be a source of health risk. Repeated-dose 90-day toxicity studies on GM crops fed to rats often produced statistically significant differences in liver and kidney biochemical parameters,[91] but it was not clear whether these differences were due to chance or to early signs of the development of chronic disease. The research group initiated the first study of the long-term toxicity of Roundup in Roundup-tolerant GM maize, that is, sprayed or not sprayed with Roundup.[92]

The first manuscript describing the main findings was published in September 2012 in the journal *Food and Chemical Toxicology*.[92] The study was exploratory and not conceived to conclude with certainty on long-term effects because the pathologies of normal aging introduce background noise that impairs statistical power with 10 animals per group. The strain of rat used, Sprague-Dawley, is particularly prone to tumors. The paper caused an explosion of media coverage because it contained graphic images of rats with large tumors.[93] The media coverage strategy used by Pr. Séralini was heavily criticized and condemned by the scientific community and journalists. This had a strong influence on public opinion about GM foods.[94] The debate even led to serious political consequences, with countries such as Kenya and Peru placing a moratorium on GM crops. I had argued against publication of the tumor pictures as misleading because 10 rats per group was insufficient to demonstrate significant incidence.

One year after publication, the editor-in-chief of the journal requested the raw data. This was examined by a panel of experts, who together with the editor-in-chief decided that the paper should be retracted because "the results presented (while not incorrect) are inconclusive."[92] Pr. Séralini raised the issues of conflict of interest and double standards, as the retraction could not be rationalized on any discernible scientific or ethical grounds.[95] This revived the debate, as many scientists protested that the retraction of a paper because it is inconclusive based on a post hoc analysis represented an erosion of scientific integrity.[96] The debate continues, as lawsuits have uncovered evidence that Monsanto, the maker of Roundup, engaged with a network of scientists to discredit the study and obtain retraction of the paper.[97]

Having finished my doctorate during this storm of criticism, I felt that the only way to retain integrity was to continue research on the topic with the most up-to-date techniques. High-throughput omics methods can be used to study biomarkers that predict the development of chronic diseases in short-term experiments. I moved to Dr. Michael Antoniou's group at King's College London to generate molecular profiles (proteomics, transcriptomics, and metabolomics) of livers and kidneys in rats to investigate further whether feeding them GM maize sprayed or not sprayed with Roundup resulted in pathology. After several years of work, we did not find any indication of altered organ function. There was no indication of adverse effects from feeding the NK603 GM maize cultivated with or without Roundup.[98]

Although there were no effects of feeding NK603, liver molecular profiles suggested the development of nonalcoholic fatty liver disease for rats exposed to Roundup in drinking water.[99] At present, and despite the serious political and economic consequences of the debate on glyphosate toxicity, Séralini's is still the only long-term toxicity study of a formulated glyphosate product. Although publicly funded projects (GMO90+ and G-TwYST) were launched to replicate and extend the arms of our study investigating the toxicity of the GM crops, the long-term toxicity of the Roundup formulation was never reproduced.[100] This situation will ultimately change with the so-called global glyphosate study, a combined chronic toxicity/carcinogenicity study performed by the Ramazzini Institute.[101] Debate on glyphosate toxicity would have been very different if a glyphosate-formulated product had been tested in this way at some point over the last 40 years of glyphosate commercialization.

References

1. Varanasi A, Prasad PVV, Jugulam M. Impact of climate change factors on weeds and herbicide efficacy. In: Sparks DL, editor. *Advances in agronomy*, vol. 135. Academic Press; 2016. p. 107–46 [chapter 3].
2. European Parliament. Regulation (EC) No 1107/2009 of the European Parliament and of the Council of 21 October 2009 concerning the placing of plant protection products on the market and repealing Council Directives 79/117/EEC and 91/414/EEC. *Off J Eur Union* 2009.
3. Butler O. Chemical, physical, and biological properties of bordeaux mixtures. *Ind Eng Chem* 1923;**15**(10):1039–41.
4. Marketsandmarkets. *Agricultural adjuvants market worth $4.7 billion by 2026*; 2019. Available at: https://www.marketsandmarkets.com/Market-Reports/adjuvant-market-1240.html.
5. Hartzell FZ. The influence of molasses on the adhesiveness of arsenate of lead. *J Econ Entomol* 1918;**11**(1):62–6.
6. Grillo R, Abhilash PC, Fraceto LF. Nanotechnology applied to bio-encapsulation of pesticides. *J Nanosci Nanotechnol* 2016;**16**(1):1231–4.
7. Hazen JL. Adjuvants—terminology, classification, and chemistry. *Weed Technol* 2000;**14**(4):773–84.

8. Gaskin RE, Steele KD. A comparison of sticker adjuvants for their effects on retention and rainfastening of fungicide sprays. *N Z Plant Prot* 2009;**62**:339–42.
9. Sandín-España P, Sevilla-Morán B, López-Goti C, Mateo-Miranda MM, Alonso-Prados JL. Rapid photodegradation of clethodim and sethoxydim herbicides in soil and plant surface model systems. *Arab J Chem* 2016;**9**:694–703.
10. Hussain M, Perschke H, Kutscher R. The effect of selected uv absorber compounds on the photodegradation of pyrethroid insecticides applied to cotton fabric screens. *Pestic Sci* 1990;**28**:345–55.
11. Apodaca MA, et al. Drift control polymers and formulation type affect volumetric droplet size spectra of propanil sprays. *J Environ Sci Health B* 1996;**31**:859–70.
12. Mueller TC, Steckel LE. Spray mixture pH as affected by dicamba, glyphosate, and spray additives. *Weed Technol* 2019;**33**:547–54.
13. Brown A, Willoughby I, Clay DV, Moore R, Dixon F. The use of dye markers as a potential method of reducing pesticide use. *Forestry: Int J Forest Res* 2003;**76**:371–84.
14. Nalewaja JD, Robert M. Optimizing adjuvants to overcome glyphosate antagonistic salts. *Weed Technol* 1993;**7**:337–42.
15. Gillette CP. Experiments with arsenites. *Bulletin* 1888;**1**(10). Article 3. Available at: https://lib.dr.iastate.edu/bulletin/vol1/iss10/3.
16. Buchanan GA. Patterns of surfactant toxicity to plant tissues. In: *Retrospective Theses and Dissertations*; 1965. https://lib.dr.iastate.edu/rtd/3905.
17. Ceafts AS, Harvey WA. Weed control. In: Norman AG, editor. *Advances in agronomy*, vol. 1. Academic Press; 1949. p. 289–320.
18. U.S. Department of Agriculture. *Technical bulletin*. U.S. Department of Agriculture; 1963.
19. Jansen LL, Gentner WA, Shaw WC. Effects of surfactants on the herbicidal activity of several herbicides in aqueous spray systems. *Weeds* 1961;**9**(3):381–405.
20. Jones GE, Anderson G. Atrazine as a foliar application in an oil water emulsion. In: *Research report north central weed control conference 20*; 1964. p. 27–9.
21. Osipow L, Snell FD, York WC, Finchler A. Methods of preparation fatty acid esters of sucrose. *Ind Eng Chem* 1956;**48**(9):1459–62.
22. Guyton KZ, Loomis D, Grosse Y, et al. Carcinogenicity of tetrachlorvinphos, parathion, malathion, diazinon, and glyphosate. *Lancet Oncol* 2015;**16**(5):490–1.
23. Mesnage R, Benbrook C, Antoniou MN. Insight into the confusion over surfactant co-formulants in glyphosate-based herbicides. *Food Chem Toxicol* 2019;**128**:137–45.
24. Defarge N, Takacs E, Lozano VL, et al. Co-formulants in glyphosate-based herbicides disrupt aromatase activity in human cells below toxic levels. *Int J Environ Res Public Health* 2016;**13**(3).
25. Tu M, Hurd C, Randall JM. The Nature Conservancy. Weed control methods handbook: tools & techniques for use in natural areas. In: *All U.S. Government Documents (Utah Regional Depository)*. Paper 533; 2001. https://digitalcommons.usu.edu/govdocs/533.
26. Willoughby IH, Stokes VJ. Mixture B new formulation adjuvant increases the rainfastness and hence effectiveness of glyphosate for rhododendron control. *Forestry* 2014;**88**(2):172–9.
27. van Os NM. *Nonionic surfactants: organic chemistry*. Taylor & Francis; 1997.
28. Koley D, Bard AJ. Triton X-100 concentration effects on membrane permeability of a single HeLa cell by scanning electrochemical microscopy (SECM). *Proc Natl Acad Sci* 2010;**107**(39):16783.
29. Groot RD, Rabone KL. Mesoscopic simulation of cell membrane damage, morphology change and rupture by nonionic surfactants. *Biophys J* 2001;**81**(2):725–[illegible]

30. Lechuga M, Fernández-Serrano M, Jurado E, Núñez-Olea J, Ríos F. Acute toxicity of anionic and non-ionic surfactants to aquatic organisms. *Ecotoxicol Environ Saf* 2016;**125**:1–8.
31. Talmage SS. *Environmental and human safety of major surfactants: alcohol ethoxylates and alkylphenol ethoxylates*. Taylor & Francis; 1994.
32. Mesnage R, Bernay B, Seralini GE. Ethoxylated adjuvants of glyphosate-based herbicides are active principles of human cell toxicity. *Toxicology* 2013;**313**(2–3):122–8.
33. Rice JR, Dunlap P, Ramaiahgari S, Ferguson F, Smith-Roe SL, DeVito M. *Effects of glyphosate and its formulations on DNA damage in HepaRG and HaCaT cell lines*; 2019. Available at: https://ntp.niehs.nih.gov/ntp/results/pubs/posters/rice_sot20190300.pdf.
34. Swartz CD, Christy NC, Sly JE, Witt KL, Smith-Roe SL. *Comparison of the genotoxicity of glyphosate, (aminomethyl)phosphonic acid, and glyphosate-based formulations using in vitro approaches*; 2019. Available at: https://ntp.niehs.nih.gov/ntp/results/pubs/posters/swartz_emgs20190919.pdf.
35. Benbrook CM. How did the US EPA and IARC reach diametrically opposed conclusions on the genotoxicity of glyphosate-based herbicides? *Environ Sci Eur* 2019;**31**(1):2.
36. Bolognesi C, Bonatti S, Degan P, et al. Genotoxic activity of glyphosate and its technical formulation roundup. *J Agric Food Chem* 1997;**45**(5):1957–62.
37. Koller VJ, Furhacker M, Nersesyan A, Misik M, Eisenbauer M, Knasmueller S. Cytotoxic and DNA-damaging properties of glyphosate and Roundup in human-derived buccal epithelial cells. *Arch Toxicol* 2012;**86**(5):805–13.
38. Peluso M, Munnia A, Bolognesi C, Parodi S. 32P-postlabeling detection of DNA adducts in mice treated with the herbicide Roundup. *Environ Mol Mutagen* 1998;**31**(1):55–9.
39. Rank J, Jensen AG, Skov B, Pedersen LH, Jensen K. Genotoxicity testing of the herbicide Roundup and its active ingredient glyphosate isopropylamine using the mouse bone marrow micronucleus test, *Salmonella* mutagenicity test, and *Allium* anaphase-telophase test. *Mutat Res* 1993;**300**(1):29–36.
40. De Almeida LKS, Pletschke BI, Frost CL. Moderate levels of glyphosate and its formulations vary in their cytotoxicity and genotoxicity in a whole blood model and in human cell lines with different estrogen receptor status. *3 Biotech* 2018;**8**(10):438.
41. Moreno NC, Sofia SH, Martinez CB. Genotoxic effects of the herbicide Roundup Transorb and its active ingredient glyphosate on the fish *Prochilodus lineatus*. *Environ Toxicol Pharmacol* 2014;**37**(1):448–54.
42. EFSA. Request for the evaluation of the toxicological assessment of the co-formulant POE-tallowamine. *EFSA J* 2015;**13**(11):4303.
43. Schwartz AM. Surface active ethylene oxide adducts. In: Schönfeldt N, editor. *Surface active ethylene oxide adducts*. Pergamon; 1969. p. xvii.
44. USDA. *Glyphosate herbicide information profile*; February 1997. Available at: https://www.fs.usda.gov/Internet/FSE_DOCUMENTS/fsbdev2_025810.pdf.
45. OEHHA. *Public hearing to consider the adoption of a regulatory amendment identifying ethylene oxide as a toxic air contaminant*; 1987. Available at: https://oehha.ca.gov/media/downloads/air/document/ethylene20oxide.pdf.
46. Filser JG, Kreuzer PE, Greim H, Bolt HM. New scientific arguments for regulation of ethylene oxide residues in skin-care products. *Arch Toxicol* 1994;**68**(7):401–5.
47. Babich R, Ulrich JC, Ekanayake EMDV, et al. Kidney developmental effects of metal-herbicide mixtures: implications for chronic kidney disease of unknown etiology. *Environ Int* 2020;**144**:106019.

48. Stajnko A, Snoj Tratnik J, Kosjek T, et al. Seasonal glyphosate and AMPA levels in urine of children and adolescents living in rural regions of Northeastern Slovenia. *Environ Int* 2020;**143**:105985.
49. Defarge N, De Vendômois JS, Séralini G. Toxicity of formulants and heavy metals in glyphosate-based herbicides and other pesticides. *Toxicol Rep* 2018;**5**:156–63.
50. Seralini G-E, Jungers G. Toxic compounds in herbicides without glyphosate. *Food Chem Toxicol* 2020;**146**:111770.
51. European Food Safety Authority. Review of the existing maximum residue levels for glyphosate according to Article 12 of Regulation (EC) No 396/2005 - revised version to take into account omitted data. *EFSA J* 2019;**17**(10):e05862.
52. Bhattacharya P, Samal AC, Majumdar J, Santra SC. Arsenic contamination in rice, wheat, pulses, and vegetables: a study in an arsenic affected area of West Bengal, India. *Water Air Soil Pollut* 2010;**213**(1):3–13.
53. Park ES, Chang SY, Hahn M, Chi SC. Enhancing effect of polyoxyethylene alkyl ethers on the skin permeation of ibuprofen. *Int J Pharm* 2000;**209**(1–2):109–19.
54. Brand RM, Mueller C. Transdermal penetration of atrazine, alachlor, and trifluralin: effect of formulation. *Toxicol Sci* 2002;**68**(1):18–23.
55. Som I, Bhatia K, Yasir M. Status of surfactants as penetration enhancers in transdermal drug delivery. *J Pharm Bioallied Sci* 2012;**4**(1):2–9.
56. Barel A, Paye M, Maibach H. *Handbook of cosmetic science and technology*. Boca Raton: CRC Press; 2014.
57. Franz R. *Evaluation of the percutaneous absorption of Roundup(R) formulation in man using an in vitro technique*. Unpublished study No UW-81-346 from University of Washington School of Medicine, Seattle, WA, USA Submitted to WHO by Monsanto Europe, SA, Brussels, Belgium; 1983.
58. European Commission. *Review report for the active substance glyphosate*. Finalised in the Standing Committee on Plant Health at its meeting on 29 June 2001 in view of the inclusion of glyphosate in Annex I of Directive 91/414/EEC. Document 6511/VI/99-final 21 January 2002; 2002.
59. Mesnage R, Teixeira M, Mandrioli D, Falcioni L, Ducarmon QR, Zwittink RD, et al. Use of shotgun metagenomics and metabolomics to evaluate the impact of glyphosate or roundup MON 52276 on the gut microbiota and serum metabolome of Sprague-Dawley rats. *Environ Health Perspect* 2021;**129**(1):17005. https://doi.org/10.1289/EHP6990.
60. Inácio ÂS, Costa GN, Domingues NS, et al. Mitochondrial dysfunction is the focus of quaternary ammonium surfactant toxicity to mammalian epithelial cells. *Antimicrob Agents Chemother* 2013;**57**(6):2631.
61. Cuhra M, Bøhn T, Cuhra P. Glyphosate: too much of a good thing? *Front Environ Sci* 2016;**4**(28).
62. Székács A, Darvas B. Forty years with glyphosate. In: Hasaneen MNAE-G, editor. Herbicides—properties, synthesis and control of weeds. Available from: *IntechOpen* 2012. https://doi.org/10.5772/32491. https://www.intechopen.com/books/herbicides-properties-synthesis-and-control-of-weeds/forty-years-with-glyphosate.
63. OECD. *Data requirements for registration of pesticides in OECD member countries: survey results*; 1996. Available at: https://www.oecd.org/env/ehs/pesticides-biocides/data-for-biopesticide-registration.htm.
64. Nielsen KM. Biosafety data as confidential business information. *PLoS Biol* 2013;**11**(3), e1001499.
65. European Food Safety Authority, Medina-Pastor P, Triacchini G. The 2018 European Union report on pesticide residues in food. *EFSA J* 2020;**18**(4):e06057.

66. Krogh KA, Vejrup KV, Mogensen BB, Halling-Sorensen B. Liquid chromatography-mass spectrometry method to determine alcohol ethoxylates and alkylamine ethoxylates in soil interstitial water, ground water and surface water samples. *J Chromatogr A* 2002;**957** (1):45–57.
67. Tush D, Maksimowicz MM, Meyer MT. Dissipation of polyoxyethylene tallow amine (POEA) and glyphosate in an agricultural field and their co-occurrence on streambed sediments. *Sci Total Environ* 2018;**636**:212–9.
68. Tush D, Meyer MT. Polyoxyethylene tallow amine, a glyphosate formulation adjuvant: soil adsorption characteristics, degradation profile, and occurrence on selected soils from agricultural fields in Iowa, Illinois, Indiana, Kansas, Mississippi, and Missouri. *Environ Sci Technol* 2016;**50**(11):5781–9.
69. Giknis MLA, Clifford CB. *Compilation of spontaneous neoplastic lesions and survival in Crl: CD® (SD) rats from control groups*. Charles River Laboratories; 2004.
70. Mesnage R, Defarge N, Rocque LM, Spiroux de Vendomois J, Seralini GE. Laboratory rodent diets contain toxic levels of environmental contaminants: implications for regulatory tests. *PLoS One* 2015;**10**(7):e0128429.
71. Fallon PG. Why does work on same mouse models give different results? *Nature* 2008;**454** (7205):691.
72. HMB4EU. *First report on the stakeholder consultation and the mapping of needs deliverable report D4.4. WP 4 - prioritisation and input to the annual workplan deadline*; April 2018. Ref Ares(2018)3724718.
73. European Commission. *D062268/02 (Draft implementing measure/act) in dossier CMTD (2019)0805*; 2019.
74. Hoppin JA, Umbach DM, Long S, et al. Pesticides are associated with allergic and non-allergic wheeze among male farmers. *Environ Health Perspect* 2017;**125**(4):535–43.
75. Carmichael SL, Yang W, Roberts EM, et al. Hypospadias and residential proximity to pesticide applications. *Pediatrics* 2013;**132**(5):e1216–26.
76. Parks CG, Santos ASE, Lerro CC, et al. Lifetime pesticide use and antinuclear antibodies in male farmers from the agricultural health study. *Front Immunol* 2019;**10**:1476.
77. Freitas EC, de Oliveira MS, Monticielo OA. Pristane-induced lupus: considerations on this experimental model. *Clin Rheumatol* 2017;**36**(11):2403–14.
78. Koutros S, Andreotti G, Berndt SI, et al. Xenobiotic-metabolizing gene variants, pesticide use, and the risk of prostate cancer. *Pharmacogenet Genomics* 2011;**21** (10):615–23.
79. Mesnage R, Defarge N, Spiroux de Vendomois J, Séralini GE. Major pesticides are more toxic to human cells than their declared active principles. *Biomed Res Int* 2014;**2014**:179691.
80. Haridy A, El AL A, Fawzi M, Al-Khafif M, El-Zemaity M. Epidemiological study of organophosphorus compounds insecticides types related to acutely intoxicated patients presented to poison control center(PCC-ASU)–Egypt. *IOSR J Environ Sci Toxicol Food Technol (IOSR-JESTFT)* 2016;**10**:2319–99.
81. EPA. *Recognition and management of pesticide poisonings*; 2013. Available at: https://www.epa.gov/pesticide-worker-safety/recognition-and-management-pesticide-poisonings.
82. Eddleston M. Patterns and problems of deliberate self-poisoning in the developing world. *QJM* 2000;**93**(11):715–31.
83. Seok SJ, Park JS, Hong JR, et al. Surfactant volume is an essential element in human toxicity in acute glyphosate herbicide intoxication. *Clin Toxicol (Phila)* 2011;**49**(10):892–9.

84. Bradberry SM, Proudfoot AT, Vale JA. Glyphosate poisoning. *Toxicol Rev* 2004;**23**(3): 159–67.
85. Bonefeld-Jørgensen EC, Long M, Hofmeister Marlene V, Vinggaard AM. Endocrine-disrupting potential of bisphenol A, bisphenol A dimethacrylate, 4-n-nonylphenol, and 4-n-octylphenol in vitro: new data and a brief review. *Environ Health Perspect* 2007;**115**(Suppl. 1):69–76.
86. Kassotis CD, Kollitz EM, Ferguson PL, Stapleton HM. Nonionic ethoxylated surfactants induce adipogenesis in 3T3-L1 cells. *Toxicol Sci* 2017;**162**(1):124–36.
87. Maher B. Lab disinfectant harms mouse fertility. *Nature* 2008;**453**(7198):964.
88. Melin VE, Potineni H, Hunt P, et al. Exposure to common quaternary ammonium disinfectants decreases fertility in mice. *Reprod Toxicol* 2014;**50**:163–70.
89. Mastropietro CW, Valentine K. Early administration of intratracheal surfactant (calfactant) after hydrocarbon aspiration. *Pediatrics* 2011;**127**(6):e1600–4.
90. Duke SO. Why have no new herbicide modes of action appeared in recent years? *Pest Manag Sci* 2012;**68**(4):505–12.
91. Séralini G-E, Mesnage R, Clair E, Gress S, de Vendômois JS, Cellier D. Genetically modified crops safety assessments: present limits and possible improvements. *Environ Sci Eur* 2011;**23**(1):10.
92. Séralini G-E, Clair E, Mesnage R, et al. Retracted: Long term toxicity of a Roundup herbicide and a Roundup-tolerant genetically modified maize. *Food Chem Toxicol* 2012;**50**(11):4221–31.
93. Butler D. Rat study sparks GM furore. *Nature* 2012;**489**(7417):484.
94. Gerasimova K. Advocacy science: explaining the term with case studies from biotechnology. *Sci Eng Ethics* 2018;**24**(2):455–77.
95. Séralini G-E, Mesnage R, Defarge N, Spiroux de Vendômois J. Conflicts of interests, confidentiality and censorship in health risk assessment: the example of an herbicide and a GMO. *Environ Sci Eur* 2014;**26**(1):13.
96. Portier CJ, Goldman LR, Goldstein BD. Inconclusive findings: now you see them, now you don't! *Environ Health Perspect* 2014;**122**(2):A36.
97. Goldman BHA. *Monsanto papers*; 2019. Available at: https://www.baumhedlundlaw.com/toxic-tort-law/monsanto-roundup-lawsuit/monsanto-secret-documents/.
98. Mesnage R, Arno M, Séralini GE, Antoniou MN. Transcriptome and metabolome analysis of liver and kidneys of rats chronically fed NK603 Roundup-tolerant genetically modified maize. *Environ Sci Eur* 2017;**29**(1):6.
99. Mesnage R, Renney G, Seralini GE, Ward M, Antoniou MN. Multiomics reveal non-alcoholic fatty liver disease in rats following chronic exposure to an ultra-low dose of Roundup herbicide. *Sci Rep* 2017;**7**:39328.
100. Schiemann J, Steinberg P, Salles B. Facilitating a transparent and tailored scientific discussion about the added value of animal feeding trials as well as in vitro and in silico approaches with whole food/feed for the risk assessment of genetically modified plants. *Arch Toxicol* 2014;**88**(12):2067–9.
101. Panzacchi S, Mandrioli D, Manservisi F, et al. The Ramazzini Institute 13-week study on glyphosate-based herbicides at human-equivalent dose in Sprague Dawley rats: study design and first in-life endpoints evaluation. *Environ Health* 2018;**17**(1):52.

Analytical strategies to measure herbicide active ingredients and their metabolites

Souleiman El Balkhi, Sylvain Dulaurent, and Franck Saint-Marcoux
University Hospital of Limoges, Limoges, France

Chapter outline

Introduction

Why should we measure herbicides?

Herbicides are the most commonly used crop-protection compounds in the world. They are chemical substances applied to agricultural soils, gardens, lawns, and plants to destroy or prevent the growth of undesirable vegetation. Selective herbicides have radically reduced the amount of labor required to manage weeds in crops. In 2007, worldwide herbicide use was estimated at 951,000 metric tons. Today, more than 130 chemical compounds are used as selective herbicides, and approximately 30 others are nonselective. They are classified into about 16 unique modes of action. They can cause adverse side effects on nontarget organisms, humans, and the environment (soil, surface water, ground water).[1] While many of the chemicals in the

Herbicides. https://doi.org/10.1016/B978-0-12-823674-1.00001-8

formulas are considered harmless, others have toxic properties or form toxic byproducts that pose risks to human and animal health[2] and the environment.[3] Therefore, assessing risks before and after authorizing a herbicide is necessary. This has led countries all over the globe to regulate these chemicals and elaborate a management system by characterizing the risks and evaluating the risk/benefit balance.

Active substances must undergo intensive evaluation to be approved by country authorities before use of a product is allowed. Recently, in many countries, including the European Union (EU) through the European Food Safety Authority (EFSA) and the United States through the Environmental Protection Agency (EPA), it was decided to periodically re-evaluate existing pesticides to ensure that they continue to meet safety standards. The risk assessment of active substances evaluates whether, when used correctly, they are likely to have direct or indirect harmful effects on human or animal health, through, for example, drinking water, food, or animal feed, or on groundwater quality. In addition, the environmental risk assessment aims to evaluate the potential impact on nontarget organisms.

The risk assessment process consists of two successive analyses:

- Hazard identification, which identifies a residue that may pose a health hazard at environmentally relevant concentrations and qualitatively describes the effects that may occur in humans.
- A dose-response assessment for humans to characterize the relationship between exposure to residues and resultant health effects.

Why is it necessary to accurately measure herbicides in food and feed?

Risk assessment is based on several fundamental toxicological concepts such as the acceptable daily intake (ADI), acute reference dose (ARfD), maximum residue level or limit (MRL), provisional tolerable weekly intake (PTWI), and theoretical maximum daily intake (TMDI) (see Fig. 5.1).[4,5] When herbicide use respects good

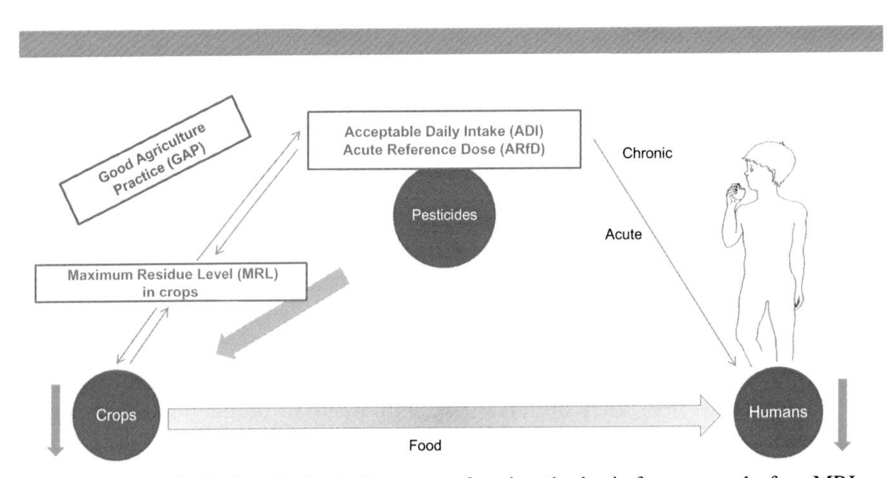

Fig. 5.1 Three principal toxicological concepts forming the basis for approval of an MRL.

agricultural practices (GAPs), residue levels should not exceed the ADI after chronic exposure or the ARfD after acute exposure.

The ADI is an estimate of the amount of a substance in food, feed, or drinking water that can be consumed daily over a lifetime without presenting an appreciable risk to health. The ARfD is an estimated intake that can be ingested over a short period of time, usually during one meal or one day, without posing a health risk.

MRLs are the upper levels of pesticide residues that are legally permissible in or on food or feed, based on GAPs and the lowest exposure necessary to protect vulnerable consumers. MRLs are derived after the assessment of the properties of the active substances and the intended uses in controlled conditions. For instance, before setting MRLs, the EFSA evaluates the behavior of the residue of the pesticide and the possible consumer health risks from residues in food. Consumer risk is evaluated by taking into account the ADI and the TMDI for chronic and acute dietary exposure and making an estimate using a calculation model developed by EFSA. If the assessment does not identify any unacceptable risks, MRLs are set and the product is authorized.[6]

MRL is determined so that the amount of a residue ingested by a given population does not exceed the ADI; that is, it is calculated for an agricultural commodity so that the TMDI of residues from a given pesticide is below its ADI. As the MRL depends on the dietary habits of an average consumer (standard 60 kg body weight) representative of a population, it varies by country and commodity. There is one MRL for each crop and for each pesticide.

The identification and quantification of hazardous effects of a pesticide residue, including herbicides, are done under controlled conditions in laboratory or field experiments by measuring how much of an individual active substance is required to kill or affect test organisms. The identification and calculation of the MRL follow strict GAPs. However, there are multiple contexts where concentrations are above the anticipated MRLs. These could include fraudulent use, failure of postharvest interval (PHI) established in the authorization procedure, inappropriate or illegal use of pesticides, or importation of food from countries where pesticide legislation differs significantly.

In real life, appropriate practices in pesticide use are not always followed and MRL violations occur. Therefore, measurement in commodities is a central part of risk assessment. In addition, the increasing public fears that were raised with the acknowledgment of health risks due to pesticides and the maximal precaution philosophy or policy applied in developed and developing countries have put down roots for a strict surveillance process.

Nearly all governments have monitoring programs in fruits, vegetables, grains, meat, poultry, seafood, beverages, eggs, nuts, tobacco, and most other agricultural commodities. Analytical determinations of herbicides are also required by manufacturers to investigate fate, environmental contamination, transport, and degradation, and to monitor compliance with governmental regulations. Fig. 5.2 shows routes by which pesticides circulate after application. Food manufacturers regularly conduct pesticide residue analysis of ingredients in their raw or transformed products. Consumer groups and research organizations also ask labs to conduct analyses for a variety of purposes.

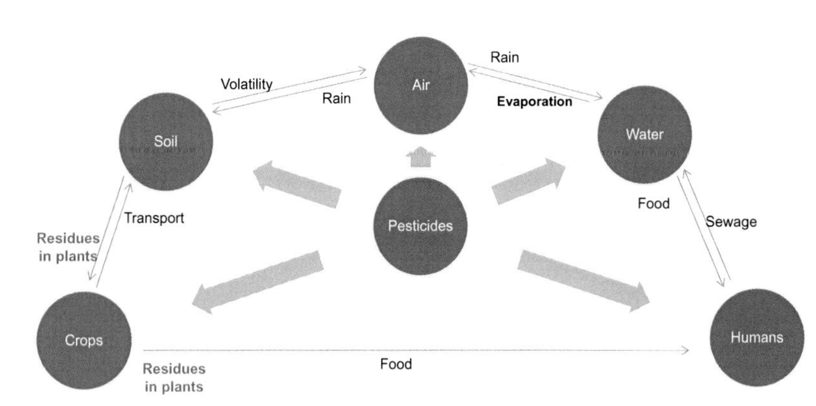

Fig. 5.2 Pesticide circulation after application.

Why is it necessary to accurately measure herbicides in human biological specimens?

Human biomonitoring (HBM)

By definition, human biomonitoring is the determination of a chemical or its metabolites in body fluids (e.g., urine, blood, saliva, breast milk), tissues (e.g., hair), or exhaled air. To identify biomarkers for the health effects of an exposure, relevant and accurate measurements reflecting this exposure must be made. For instance, measuring the concentration of a herbicide in urine can provide valuable information on environmental exposure and, in some cases, identify potential health risks. HBM allows the assessment of whether and to what extent chemical substances have entered the body, and how exposure has changed during a specific period. HBM is complementary to environmental monitoring. It can accurately quantify a herbicide and its metabolites in a suitable sample of biological material obtained from an individual. In the same way as for environmental monitoring, the aim is to relate the result to an internal dose (body burden) of a substance able to cause toxic effects.[7]

HBM is rather simple for occupational situations but much more complex for the general population, where exposure may occur via multiple pathways, routes of entry, and media. Once a herbicide is absorbed, it is distributed in the different tissues of the organism, potentially reaching key organs where toxicity can occur. This is influenced by a multiplicity of factors, including the route of exposure, the duration of exposure, and the physicochemical properties of the active molecule. Most chemicals, including herbicides, undergo extensive metabolism leading to the excretion in urine of very low concentrations of the unchanged molecules, which makes difficult the quantification of the compound of interest at low levels of exposure. Elimination can be slow or rapid and is subject to intra- and interindividual variations. Therefore, metabolism and its variations, the extent of bioaccumulation, and elimination kinetics are important elements in the correct performance of HBM. Phase I and II metabolism lead to the production of a low-toxicity, water-soluble product suitable for excretion in urine.[8] Consequently, most HBM using urine focuses on the measurement of metabolite levels, which can be very challenging for some herbicides because the structures of the metabolites may be unknown. Also, when a structure is known, obtaining a pure standard can be problematic.

However, great advances in HBM in general (identification of pertinent biomarkers) and particularly in analytical measurements now allow detection and quantitation at increasingly lower levels of hundreds of chemicals, including metabolites, in biological samples. Modern analytical methods are accessible to many labs, and HBM can be applied to monitor combined or mixed exposure, an issue of increasing concern in risk assessment.[9] What remains challenging is obviously the interpretation of the measured concentrations. Knowledge about biological effects has not developed as fast as the analytical technology.[10] Unfortunately, the results of human toxicity studies remain rarely available, and animal studies can be difficult to obtain.[11]

HBM might not be the most accurate tool for assessing past exposure. Addressing past exposure is still challenging because of a lack of reliable analytical tools. Although hair analysis has been proposed, it has limits. The presence of herbicides in hair could be due to very recent environmental exposure through particle deposition, or to direct exposure during pesticide application (spray drift and splashing), making the differentiation of recent from past exposure impossible. Unlike in the case of drugs, the presence of a herbicide metabolite does not necessary indicate a systemic passage.[12]

Occupational exposure

Occupational exposure can be encountered at all stages of the life of a herbicide: manufacturing, transport, storage, preparation, application, and afterward. Exposure of farmers and their children and spouses can be significant where a combination of para-occupational and environmental exposures can occur.

Exposure varies by crop, climate, application method, personal protective equipment (PPE), and clothing.[13–16] In this context, herbicides are prevalently absorbed through the skin and airways. Exposure through inhalation of aerosol or vapor is lower than by the cutaneous route. Exposure varies between countries according to the climate and the number of applications per year.

HBM techniques, such as the measurement of herbicides and their metabolites in urine, provide a direct and objective means of assessing exposure. Occupational studies of pesticide workers in general have appeared steadily in the published literature over the past 15–20 years, but studies of herbicide workers in particular are relatively scarce and typically investigated only one or two agents. Recent examples include 2,4-D,[17] atrazine,[18] and 2,4-D and triclopyr.[19] A comprehensive review of analytical methods for pesticides in general was published by Yusa et al. in 2015.[20]

Several good examples of how HBM helped to refine risk assessment have been published.[21] Hays et al. used the concept of the biomonitoring equivalent (BE) instead of the reference dose to a concentration of a compound in blood or urine. This provided an alternative way to interpret HBM data. The suggested BE for 2,4-D was exploited in the United States and Canada using new data for HBM.[22] A comparison of conventional exposure assessment to 2,4-D using environmental monitoring with extrapolation to estimate exposure doses with an HBM-based approach was reported recently.[17] An interesting pilot study for HBM of atrazine-exposed workers has been reported[18] where pre-exposure, postexposure, and 12-hour postexposure spot urine samples were taken. The results indicated that a specific metabolite was sufficiently sensitive to indicate exposure.

Environmental exposure

The general population is exposed to herbicides through residues present in food, water, and the personal environment.[23] Domestic use to eliminate weeds in gardens is also an important source.[24] However, in these contexts, herbicides are present in the biological matrix at only trace levels. Therefore, highly sensitive, specific, and selective analytical methods for the extraction, separation, and quantification of these chemicals must be developed to fulfill HBM study objectives.[25]

Accidental and nonaccidental poisoning

Most published cases involve paraquat.[26] It is one of the most commonly used lethal herbicides. Even a small amount can kill.[27] Poisoning is common in agriculturally based countries, particularly in Asia.[28] Other cases of accidental and nonaccidental poisoning have been reported for glyphosate,[29,30] 2,4-D,[31] and atrazine.[32]

The determination of herbicides in biological fluids is often needed to confirm the intoxicant and exclude other sources of clinical symptoms. Identification of the herbicide is relevant and sometimes necessary for medical care. Measurement of the agent and its metabolites in blood and urine allows an accurate evaluation of the elimination capacity, especially in a case of renal or hepatic failure, if extracorporeal filtration of blood is indicated.[33] In acute intoxication, metabolic pathways are often saturated, leading to low metabolism; unchanged herbicide would be more concentrated in urine and blood than corresponding metabolites.[34] Table 5.1 summarizes the approach.

Matrices of interest

Several types of samples—including blood, urine, hair, saliva, exhaled breath, nails, feces, breast milk, and umbilical cord blood—have been proposed to explore exposure.[8] Whatever the context, the choice of matrix should be driven by the pharmacokinetics of the targeted molecules and their physicochemical properties. Blood and urine are the major samples. Other types are more used in research studies. Knowledge of the pharmacokinetics and the context of exposure is necessary to select the most appropriate matrix and to define the time of sampling. Good knowledge of the phases depicted in Fig. 5.3—absorption, distribution, metabolism, and excretion—is important in the selection of matrices, the biomarkers to monitor, and the interpretation of measured concentrations.[7,11,14,16,35]

Table 5.1 Context, matrices, and procedures used in the determination of herbicide exposure.

Context	Available matrices	Compounds to investigate	Obtained concentrations	Use of matrices	Analytical methods
Acute intoxication	Blood, urine, hair, gastric contents	Herbicide, metabolites	mg/L to g/L	Identification: blood, urine, gastric contents Interpretation: blood, urine	Qualitative (screening) for identification Quantitative (specific) for interpretation with diluted sample
Occupational exposure	Blood, urine	Metabolites	$\leq$1 mg/L	Blood, urine	Quantitative for interpretation
Environmental exposure	Blood, urine	Metabolites	$\leq$0.1 mg/L	Blood, urine	Quantitative for interpretation

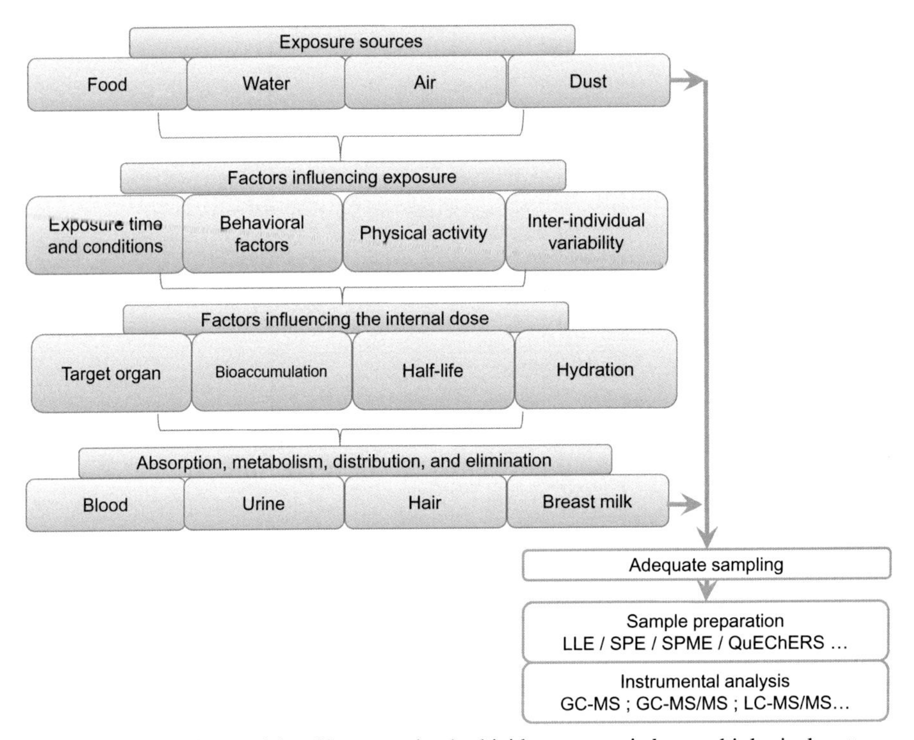

Fig. 5.3 Factors to be considered in measuring herbicide exposure in human biological systems.

Absorption occurs through the oral, dermal, and inhalation routes. Dermal exposure is most common in work situations where workers mix, load, or apply herbicides. It also occurs from touching an item or person contaminated with residue. The absorbed molecule is distributed in the tissues, where it can accumulate, and is metabolized mostly by the liver before elimination in urine or feces.[3,36] The elimination rate can be rapid or slow, leading to a short or long elimination half-life. Therefore, blood samples are used because blood reflects the level that is bioavailable to potentially cause adverse effects. The principal disadvantage of blood sampling is invasiveness. Urine samples are easier to obtain and are preferred; they are used for exposure evaluation in the general population and in occupationally exposed workers. HBM typically relies on a spot urine sample, which introduces uncertainty due to the variation of hydration. Spot samples are often corrected for hydration level by using the ratio of the molecule concentration to creatinine concentration.[37] Urine is composed of more than 95% water, plus sodium, ammonia, phosphate, sulfate, urea, creatinine, proteins, and products processed by the kidney and liver, including drugs and metabolites. Spot samples should be standardized because urine is slightly acidic in the morning (pH 6.5–7.0) and more alkaline in the evening (pH 7.5–8.0). This could influence the elimination rate of herbicides and their metabolites.[38] Among the analytical advantages of urine over serum are that the urine can be obtained in large quantities by noninvasive sampling, and that it requires less complex sample preparation because of lower protein content. However, the number of conjugated herbicides can be higher than in serum, which might require a supplementary deconjugation step.[39]

Hair testing for herbicides showed utility in confirming past exposure in an intoxication context.[40] It is also useful to highlight cumulative exposure to herbicides and other pesticides.[41] The reliability of hair testing in biomonitoring studies is now supported by an increasing number of datasets.[42,43] The actual analytical methods are specific and sensitive enough to measure most herbicides, but some molecules, such as glyphosate, cannot be detected in hair samples. There is an ongoing debate about the use of hair for biomonitoring because of the difficulty in discriminating internal from external exposure. In our opinion, knowing whether the presence of a herbicide is due to direct deposition or accumulation in hair after absorption in not a key question.[12] Testing segments of a strand of hair to date past exposure is known to be erroneous.

Analytical strategies

Because the exposure of the general population and of workers involves numerous molecules, it is common sense for a lab to develop analytical methods allowing the simultaneous determination of as many compounds as possible. The objective of such "multimethods" is the fast generation of quantitative data for a large number of analytes in a single analysis. This is very useful in HBM epidemiological studies as well as for food and feed analysis when many samples and multiple classes of pesticides have to be analyzed.

Whatever the analytical method, the steps in Fig. 5.4 must be defined, optimized, and validated. For each, several guidelines are available to help analysts produce robust and reproducible results. For instance, in food analysis, the SANTE/11813/2017 guidelines describe the means of validation and the analytical quality control requirements to support the validity of data used for checking compliance with MRL, enforcement actions, or assessment of consumer exposure to pesticides. These guidelines are currently the most commonly used in Europe.

Multiclass, multiresidue analysis of pesticides in food and feed is a major challenge to the chemist because of the sheer number and diversity of analytes and sample types. For herbicides, the challenge is even greater because of the wide chemical spectrum.

Fig. 5.4 The steps in an analytical method.

Sample preparation for food and feed

For any quantitative analysis, it is necessary to guarantee that a representative and homogeneous sample is taken. All processes must be designed to avoid under- or over-estimation and any loss of the targeted analyte. Every step can be a source of error.

First, the storage conditions between sampling and analysis must not degrade the molecule. Stability must be tested after sampling. For example, paraquat, diquat, and glyphosate are known to interact with glass tubes, which makes the use of plastic highly recommended (EU Reference Laboratory for pesticides requiring single residue methods (EURL-SRM)).

A simple process such as comminution alone can affect residue concentrations if not properly performed. Sample preparation at too high or room temperature can lead to residue loss. Therefore, the sample should consist of the whole units or a part of the analyzed commodity at low temperature when necessary. If a simple approach such as chopping and blendering can be used to homogenize the sample, subsequent steps such as the use of an ultrasonic probe can have added value.[44] This approach facilitates homogenization and significantly shortens the time necessary for sample preparation.

The last step in the preparation process is clean up to minimize the introduction of undesirable compounds into the analytical system and to reduce interference. Cleanup techniques take advantage of the difference in physicochemical proprieties (polarity, solubility, molecular size) between the molecule of interest and the compounds present in the analyzed matrix. The golden rule for extraction is: the greater the selectivity of the sample preparation and the extraction steps, the cleaner will be the samples and the more robust will be the analysis. With cleaner extracts, fewer matrix effects will be present, which reduces potential interference and background noise. Clean extracts improve the selectivity of the method by reducing false positives, which increases the sensitivity with lower limits of detection and improves the accuracy of the results.

Sample preparation and extraction can become a bottleneck if not properly optimized. Insufficient cleaning or extraction is always associated with analytical errors, whatever the performance of the system used for the determination. Each kind of fruit has to be considered as a particular matrix with its own sample preparation and cleanup steps.[45] Families can be divided roughly as follows: high water content, high acid and high water content, high sugar and low water content, high oil and very low water content, high oil and intermediate water content, and high starch and/or protein and low water content.

Fruits have multiple natural pigments, lipids, proteins, and carbohydrates. These can be present at high concentrations and are responsible for the matrix interferences that can mask the presence or modify the quantitation of a pesticide. These troublesome components can damage or shorten the lifetime of laboratory instrumentation. Therefore, sample preparation for matrix interference removal in food is extremely important. Clean up should eliminate most interfering peaks and allow good recovery at low levels.

The physicochemical properties of the targeted analyte determine the approach that leads to a high-performance extraction procedure. One of the most important aspects is the polarity of the analyte, which is indicated by its octanol/water partitioning coefficient ($K_{o/w}$). For herbicide analysis, pH and ionic strength are critical, as most of them are weak acids and bases. Other important factors are volatility and stability. These properties indicate which precautions must be made to avoid a loss of analyte.

The most important chemical properties for many pesticides, including herbicides, are found in databases from the USDA Agricultural Research Service (https://www.ars.usda.gov) and on Alan Wood's website (http://www.alanwood.net). The polarity range of the molecules recovered by an analytical method can help during the analytical development phase, especially in multiresidue methods, and can set the scope of analysis and number of pesticides that can be determined at the same time.

Extraction

Liquid-liquid extraction (LLE)

Based on polar to apolar solvents, LLE has been used historically for the determination of most herbicides. It is cheap, relatively quick to perform, and fairly simple. It has a short method development time and requires no expensive equipment or skills and no consumable products. The analyst must answer two main questions beforehand: what solvents will the target herbicide most likely solubilize (i.e., what is the log *P* of the target analytes), and are the extraction solvents compatible with the analytical approach? The efficiency of herbicide extraction methods using solvents such as acetonitrile (ACN), hexane, dichloromethane (DCM), acetone, petroleum ether, ethyl acetate, cyclohexane, toluene, and methanol (MeOH) has been extensively studied. These solvents play different roles and allow good recovery for a wide range of molecules.[46] The *n*-hexane extraction selectively yields nonpolar pesticides while the DCM extraction covers a wider polarity range, though it is less efficient at eliminating matrix interference. The drawbacks of LLE that make it less popular today are the large quantities of solvent consumed, the multiple operational steps, the necessity for preconcentration of the extract before analysis, and interference from compounds that are likely to be coextracted.

Solid-phase extraction (SPE)

SPE is one of the most selective techniques used in food safety testing. Its selectivity comes from the possibility to choose the sorbent that selectively interacts with the analyte of interest or with the matrix-interfering compounds. The intermolecular interaction between a solid stationary phase and the target analyte helps by removing the undesirable contaminants while the targeted analyte is concentrated. SPE addresses the three primary goals: analyte extraction, concentration, and solvent switching. It can be used for matrix removal or for enrichment of the targeted compounds. The development of an SPE procedure requires defining the chemical nature of the sorbent and the buffer conditions used during the loading, washing, and elution stages. SPE can be highly selective and robust, and it produces high recoveries and provides repeatable results. However, there are some disadvantages that prevent its systematic use. It requires special equipment and is generally much more expensive than other methods. A large variety of sorbents with different chemistries are available, including chemically bonded silica (C18, C8, C4), graphitized carbon, ion-exchange materials, polymeric materials, and mixed-mode products. Different mixed-mode sorbents are available, including weak and strong cation and anion exchangers. Mixed-mode sorbents are regularly used for herbicide analysis, as they are expected to retain the analytes with their wide range of physicochemical properties by combining hydrophobic interaction and ion exchange. Fractioned elution using SPE could be interesting for multiclass, multiresidue methods where methanol, for example, could be used to elute compounds retained by hydrophobic interaction followed by acidified or alkalinized methanol for compounds retained by ionic interaction. This fractionation can also help to reduce matrix effects during ionization if matrix compounds and analytes become separated.[47]

QuEChERS

QuEChERS is an acronym for "quick, easy, cheap, effective, rugged, and safe." It has been widely used since appearing in an article on the determination of pesticide residues in fruits and vegetables by Anastassiades et al. in 2003.[48] Roughly speaking, QuEChERS consists of adding salts to the matrix that has been previously mixed in a polar solvent (ACN). Originally, QuEChERS extraction required two steps: (1) extraction partitioning, where the matrix was mixed with ACN before adding anhydrous $MgSO_4$ and NaCl (with the objective of drying the organic phase to allow the separation of the two phases), and (2) a dispersive SPE clean up, where the remaining impurities (contained in fruits and vegetables) are removed by adsorbents such as primary and secondary amines (PSA). Before the extraction-partitioning phase, the sample must be homogenized, as the contact surface between the sample and the extracting solvent could influence the efficiency of the extraction. Numerous papers reporting the use of this kind of preparation procedure have been published in the field of environmental toxicology.[49] Between the original article and the end of 2018, more than 1400 articles dealing with QuEChERS extraction were published. Among these, about 100 concerned herbicides.

Several modified QuEChERS procedures have been successfully used to measure herbicides in mono- or multiclass methods.[50] The modifications concern the use of acetonitrile as a solvent without a dispersive phase clean-up step,[51] the use of acetonitrile saturated with hexane for extraction of PSA in the clean-up step,[52] and the use of acidified acetonitrile to extract acid herbicides.[50]

QuPPe

QuPPe (for "quick polar pesticides") was developed by Anastassiades et al. for very polar pesticides that are not amenable to QuEChERS. It involves extraction with acidified methanol and LC-MS/MS measurement using labeled internal standards for quantification. Glyphosate can be extracted and analyzed by this approach without any derivatization. Amitrole, bialaphos, difenzoquat, and glufosinate also can be extracted.

Preparation of biological samples

Pretreatment is required to remove substances that might interfere with analysis, to concentrate analytes, and to hydrolyze conjugated forms of the targeted biomarkers.[53,54] A hydrolysis step could be necessary when the herbicide's phase I metabolite undergoes phase II metabolism and is conjugated to a glucuronide or sulfate. For example, all ingested 2,4-D is excreted as either the parent compound or a conjugate of it in urine, and analytical methods used for urinary biomonitoring recover both components.[22] Deconjugation is usually done by an enzymatic hydrolysis treatment with beta-glucoronidase or sulfatase or by acid hydrolysis.[53,54]

Although all the preparation procedures cited above for food and feed can be used for biological fluids, it is often more challenging, especially in the HBM context. The limits of detection and quantification needed are much lower and the sample volumes available are often much smaller than in food analysis.

Several sample preparation techniques have been developed to avoid the drawbacks of LLE and SPE. Requirements include shortening the sample preparation time, reducing the number of steps, and decreasing the sample and solvent consumption. The objective is to reduce the sample preparation cost and volume necessary to perform the analysis. However, the techniques to accomplish this are more sophisticated and need highly skilled manpower and special equipment. This restrains their ability to replace conventional extractions. Solid-phase microextraction (SPME), stir-bar sorptive extraction (SBSE), microextraction by packed sorbent (MEPS), disposable pipette tip extraction (DPX), single-drop microextraction (SDME), hollow-fiber liquid phase microextraction (HF-LPME), and dispersive liquid-liquid microextraction (DLLME), among others, could be considered.[55] The most popular approach for day-to-day application in a lab measuring herbicides is online extraction, where an extraction system is integrated with a chromatographic system and operates before the chromatographic separation.

Separation and detection

The determination of herbicides in food and human samples is done mainly by liquid chromatography coupled with tandem mass spectrometry (LC-MS/MS) and gas chromatography coupled with single (GC-MS) or tandem mass spectrometry (GC-MS/MS) (Fig. 5.5). The mass analyzer is a triple quadrupole in LC-MS/MS and GC-MS/MS systems.

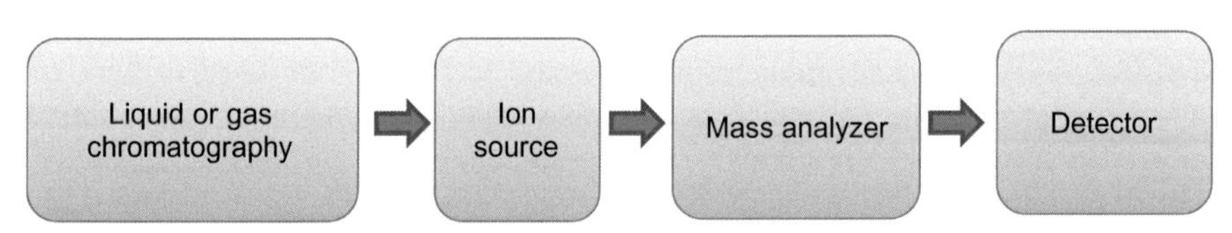

Fig. 5.5 Basic components of an LC-MS or GC-MS system.

Separation techniques

Various chromatographic techniques such as GC, LC, thin layer (TLC), high-performance LC (HPLC), and ultraperformance LC (UPLC) are used in analytical quality control and environmental sciences. Chromatography is applied for the residue analysis of xenobiotics in air, ground and surface water, sludge, soil matrices, food and food products, and human and veterinary health care.

Some herbicides such as amides, thiocarbamates, and triazines are amenable to GC-MS and GC-MS/MS, and many other compounds such as phenoxycarboxylic acids are made amenable to GC-MS by simple derivatization reactions.[56] However, most of the time, metabolites and degradation products, quaternary ammonium salts, and ureas are not amenable to GC, which is also the case with many compounds with complex structures and several functional groups.[57] Therefore, LC-MS/MS multi-methods have become increasingly popular. This includes methods for analyzing food and drinking, surface, and waste water[58,59] as well as biological fluids such as urine[60] and blood[61] where hundreds of molecules, including insecticides, herbicides, fungicides, and their breakdown products might be analyzed in a single run.[47]

A typical LC-MS/MS system for herbicides is composed of one or two pumps able to dispense mobile phases in a chromatographic column coupled to a quadripolar mass spectrometer. The length of the column is generally between 5 and 15 cm. Reversed phase chromatography is the most used mode, receiving polar mobile phases composed of a buffer with salts and acids in water (ammonium formate or acetate and formic or acetic acid) and methanol or acetonitrile. The chemistry of the apolar stationary phase is mainly based on the use of an octadecylsilyl group (C18), but other chemistries such as biphenyl and pentafluorophenyl are used as chemically bonded groups.

The use of GC-MS or GC-MS/MS could be complementary to the use of LC-MS/MS in most labs, the two technologies being necessary to cover a majority of herbicides. GC-MS and MS/MS are useful when a compound is not ionizable by LC-MS/MS, and for apolar compounds. Compounds are eluted by a carrier gas, usually helium. A capillary column of 10–30 m in length with a 5% phenyl and 95% dimethylpolysiloxane chemistry is common; this is the most often used apolar phase for organic compounds. However, this kind of chemistry is hardly compatible with the analysis of polar compounds without sample pretreatment. Indeed, the remaining polar function could be at the origin of hydrogen bonds with compounds of interest, leading to bad peak shape (nonsymmetric peak). For example, to analyze 2,4-D, a derivatization is required to modify its carboxylic acid functional group and to obtain a correct elution without hydrogen bonds. Using GC-MS or GC-MS/MS, the ionization step can be performed with two different approaches: electronic impact and chemical ionization.

ELISA (enzyme-linked immunosorbent assay) techniques are used to measure some herbicides.[62] The advantage is that no sample preparation or separation is needed. However, ELISA has drawbacks in pesticide residue analysis and has failed to fulfill the initial promising claims. Although developed ELISA antibodies showed a high degree of selectivity and sensitivity, matrix effects and cross-reactivities in complex samples lead to excessive false positives and false negatives in use. The technique is better with water-based applications.[63–65]

Mass spectrometry

Importance of ion source

The ion source is used for the vaporization and ionization of the target molecules after LC separation. Usually, two different probes or sources can be used for the ionization of organic compounds: electrospray ionization (ESI) and atmospheric pressure chemical ionization (APCI). Analytical systems are often designed with the possibility to install either source, but ESI is generally favored. It is more useful for very polar compounds, with APCI for nonpolar compounds. ESI allows lower limits of quantification and exhibits less matrix effect.[66,67] APCI provides higher sensitivity for neutral and basic herbicides (phenylureas, triazines)[68]; ESI (especially negative ion) is more sensitive for cationic and anionic herbicides (bipyridylium ions, sulfonic acids).[69]

The electronic impact used in GC-MS is universal and reproducible (low interlaboratory variability), and the spectra obtained are very rich because this ionization process involves an intense fragmentation of the molecules when used at an energy of 70 eV. Consequently, the molecular ion peak is very often absent, which can be problematic for identification. Chemical ionization, by contrast, is "soft," leading to a high molecular ion peak with weak fragmentation.[67] This approach could be interesting to identify or confirm a molecular ion.

Mass analyzers

Mass analyzers are used to separate by mass-to-charge ratio (*m*/*z*) the gas phase ions produced in the ionization source. Two types of mass analyzers have been historically employed: quadrupoles and ion traps. They can be arranged in tandem, such as a triple quadrupole mass spectrometer. These systems have good sensitivity but limited resolving power. Recently, higher-resolution instruments such as time of flight (TOF), Fourier transform ion cyclotron resonance (FTICR), and orbitrap have been launched. Different types of analyzers can be combined to form hybrids such as quadrupole-TOF (Q-TOF) and ion trap-orbitrap.

The triple quadrupole mass spectrometer (LC-MS/MS, Fig. 5.6) has been the workhorse for a decade in the quantitation of small molecules and metabolites of herbicides, as it offers sufficient sensitivity and reproducibility and a broad dynamic range (i.e., capability of measuring a wide range of concentrations).[70] The most used approach for the systematic determination of herbicides in a multiresidue context is the MRM or scheduled MRM mode (acronym for multiple reaction monitoring). With such an approach, ions are produced in the ion source; then the first mass analyzer conditions are set to select only the ions corresponding to the compounds of interest. The parent ions are then broken into daughter ions in a collision cell—the transition from parent to daughter—and the downstream mass analyzer follows these daughter ions.

For a multiresidue approach, dozens of transitions are considered, but an alternative has been developed that consists of following a transition only around its expected retention time. This is scheduled MRM. It improves the limits of quantification two- to sixfold.[71] However, compounds can move out of the fixed window due to slight

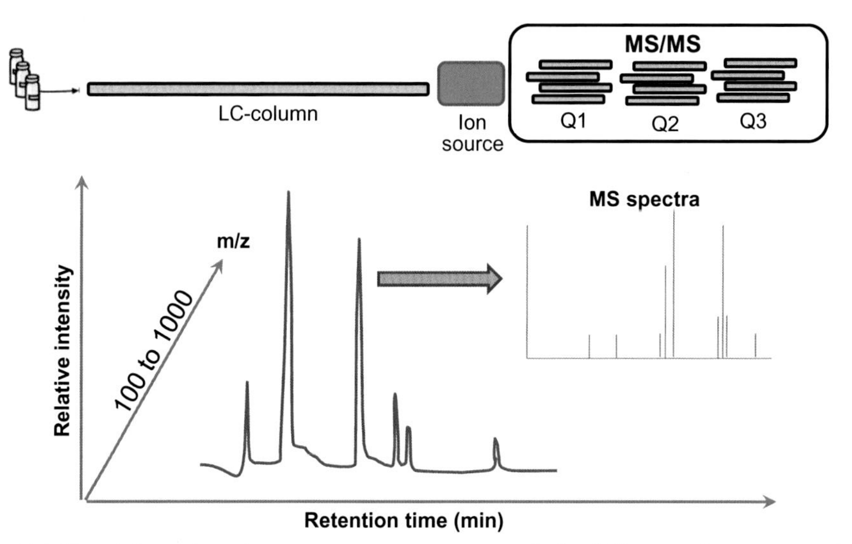

Fig. 5.6 The triple quadrupole mass spectrometer, composed of an LC part, an ion source, and a triple quadrupole analyzer (Q1, Q2, Q3). LC-MS analysis is defined by a retention time, a relative intensity (arbitrary units), and the ions detected (*m*/*z*).

changes in the mobile phase pH or to the aging of the chromatographic column, which can require the tedious and frequent readjustment of parameters.

For the determination of a compound, at least two transitions are commonly considered: one for quantification and one for qualification (i.e., confirmation). The area ratio between these two transitions should be stable between calibration standards or unknown samples and helps to draw a conclusion on the presence or not of a compound.

The main limit of these approaches is that they are, by definition, targeted. The method looks for a predefined list of compounds and in no way can it detect the presence of an unexpected compound.

In cases where a low-resolution system is insufficient to identify a compound or its metabolites, a high-resolution system is used. A frequent situation is that of isobaric coeluting of matrix components with similar calculated masses. For example, when acetochlor and alachlor are in the same sample, they can coelute. They have the same molecular mass but produce some different fragment ions. LC-TOF can discriminate between the target species and "isobaric" interference within 0.05 Da of mass difference (using 350 *m*/*z* as an example). Other authors used high-resolution technology to determine the proposed structures of transformation products of bromacil following water chlorination.[72] The main drawback of this technology is the lower dynamic range compared with quadrupole mass analyzers. Therefore, reluctance persists because these technologies have the reputation of being less useful for the quantification of several hundreds of molecules in a matrix. Recent reviews dealing with the use of TOF or orbitrap-MS reported the possibility of determining a large number of pesticides in vegetables and fruits with Q-TOF.[73–75]

Chemical particularities

The choice of sample preparation and analytical methods is driven by the physicochemical characteristics of the targeted analytes.[76] Herbicides can be extracted and analyzed with a multiresidue method when they have neutral properties, but will need specific developments and dedicated methods when they are highly acidic or alkaline. Examples of such orphan molecules are detailed in the following paragraphs.

Phenoxy herbicides such as 2,4-D, mecoprop, and dichlorprop are acidic and difficult to extract from aqueous matrices with multiresidue methods that do not have pH conditions that fit with the pK_a of these compounds. They can be extracted using acidic conditions and organic solvents for LLE. Alternatively, two approaches using SPE are possible: a reverse phase mechanism using apolar sorbents and neutral herbicides, and ionic retention with anionic compounds and cationic sorbents.[46,77,78]

Chlorotriazine herbicides such as atrazine, simazine, and terbuthylazine as well as substituted urea herbicides such as isoproturon, rimsulfuron, and diuron are all basic compounds (due to the presence of nitrogen atoms). As for phenoxy herbicides, they are potentially difficult to extract from aqueous matrices except when using basic conditions and organic solvents if LLE is chosen. Where SPE is preferred, two solutions are possible: a reverse phase mechanism using apolar sorbents and pH conditions if the pesticides are not ionized, and ionic retention with cationic compounds and anionic sorbents.[79,80]

Paraquat is a quaternary ammonium that is ionized whatever the pH conditions. Consequently, it is very water-soluble. Extraction and chromatographic separation usually involve ion pairing reagents and often LC coupled with UV detection. Ion pairing is used to enhance the retention in reversed phase conditions, but it causes significant ion suppression, rendering LC-MS unsuitable for detection and quantification.

Development strategies

Most food testing labs follow approved standard methods with validated sample preparation procedures. In some cases, when special needs must be met, analysts are granted the flexibility to deviate from the official method and can develop a more appropriate sample preparation. When a lab needs to develop a method to quantify a herbicide molecule and its metabolites in several matrices (e.g., fruits, urine, blood), there are several steps that can require extensive investigation. The ideal scenario is as follows:

- The active molecules and their metabolites are available from reliable suppliers as soluble salts in common solvents (methanol, ACN, toluene, ethyl-acetate) or liquid form allowing the preparation of a stock solution usually at 1 g/L.
- An isotopic derivative of the molecule is available as a standard for internal calibration.
- The molecules present a chemical function that could be easily ionized: the presence of an amine function can lead to ionization in positive mode after the addition of a hydrogen (H) atom. A quaternary ammonium ion (pseudomolecular ion $[M+H]^+$)

is then produced. In molecules where carboxylic acid or alcohol functions are present, negative ionization is possible with the loss of a hydrogen atom (pseudomolecular ion $[M-H]^-$).

- The pseudomolecular ions are rapidly identified by the LC-MS/MS or GC-MS/MS system that is able to drive the optimization and fragment them, and to select more than two transitions. One or more of these transitions will be implemented in the final acquisition method. This allows the selection of optimum parameters such as orifice voltage and collision energy.
- The injection on the chromatographic column shows an acceptable retention time that is distant from the void volume and compatible with the retention time of other molecules in case of a multiresidue method. If coelution occurs, the mass spectrometer can discriminate the coeluted molecules.
- The new compound is extracted by an easy sample preparation approach such as QuEChERS and yields satisfactory recovery and no matrix effects.
- The calibration curve is adapted to the targeted concentration with a large dynamic range between the lower and upper limits of quantification and covering easily three- to fivefold in the concentration range (e.g., 0.1–100 μg/L).

This ideal scenario rarely goes as smoothly as described. Several issues can occur, and very often analysts will have to test multiple approaches at the different stages. In short, method development can be very tedious. The analyst has to clearly define needs beforehand and stratify the compromises that must be accepted depending on the research objectives and the field of application. As stated by Steven J. Lehotay and Katerina Mastovska, "The good analytical chemist provides results to meet the needs of the analysis. The great analytical chemist provides results to meet the needs of the analysis quickly and efficiently."[81]

For example, a few years ago the analytical standards of many metabolites were not commercially available, which necessitated synthesis of the molecules in collaboration with organic labs and chemical evaluation of the new standards. Unequivocal identification of the detected metabolites is necessary. Unless a new herbicide is used, this obstacle is less likely to be faced nowadays, as almost all metabolites of interest are available.

If low ionization of the target molecule in the ionization source is an issue, it could be solved by using GC-MS or GC-MS/MS as an alternative to LC-MS/MS. Other ion sources such as ACPI and APPI or other ionization modes could be used, if available. When the retention time is not suitable, other stationary [unusual reversed phase such as hydrophilic interaction liquid chromatography (HILIC)] or mobile (e.g., basic conditions) phases could be tested to increase the retention of compounds of interest. Or other extractions such as SPE and LLE using particular solvents or buffers could be used.

Derivatization can be a solution to several analytical problems during the development of a method, including those listed above. However, derivatization cannot be applied to multiclass, multimolecule methods. It is generally limited to methods analyzing one or a few molecules. It can be used for LC-MS/MS and GC-MS or MS/MS. The objectives of derivatization are different in LC and GC. For example, glyphosate and one of its metabolites, aminomethylphosphonic acid (AMPA), can be derivatized in LC-MS/MS to increase their retention time and molecular weight, which increases the specificity of selected MRM transitions. In GC-MS and MS/MS, it is necessary to eliminate hydrogen bonds between the apolar stationary phase and the polar functions of the herbicide of interest to allow a correct elution. For example, the carboxylic acid function of 2,4-D can be derivatized using diazomethane and its analogs[82] or pentafluorobenzylbromide.[83] The amine function of atrazine can be derivatized with PFPA (pentafluoropropionic acid).[84]

For some applications, method development is a very large task. The work reported by Kuklenyik et al. in 2012 for the determination of atrazine and its metabolites is a good example. Because of the wide range of pK_a and the polarity of the analytes, a two-level separation was used by developing a two-dimensional HPLC method with isotope dilution to measure atrazine along with the 11 atrazine metabolites and hydrolysis products in urine. The 2D-HPLC system incorporated strong cation exchange and reverse phase separation modes and was preceded by an automated offline solid phase extraction. On the basis of theoretical predictions and their experimental observations, a 2D-HPLC timetable was designed to capture all atrazine metabolites from the urine matrix on a first cation exchange column. Then the chloro and mercapto analogs were eluted from the first cation exchange column and sent to a reverse phase column for analytical separation. The analogs with hydroxyl functions (OH) were then eluted from the first cation exchange column to a second one. To obtain a method suitable for exploring environmental exposure to atrazine (limit of detection of 0.1–0.5 mg/mL), these authors had to optimize a huge number of parameters, including the offline SPE purification and preconcentration, the separation columns, the quaternary pump gradient programs, the tubing configurations, the timetables, the MS ionization conditions, the pH of the gradient, and the solvent percentages. Their optimized method was in the end suitable for the analysis to characterize environmental atrazine exposure for risk assessment with good sensitivity using only 500 μL of urine.[85]

Method validation

Once the analytical conditions for the target herbicide have been defined, the next step is to verify that these conditions lead to a robust method that delivers reliable results in the long term. To ensure that, a full validation of the whole procedure is mandatory. Many guidelines describing the process have been published. They share the following elements:

- Selectivity: checking for the absence of interfering signals, studied with at least six different blank samples.
- Accuracy (bias) and precision (repeatability, intermediate precision): studied with quality controls at low, medium, and high concentrations within the calibration range; analysis of several repetitions ($n = 6$ is a minimum).
- Calibration model (linearity): at least 4–6 concentration levels, usually 1 concentration more than the factor in the calibration equation (e.g., 4 concentration levels for a quadratic model).
- Lower limit of quantification: the lowest concentration at which accuracy and precision meet predefined criteria (e.g., bias less than 15%).

- Recovery: comparison of the signal of a spiked sample extracted to a blank sample previously extracted and then spiked.
- Matrix effect evaluation: several approaches can be used. The ion suppression phenomenon is studied by evaluating the decrease of the signals after continuous infusion of a standard solution containing the compounds of interest into the MS system followed by a simultaneous LC flow containing either a pure mobile phase or the matrix extract (matrices from six blank samples) introduced through a T-union coupling system. The evolution of the signal of the transitions at the retention times of the corresponding analytes and the internal standard is studied to evaluate the intensity of ion suppression. Matrix effects can also be assessed by comparing the peak area of analytes of interest in the presence and absence of the matrix. Concretely, the peak area of a blank matrix spiked after extraction is compared to the peak area of the analytes in the absence of a matrix. The difference should not exceed 20%.

Fig. 5.7 shows an example of steps to follow for the validation of a molecule determination analysis in a given matrix. This scheme conforms to the validation items above and is designed to reduce the number of tests required. In the case of a multi-analyte method, the spiking of the calibration points and controls should be performed with one or several stock solutions containing the target molecules.

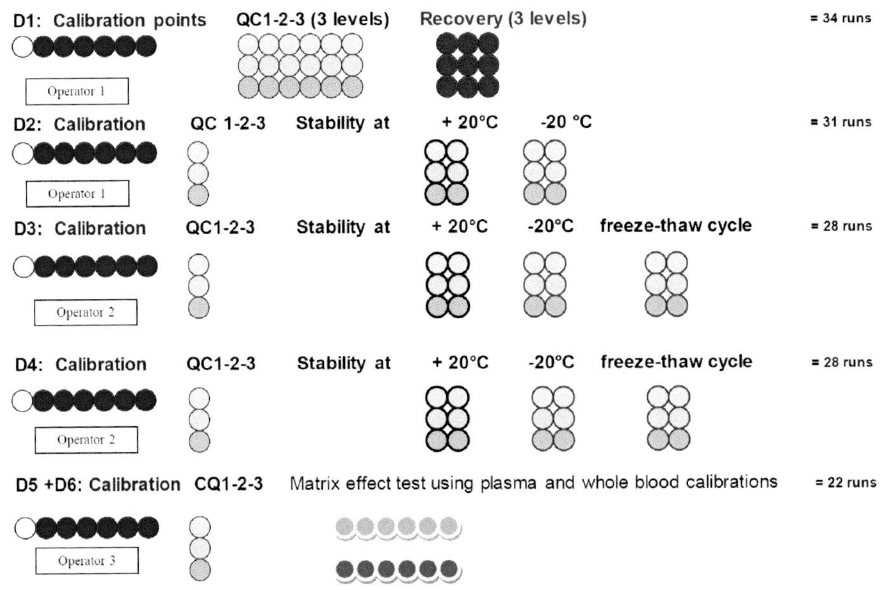

Fig. 5.7 Scheme for the validation of an analytical method in urine. ● Calibration points on a spiked pool of urine. ○○○ Three levels of quality controls prepared at concentrations covering the calibration curve; these are repeatedly used to test the accuracy and precision of the method. ● Calibration points spiked in water or the mobile phase injected directly without extraction to test the recovery rate of the extraction. The matrix effect could be tested by fortifying plasma or whole blood if needed to test the impact of the matrix on signal intensities.

For validation in fruit analysis, SANTE/11813/2017 and LAB-GTA26 guidelines classify the fruits into groups, each with several categories, each category represented by several commodities (Table 5.2). To validate the method for one group, the recommendation is to perform a complete validation for one matrix of the group, then partial validations for other matrices of the group. For example, a full validation of the method for analyzing atrazine in raspberries added to a partial validation of this method in oranges is sufficient to declare the method validated for the determination of atrazine in the group of high acid content and high water content. A full validation should include selectivity, linearity, matrix effect, limit of quantification, specificity, accuracy, precision, robustness, ion ratio, and retention time. A partial validation includes only the limit of quantification, specificity, accuracy, and precision. For a method covering 20 herbicides, the validation of a single group can represent more than 1000 tests. Analysis of the same 20 herbicides in another group will necessitate another complete validation. Fig. 5.8 shows validation steps in a flowchart.

To assess the performance of an analytical method on a day-to-day basis, several experiments must be performed using internal and external quality controls (IQCs):

- IQCs are proposed by a supplier or can be homemade. Those samples for which the concentration is known are used to verify performance. Typically, three IQC levels are run before starting the analysis of samples, and IQCs are measured every 10 samples and at the end of a batch. If the expected concentrations are not obtained, no result can be delivered.
- The lab must belong to a proficiency testing program that performs an external quality assessment (EQA). The World Health Organization proposed a definition in 2009: "EQA is defined as a system for objectively checking the laboratory's performance using an external agency or facility." The principle is to compare the results of a lab to those of other labs.
- Quality assurance under an ISO 15189 environment is needed. The ISO website proposed this definition: "ISO 15189:2012 can be used by medical laboratories in developing their quality management systems and assessing their own competence. It can also be used for confirming or recognizing the competence of medical laboratories by laboratory customers, regulating authorities and accreditation bodies."

Table 5.2 Example of groups, categories, and representative commodities to consider for analytical method validation.

Group	Category	Representative commodities
High water content	Pome fruits	Apples, pears
	Stone fruits	Apricots, peaches, cherries
	Other fruits	Bananas
High acid content and high water content	Citrus fruits	Lemons, oranges
	Small fruits and berries	Raspberries, strawberries, grapes
	Other fruits	
High sugar and low water content	Honey, dried fruits	Honey, raisins, fruit jams

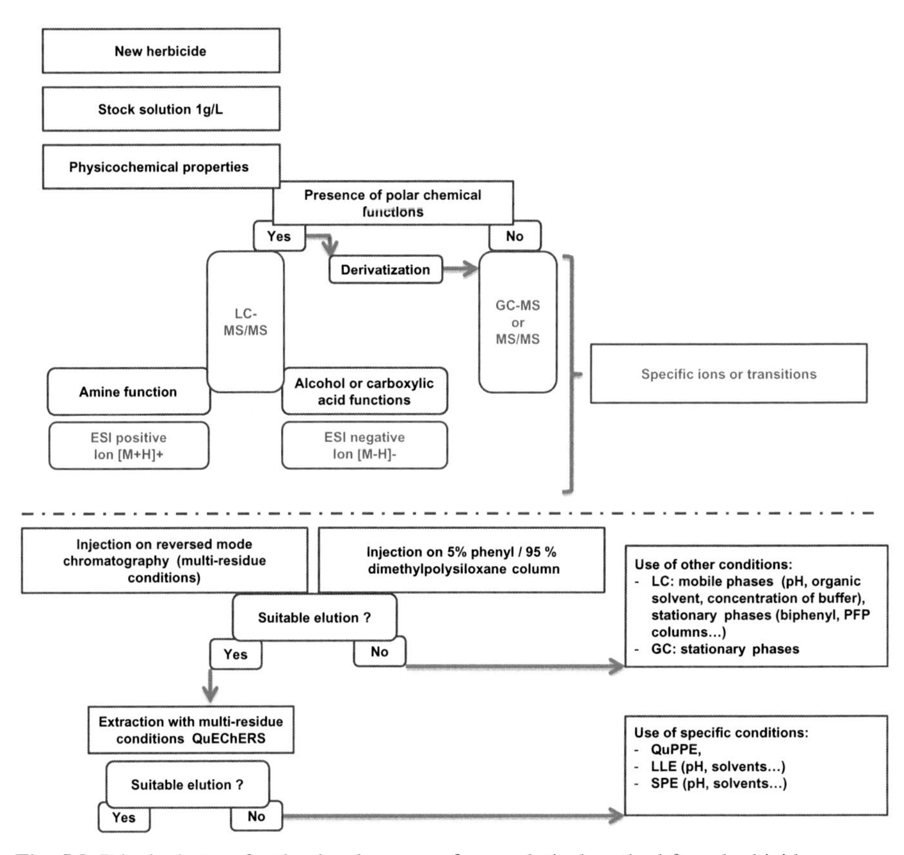

Fig. 5.8 Principal steps for the development of an analytical method for a herbicide.

Glyphosate: A typical application in a lab analyzing herbicides

Glyphosate [(*N*-(phosphonomethyl)glycine)] is a nonselective herbicide. It was first sold in 1974 and has become the most commonly and intensively used herbicide worldwide. This has increased the spread of glyphosate-tolerant or glyphosate-resistant weeds, which has in turn increased the need for higher concentrations and more frequent applications. A direct consequence of this misuse is that the compound may now be widely present in food, drinking water, and dust, suggesting that anyone in the relevant occupational and environmental settings could be exposed. MRLs of 10–20 mg/kg have been established by the EFSA for different crops. Additionally, the EU has defined analytical limits of quantification of 0.05–0.1 mg/kg for most plant and animal commodities.[86]

The greatest reason why glyphosate use is so controversial is the unsettled question of carcinogenicity. Despite glyphosate's widespread use, very few studies have explored human exposure. In 2019, Gillezeau et al.[87] conducted a review of scientific publications on glyphosate levels in humans and found fewer than 20, of which a dozen documented exposure in the general population. They estimated that levels had been measured in no more than 3000 subjects from the general population around the world and pointed out "the paucity of data on glyphosate levels among individuals exposed occupationally, para-occupationally, or environmentally." Thus, there is a need to develop sensitive and robust analytical methods for the determination of glyphosate and its metabolites in biological fluids, and to improve the performances of analytical methods dedicated to feed and food to better understand the sources of exposure.

Glyphosate rapidly degrades to AMPA, the main environmental degradant. It must be analyzed together with glyphosate when evaluating exposure.[88] Glyphosate and AMPA have physicochemical properties that make them difficult to analyze with simple tools: high hydrophilicity, nonvolatility, low molecular weight (169.1 g/mol for glyphosate), and the absence of chromophore groups. If GC-MS is theoretically possible, this physicochemical profile requires a derivatization procedure that is time-consuming and uses hazardous chemicals. Due to high polarity, glyphosate and AMPA are better candidates for MS or tandem MS using ESI, that is, LC-MS/MS.

The literature reports many methods to measure glyphosate and AMPA in food of plant origin and in feed. Analysis in nonhuman matrices is less challenging when targeting the requirements of the European Commission Santé. Our lab uses a method based on QuPPe: 10 g samples are mixed in 10 mL of 1% formic acid in methanol for 10 minutes. After centrifugation, the supernatant is filtered using UptiDisc Nylon (Interchim) and injected in the LC-MS system. A 6-point calibration curve from 0.005 to 0.25 mg/kg is used with $^{13}C_2{}^{15}N$ glyphosate as the internal standard. This method is quite similar to those published in the literature that are usually based on 1–10 g samples diluted 10–20 times with water or acidified water (e.g., 1% formic acid in water) with or without methanol before centrifugation. Depending on the matrix and in cases where proteins and lipids are present at significant levels, a matrix effect could be significant, so SPE or filtration has to be added. This supplementary cleanup is important to prevent ion suppression or enhancement. This phenomenon is present when a competition among ions occurs in the ion source, implying heavy cleanup of samples, dilution, and optimized chromatographic separation. Due to their short retention times and low molecular weights, glyphosate and AMPA are very much affected by ion suppression and enhancement. Even if the addition of stable isotope labeled internal standards can reduce the impact of this phenomenon, cleanup is strongly recommended.

As mentioned above, there are far fewer analytical methods dedicated to the measurement of glyphosate in human matrices, especially urine, the matrix of choice to explore environmental, occupational, and para-occupational exposure to glyphosate. Indeed, the measurement of glyphosate and AMPA is challenging for this matrix as well as for blood or milk, which are also sometimes considered. Multiple studies have reported the usefulness of immunoassay approaches for the rapid determination of glyphosate in nonhuman matrices. But it can be fairly said that these immunoassays should not be recommended for complex human matrices such as urine. The specificity of this approach has not been demonstrated for urine. In normal conditions, urine has a low content of proteins and lipids but contains numerous molecules capable of interfering when measuring glyphosate and AMPA. Among them are endogenous compounds such as creatinine, urea or urobilin, and potentially water-soluble exogenous compounds that are eliminated by the body. There is no consensually recommended analytical procedure for the measurement of glyphosate and AMPA in human matrices.

As with most published methods for glyphosate and AMPA, our lab had to choose a procedure with a derivatization to achieve high sensitivity. 1 mL of urine is mixed with 1 mL of pH 9 tetraborate sodium buffer and 3 mL of 9-fluorenylmethylchloroformate (FMOC). This mixture is kept out of the light for 30 minutes. Then two clean-up steps with ethyl acetate and diethyl ether after acidification are necessary before injection into the LC-MS system. A 7-point calibration curve from 0.05 to 10 μg/L is used with $^{13}C_2{}^{15}N$ glyphosate as the internal standard. Obviously, this kind of method with derivatization and LLE is time-consuming, but it produces good results. Several LC-MS/MS methods without derivatization in various human matrices have been published.[88] However, whatever the approach, none can easily solve the problem of a molecule having very high polarity and low molecular weight. High polarity complicates chromatographic separation, and molecular weight in the range where interferences are most frequent is associated with poor specificity (high risk of false positives). Meanwhile, with glyphosate being more frequently assessed for environmental exposure (not to mention the key question of carcinogenicity), the limits of quantification of the methods must be continuously improved.

References

1. Burke IC, Bell JL. Plant health management: herbicides. In: Van Alfen NK, editor. *Encyclopedia of agriculture and food systems.* 2nd ed. Academic Press/Elsevier; 2014. p. 425–40.
2. Mostafalou S, Abdollahi M. Pesticides and human chronic diseases: evidences, mechanisms, and perspectives. *Toxicol Appl Pharmacol* 2013;**268**(2):157–77.
3. Damalas CA, Eleftherohorinos IG. Pesticide exposure, safety issues, and risk assessment indicators. *Int J Environ Res Public Health* 2011;**8**(5):1402–19.
4. Heberer T, Lahrssen-Wiederholt M, Schafft H, et al. Zero tolerances in food and animal feed—are there any scientific alternatives? A European point of view on an international controversy. *Toxicol Lett* 2007;**175**(1–3):118–35.
5. Brandon EFA, Bulder AS, van Engelen JGM, et al. Does EU legislation allow the use of the benchmark dose (BMD) approach for risk assessment? *Regul Toxicol Pharmacol* 2013;**67**(2):182–8.
6. Guidance of the Scientific Committee on a request from EFSA on the use of the benchmark dose approach in risk assessment. *EFSA J* 2009;**1150**:1–72.
7. Angerer J, Ewers U, Wilhelm M. Human biomonitoring: state of the art. *Int J Hyg Environ Health* 2007;**210**(3–4):201–28.
8. Esteban M, Castano A. Non-invasive matrices in human biomonitoring: a review. *Environ Int* 2009;**35**(2):438–49.
9. Silins I, Hogberg J. Combined toxic exposures and human health: biomarkers of exposure and effect. *Int J Environ Res Public Health* 2011;**8**(3):629–47.

10. Bevan R, Angerer J, Cocker J, et al. Framework for the development and application of environmental biological monitoring guidance values. *Regul Toxicol Pharmacol* 2012;**63**(3):453–60.
11. Angerer J, Aylward LL, Hays SM, Heinzow B, Wilhelm M. Human biomonitoring assessment values: approaches and data requirements. *Int J Hyg Environ Health* 2011;**214**(5):348–60.
12. Lehmann E, Oltramare C, de Alencastro LF. Development of a modified QuEChERS method for multi-class pesticide analysis in human hair by GC-MS and UPLC-MS/MS. *Anal Chim Acta* 2018;**999**:87–98.
13. Aprea C, Sciarra G, Sartorelli P, Desideri E, Amati R, Sartorelli E. Biological monitoring of exposure to organophosphorus insecticides by assay of urinary alkylphosphates: influence of protective measures during manual operations with treated plants. *Int Arch Occup Environ Health* 1994;**66**(5):333–8.
14. Aprea MC. Environmental and biological monitoring in the estimation of absorbed doses of pesticides. *Toxicol Lett* 2012;**210**(2):110–8.
15. Thomas KW, Dosemeci M, Coble JB, et al. Assessment of a pesticide exposure intensity algorithm in the agricultural health study. *J Expo Sci Environ Epidemiol* 2010;**20**(6):559–69.
16. Thomas KW, Dosemeci M, Hoppin JA, et al. Urinary biomarker, dermal, and air measurement results for 2,4-D and chlorpyrifos farm applicators in the agricultural health study. *J Expo Sci Environ Epidemiol* 2010;**20**(2):119–34.
17. Hays SM, Aylward LL, Driver J, Ross J, Kirman C. 2,4-D exposure and risk assessment: comparison of external dose and biomonitoring based approaches. *Regul Toxicol Pharmacol* 2012;**64**(3):481–9.
18. Mendas G, Vuletic M, Galic N, Drevenkar V. Urinary metabolites as biomarkers of human exposure to atrazine: atrazine mercapturate in agricultural workers. *Toxicol Lett* 2012;**210**(2):174–81.
19. Zhang X, Acevedo S, Chao Y, et al. Concurrent 2,4-D and triclopyr biomonitoring of backpack applicators, mixer/loader and field supervisor in forestry. *J Environ Sci Health B* 2011;**46**(4):281–93.
20. Yusa V, Millet M, Coscolla C, Roca M. Analytical methods for human biomonitoring of pesticides. A review. *Anal Chim Acta* 2015;**891**:15–31.
21. Bevan R, Brwon T, Matthies F, et al. *Human biomonitoring data collection from occupational exposure to pesticides*; 2017. https://efsa.onlinelibrary.wiley.com/doi/abs/10.2903/sp.efsa.2017.EN-1185. [Accessed February 2021].
22. Aylward LL, Hays SM. Interpreting biomonitoring data for 2,4-dichlorophenoxyacetic acid: update to biomonitoring equivalents and population biomonitoring data. *Regul Toxicol Pharmacol* 2015;**73**(3):765–9.
23. Dereumeaux C, Fillol C, Quenel P, Denys S. Pesticide exposures for residents living close to agricultural lands: a review. *Environ Int* 2020;**134**:14.
24. Deziel NC, Colt JS, Kent EE, et al. Associations between self-reported pest treatments and pesticide concentrations in carpet dust. *Environ Health* 2015;**14**:11.
25. LaKind JS, Burns CJ, Naiman DQ, et al. Critical and systematic evaluation of data for estimating human exposures to 2,4-dichlorophenoxyacetic acid (2,4-D)—quality and generalizability. *J Toxicol Environ Health B Crit Rev* 2017;**20**(8):423–46.
26. Elenga N, Merlin C, Le Guern R, et al. Clinical features and prognosis of paraquat poisoning in French Guiana a review of 62 cases. *Medicine (Baltimore)* 2018;**97**(15):7.
27. Honore P, Hantson P, Fauville JP, Peeters A, Mahieu P. Paraquat poisoning—state-of-the-art. *Acta Clin Belg* 1994;**49**(5):220–8.

28. Sikary AK. Homicidal poisoning in India: a short review. *J Forensic Leg Med* 2019;**61**: 13–6.
29. Moon JM, Chun BJ. Predicting acute complicated glyphosate intoxication in the emergency department. *Clin Toxicol (Phila)* 2010;**48**(7):718–24.
30. Planche V, Vergnet S, Auzou N, Bonnet M, Tourdias T, Tison F. Acute toxic limbic encephalopathy following glyphosate intoxication. *Neurology* 2019;**92**(11):534–6.
31. Pannu AK, Saroch A, Agrawal J, Sharma N. 2,4-D poisoning: a review with illustration of two cases. *Trop Doct* 2018;**48**(4):366–8.
32. Yuan G, Zhang R, Chen X, Wang B, Guo R. A simple and economical method of gas chromatography-mass spectrometry to determine the presence of 6 pesticides in human plasma and its clinical application in patients with acute poisoning. *Biosci Trends* 2018;**12**(2):201–7.
33. Lin GD, Long JH, Luo Y, Wang YG, Qiu ZW. Continuous venovenous hemofiltration in the management of paraquat poisoning a meta-analysis of randomized controlled trials. *Medicine (Baltimore)* 2017;**96**(20):7.
34. Reisinger A, Rabensteiner J, Hackl G. Diagnosis of acute intoxications in critically ill patients: focus on biomarkers—part 2: markers for specific intoxications. *Biomarkers* 2020;**25**(2):112–25.
35. Yusa V, Millet M, Coscolla C, Pardo O, Roca M. Occurrence of biomarkers of pesticide exposure in non-invasive human specimens. *Chemosphere* 2015;**139**:91–108.
36. Lushchak VI, Matviishyn TM, Husak VV, Storey JM, Storey KB. Pesticide toxicity: a mechanistic approach. *EXCLI J* 2018;**17**:1101–36.
37. Barr DB, Wilder LC, Caudill SP, Gonzalez AJ, Needham LL, Pirkle JL. Urinary creatinine concentrations in the US population: implications for urinary biologic monitoring measurements. *Environ Health Perspect* 2005;**113**(2):192–200.
38. Udy AA, Jarrett P, Stuart J, et al. Determining the mechanisms underlying augmented renal drug clearance in the critically ill: use of exogenous marker compounds. *Crit Care* 2014;**18** (6):9.
39. Fernandez-Peralbo MA, de Castro MDL. Preparation of urine samples prior to targeted or untargeted metabolomics mass-spectrometry analysis. *TrAC Trends Anal Chem* 2012;**41**:75–85.
40. Dulaurent S, Gaulier JM, Baudel JL, Fardet L, Maury E, Lachatre G. Hair analysis to document non-fatal pesticide intoxication cases. *Forensic Sci Int* 2008;**176**(1):72–5.
41. Iglesias-Gonzalez A, Hardy EM, Appenzeller BMR. Cumulative exposure to organic pollutants of French children assessed by hair analysis. *Environ Int* 2020;**134**:16.
42. Appenzeller BMR, Tsatsakis AM. Hair analysis for biomonitoring of environmental and occupational exposure to organic pollutants: State of the art, critical review and future needs. *Toxicol Lett* 2012;**210**(2):119–40.
43. Appenzeller BMR, Hardy EM, Grova N, et al. Hair analysis for the biomonitoring of pesticide exposure: comparison with blood and urine in a rat model. *Arch Toxicol* 2017;**91** (8):2813–25.
44. Gil Garcia MD, Martinez Galera M, Ucles S, Lozano A, Fernandez-Alba AR. Ultrasound-assisted extraction based on QuEChERS of pesticide residues in honeybees and determination by LC-MS/MS and GC-MS/MS. *Anal Bioanal Chem* 2018;**410**(21):5195–210.
45. Fernandes VC, Domingues VF, Mateus N, Delerue-Matos C. Determination of pesticides in fruit and fruit juices by chromatographic methods. An overview. *J Chromatogr Sci* 2011;**49** (9):715–30.
46. Mahara BM, Borossay J, Torkos K. Liquid-liquid extraction for sample preparation prior to gas chromatography and gas chromatography mass spectrometry determination of herbicide and pesticide compounds. *Microchem J* 1998;**58**(1):31–8.

47. Schlittenbauer L, Seiwert B, Reemtsma T. Matrix effects in human urine analysis using multi-targeted liquid chromatography-tandem mass spectrometry. *J Chromatogr A* 2015;**1415**:91–9.
48. Anastassiades M, Lehotay SJ, Stajnbaher D, Schenck FJ. Fast and easy multiresidue method employing acetonitrile extraction/partitioning and "dispersive solid-phase extraction" for the determination of pesticide residues in produce. *J AOAC Int* 2003;**86**(2): 412–31.
49. Lehotay SJ. QuEChERS sample preparation approach for mass spectrometric analysis of pesticide residues in foods. In: Zweigenbaum J, editor. *Mass spectrometry in food safety: methods and protocols.* vol. 747. Totowa: Humana Press Inc; 2011. p. 65–91.
50. Sack C, Vonderbrink J, Smoker M, Smith RE. Determination of acid herbicides using modified QuEChERS with fast switching ESI(+)/ESI(−) LC-MS/MS. *J Agric Food Chem* 2015;**63**(43):9657–65.
51. Roca M, Leon N, Pastor A, Yusa V. Comprehensive analytical strategy for biomonitoring of pesticides in urine by liquid chromatography-orbitrap high resolution mass spectrometry. *J Chromatogr A* 2014;**1374**:66–76.
52. Wittsiepe J, Nestola M, Kohne M, Zinn P, Wilhelm M. Determination of polychlorinated biphenyls and organochlorine pesticides in small volumes of human blood by high-throughput on-line SPE-LVI-GC-HRMS. *J Chromatogr B Analyt Technol Biomed Life Sci* 2014;**945–946**:217–24.
53. Berube R, Belanger P, Bienvenu JF, et al. New approach for the determination of ortho-phenylphenol exposure by measurement of sulfate and glucuronide conjugates in urine using liquid chromatography-tandem mass spectrometry. *Anal Bioanal Chem* 2018;**410** (28):7275–84.
54. Toshima H, Yoshinaga J, Shiraishi H, Ito Y, Kamijima M, Ueyama J. Comparison of different urine pretreatments for biological monitoring of pyrethroid insecticides. *J Anal Toxicol* 2015;**39**(2):133–6.
55. Niu ZL, Zhang WW, Yu CW, Zhang J, Wen YY. Recent advances in biological sample preparation methods coupled with chromatography, spectrometry and electrochemistry analysis techniques. *TrAC Trends Anal Chem* 2018;**102**:123–46.
56. Crespo-Corral E, Santos-Delgado MJ, Polo-Diez LM, Soria AC. Determination of carbamate, phenylurea and phenoxy acid herbicide residues by gas chromatography after potassium tert-butoxide/dimethyl sulphoxide/ethyl iodide derivatization reaction. *J Chromatogr A* 2008;**1209**(1–2):22–8.
57. Budde WL. Analytical mass spectrometry of herbicides. *Mass Spectrom Rev* 2004;**23**(1): 1–24.
58. Reemtsma T, Alder L, Banasiak U. A multimethod for the determination of 150 pesticide metabolites in surface water and groundwater using direct injection liquid chromatography-mass spectrometry. *J Chromatogr A* 2013;**1271**(1):95–104.
59. Rodil R, Quintana JB, Lopez-Mahia P, Muniategui-Lorenzo S, Prada-Rodriguez D. Multi-residue analytical method for the determination of emerging pollutants in water by solid-phase extraction and liquid chromatography-tandem mass spectrometry. *J Chromatogr A* 2009;**1216**(14):2958–69.
60. Cazorla-Reyes R, Fernandez-Moreno JL, Romero-Gonzalez R, Frenich AG, Vidal JL. Single solid phase extraction method for the simultaneous analysis of polar and non-polar pesticides in urine samples by gas chromatography and ultra high pressure liquid chromatography coupled to tandem mass spectrometry. *Talanta* 2011;**85**(1):183–96.
61. DeArmond PD, Brittain MK, Platoff Jr GE, Yeung DT. QuEChERS-based approach toward the analysis of two insecticides, methomyl and aldicarb, in blood and brain tissue. *Anal Methods* 2015;7(1):321–8.

62. Argarate N, Arestin M, Ramon-Azcon J, et al. Evaluation of immunoassays as an alternative for the rapid determination of pesticides in wine and grape samples. *J AOAC Int* 2010;**93**(1):2–11.
63. Byer JD, Struger J, Klawunn P, Todd A, Sverko E. Low cost monitoring of glyphosate in surface waters using the ELISA method: an evaluation. *Environ Sci Technol* 2008;**42**(16):6052–7.
64. Hall JC, Vandeynze TD, Struger J, Chan CH. Enzyme-immunoassay based survey of precipitation and surface-water for the presence of atrazine, metolachlor and 2,4-D. *J Environ Sci Health B* 1993;**28**(5):577–98.
65. Rubio F, Veldhuis LJ, Clegg BS, Fleeker JR, Hall JC. Comparison of a direct ELISA and an HPLC method for glyphosate determinations in water. *J Agric Food Chem* 2003;**51**(3):691–6.
66. De O, Silva R, De Menezes MGG, De Castro RC, De A, Nobre C, et al. Efficiency of ESI and APCI ionization sources in LC-MS/MS systems for analysis of 22 pesticide residues in food matrix. *Food Chem* 2019;**297**:124934.
67. Lipok C, Uteschil F, Schmitz OJ. Development of an atmospheric pressure chemical ionization Interface for GC-MS. *Molecules* 2020;**25**(14):14.
68. Olsson AO, Baker SE, Nguyen JV, et al. A liquid chromatography-tandem mass spectrometry multiresidue method for quantification of specific metabolites of organophosphorus pesticides, synthetic pyrethroids, selected herbicides, and DEET in human urine. *Anal Chem* 2004;**76**(9):2453–61.
69. Thurman EM, Ferrer I, Barcelo D. Choosing between atmospheric pressure chemical ionization and electrospray ionization interfaces for the HPLC/MS analysis of pesticides. *Anal Chem* 2001;**73**(22):5441–9.
70. Feng C, Xu Q, Qiu X, et al. Comprehensive strategy for analysis of pesticide multi-residues in food by GC-MS/MS and UPLC-Q-Orbitrap. *Food Chem* 2020;**320**:126576.
71. Fillatre Y, Rondeau D, Jadas-Hecart A, Communal PY. Advantages of the scheduled selected reaction monitoring algorithm in liquid chromatography/electrospray ionization tandem mass spectrometry multi-residue analysis of 242 pesticides: a comparative approach with classical selected reaction monitoring mode. *Rapid Commun Mass Spectrom* 2010;**24**(16):2453–61.
72. Ibanez M, Sancho JV, Pozo OJ, Hernandez F. Use of quadrupole time-of-flight mass spectrometry to determine proposed structures of transformation products of the herbicide bromacil after water chlorination. *Rapid Commun Mass Spectrom* 2011;**25**(20):3103–13.
73. Elbashir AA, Aboul-Enein HY. Application of gas and liquid chromatography coupled to time-of-flight mass spectrometry in pesticides: multiresidue analysis. *Biomed Chromatogr* 2018;**32**(2).
74. Saito-Shida S, Hamasaka T, Nemoto S, Akiyama H. Multiresidue determination of pesticides in tea by liquid chromatography-high-resolution mass spectrometry: comparison between Orbitrap and time-of-flight mass analyzers. *Food Chem* 2018;**256**:140–8.
75. Wang J, Chow W, Wong JW, Leung D, Chang J, Li M. Non-target data acquisition for target analysis (nDATA) of 845 pesticide residues in fruits and vegetables using UHPLC/ESI Q-Orbitrap. *Anal Bioanal Chem* 2019;**411**(7):1421–31.
76. Raina-Fulton R. A review of methods for the analysis of orphan and difficult pesticides: glyphosate, glufosinate, quaternary ammonium and phenoxy acid herbicides, and dithiocarbamate and phthalimide fungicides. *J AOAC Int* 2014;**97**(4):965–77.
77. Dulaurent S, Moesch C, Marquet P, Gaulier JM, Lachatre G. Screening of pesticides in blood with liquid chromatography-linear ion trap mass spectrometry. *Anal Bioanal Chem* 2010;**396**(6):2235–49.

78. Caldas SS, Demoliner A, Primel EG. Validation of a method using solid phase extraction and liquid chromatography for the determination of pesticide residues in groundwaters. *J Braz Chem Soc* 2009;**20**(1):125–32.
79. Dopico MS, Gonzalez MV, Castro JM, et al. Determination of triazines in water samples by high-performance liquid chromatography with diode-array detection. *J Chromatogr Sci* 2002;**40**(9):523–8.
80. Dopico MS, Gonzalez MV, Castro JM, et al. Determination of chlorotriazines, methylthiotriazines and one methoxytriazine by SPE-LC-UV in water samples. *Talanta* 2003;**59**(3):561–9.
81. Lehotay SJ, Mastovska K. Determination of pesticide residues. In: Otles S, editor. *Methods of analysis of food components and additives*. CRC Press; 2011.
82. Ranz A, Korpecka J, Lankmayr E. Optimized derivatization of acidic herbicides with trimethylsilyldiazomethane for GC analysis. *J Sep Sci* 2008;**31**(4):746–52.
83. Garcia-Reyes JF, Ferrer C, Thurman EM, Fernandez-Alba AR, Ferrer I. Analysis of herbicides in olive oil by liquid chromatography time-of-flight mass spectrometry. *J Agric Food Chem* 2006;**54**(18):6493–500.
84. Lucas AD, Jones AD, Goodrow MH, et al. Determination of atrazine metabolites in human urine: development of a biomarker of exposure. *Chem Res Toxicol* 1993;**6**(1):107–16.
85. Kuklenyik Z, Panuwet P, Jayatilaka NK, Pirkle JL, Calafat AM. Two-dimensional high performance liquid chromatography separation and tandem mass spectrometry detection of atrazine and its metabolic and hydrolysis products in urine. *J Chromatogr B Analyt Technol Biomed Life Sci* 2012;**901**:1–8.
86. Review of the existing maximum residue levels for glyphosate according to Article 12 of Regulation (EC) No 396/2005. *EFSA J* 2018;**16**(5).
87. Gillezeau C, van Gerwen M, Shaffer RM, Rana I, Zhang L, Sheppard L, et al. The evidence of human exposure to glyphosate: a review. *Environ Health* 2019;**18**(1):2.
88. Jensen PK, Wujcik CE, McGuire MK, McGuire MA. Validation of reliable and selective methods for direct determination of glyphosate and aminomethylphosphonic acid in milk and urine using LC-MS/MS. *J Environ Sci Health B* 2016;**51**(4):254–9.

Mammalian toxicity of herbicides used in intensive GM crop farming

Robin Mesnage and Michael Antoniou
King's College London, London, United Kingdom

Chapter outline

Herbicides. https://doi.org/10.1016/B978-0-12-823674-1.00007-9

Introduction

Human exposure to pesticides, including herbicides, has been implicated in the development of various pathologies after acute intoxication or repeated occupational exposure. In agricultural workers, this can increase the risk of developing cancer,[1] diabetes,[2] infertility,[3] and neurodegenerative disorders such as Parkinson's disease.[4] Populations residing in farming areas can also be affected, as residential proximity to pesticide use is associated with neurodevelopmental effects.[5,6] Although acute and occupational health effects of herbicides are well characterized,[7] whether current exposure to environmental levels of herbicide residues causes adverse health effects is less clear.

Herbicide use patterns are changing rapidly to improve the efficiency of weed control, including attempts to cope with the development of herbicide resistance. The increasing reliance on glyphosate has resulted in the widespread growth of glyphosate-resistant weeds.[8] Since the introduction of genetically modified (GM) crops and the massive increase in glyphosate use, 17 plant species were known to have evolved resistance in 32 of the 50 United States in 2016.[9] Resistance has major impacts: reduction of crop yield with consequent economic damage, and a steady increase in herbicide use. As the Earth warms due to man-made climate change, wintertime pest control effects will diminish, increasing pesticide use even further.[10] The demand for pesticides is thus expected to increase in the coming years.

The application of genetic engineering to modify edible crops is often advocated as one of the most important advances to improve the sustainability of farming systems and to feed the world.[11] It has resulted in the creation of a range of traits, including adaptation to abiotic stress,[12] resistance to pathogens,[13] longer shelf life,[14] and enhanced nutritional properties. For example, some edible plants have been engineered to produce omega-3 fish oils,[15] others to contain elevated concentrations of antioxidants with cancer prevention properties[16] and vitamins.[17] However, more than two decades since their introduction, virtually all GM crops cultivated on a large scale have been engineered to tolerate being sprayed with, and thus accumulate, a herbicide, to produce their own (Bt toxin) insecticide,[9] or both. The vast majority have been engineered to tolerate the application of glyphosate-based herbicides (GBHs), the use of which has increased approximately 100-fold since the 1970s.[18] Concomitantly, the use of other herbicides such as alachlor and metolachlor has decreased.[19]

With the widespread emergence of glyphosate-resistant weeds, the agricultural biotechnology industry has responded by creating a new generation of GM food crops that are tolerant to multiple herbicides (Fig. 6.1), which are either federally approved or awaiting approval for commercial release (Table 6.1[20]). GM maize and soy tolerant to glyphosate in combination with either dicamba or 2,4-D have been grown commercially in the United States for the last several years. The maize variety MON 87429 combines tolerance to dicamba, glufosinate, 2,4-D, and quizalofop, and also has male sterility inducible by glyphosate.[21] However, a recent study has shown that the application of multiple herbicides can lead to the evolution of generalized resistance.[22] This observation calls into question whether the introduction of GM crops tolerant to multiple herbicides constitutes an effective long-term weed management strategy.

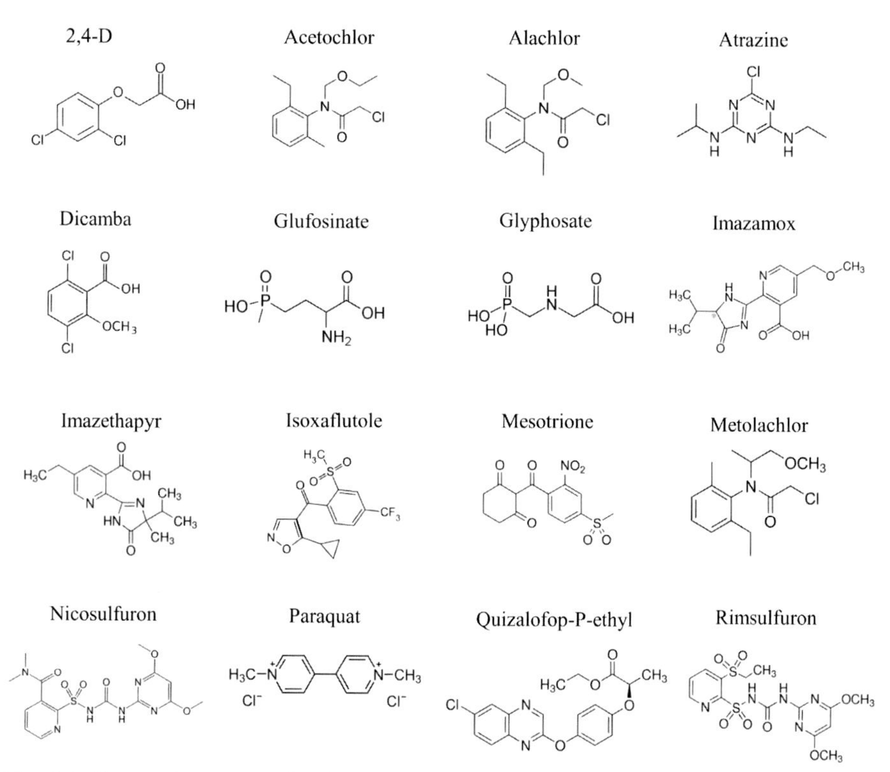

Fig. 6.1 Chemical structure of major herbicides.

Table 6.1 Herbicides tolerated by GM crops.

Product	Mode of action	Genes	Crops
Glyphosate	EPSPS inhibitor	cp4 epsps (aroA:CP4), gat4601, gat4621, goxv247, mepsps, 2mepsps	Alfalfa, canola, cotton, maize, soybean, creeping bentgrass, potato, sugar beet, wheat
Glufosinate	Glutamine synthetase inhibitor	bar, pat (syn), pat	Canola, chicory, cotton, maize, soybean, rice, sugar beet
Dicamba	Synthetic auxin	dmo	Cotton, maize, soybean
2,4-D		aad-12, aad-1	Cotton, maize, soybean
Isoxaflutole	Carotenoid biosynthesis inhibitor	hppdPF W336	Cotton, soybean
Mesotrione		avhppd-03	Soybean
Quizalofop	Inhibition of acetyl CoA carboxylase	aad-1, aad-12	
Sulfonylurea	Inhibition of acetolactate synthase	gm-hra, zm-hra, als, surB; csr1-2	Carnation, cotton, flax, maize, soybean
Imazamox		AtAHAS	Canola

Data extracted from the ISAAA GM approval database.

Nevertheless, what is clear is that this shift in herbicide use will continue to take shape over the coming years. It will result in an increase in environmental, animal, and human exposure to a mixture of compounds at unprecedented levels.

Fig. 6.2 plots the worldwide use of major herbicides by year and crop over the past three decades. An examination of the trends indicates a rise in use of some products from the introduction of glyphosate-tolerant GM crops starting in 1996, and another rise concurrent with the rapid spread of glyphosate-resistant weeds in the last decade. We have focused primarily on herbicides used in conjunction with major agricultural systems employing GM crops (soybeans, maize, canola, cotton, alfalfa, sugar beet) because their use is expected to increase globally. We also present the toxicity of other top herbicides applied in the United States and the European Union (EU) (paraquat, atrazine, metolachlor, acetochlor). Although the most toxic of these active substances are banned in the EU, they continue to be used in the United States and many third world countries.

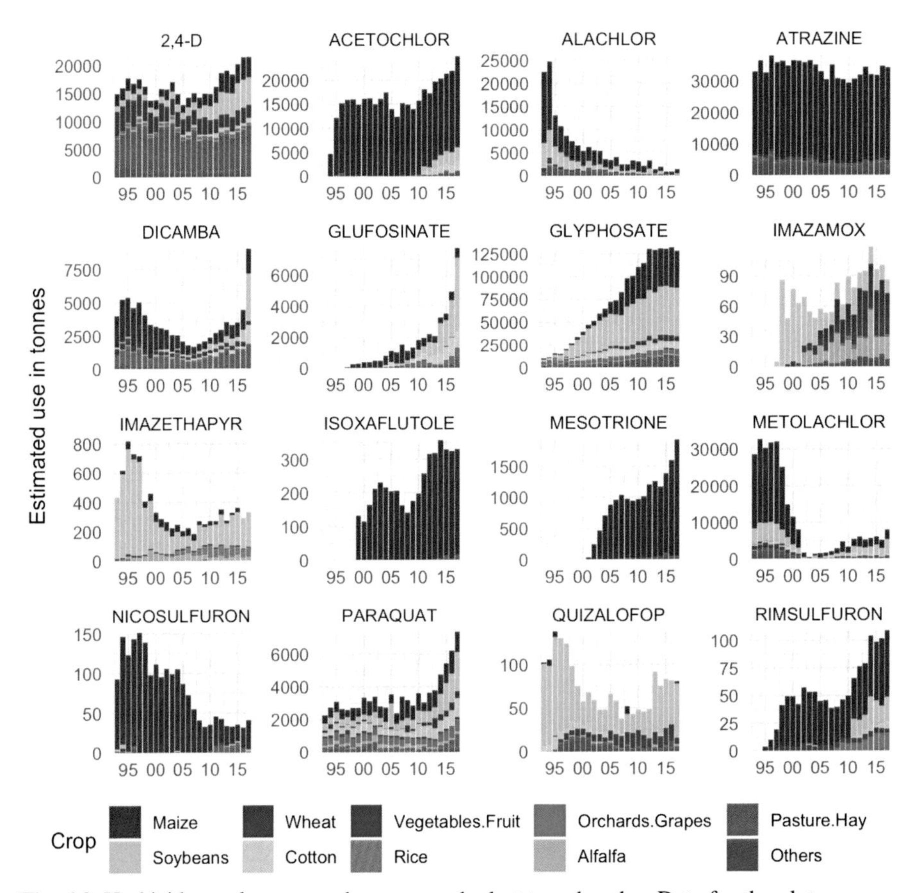

Fig. 6.2 Herbicide use by year and crop over the last two decades. Data for the plots were extracted from the website of the US Geological Survey.[23]

Health risk assessment of herbicide toxicity

Different approaches are used to evaluate health risks depending on the exposure scenario. In most cases, the evaluation for consumers or workers is done by testing herbicides in laboratory animals according to standard methods described in OECD Test Guidelines for Chemicals, although risk evaluations are increasingly based on mechanistic data (Fig. 9.3).

For acute hazards generally resulting from a single dose or short-term repeated exposures, health risks are evaluated by measuring lethality or acute clinical signs such as effects on gastrointestinal, cardiovascular, or respiratory systems. Other health outcomes that are likely to result from acute exposure include developmental effects, neurotoxicity, and hormonal alterations. These toxic effects are covered by the implementation of a guidance value called the acute reference dose.[24] For occupational settings, an acceptable operator exposure level (AOEL) is calculated to account for different routes of exposure in mixers, loaders, applicators, and re-entry workers.[25] Acute effects are generally well estimated by the current battery of tests. Protective measures are ultimately efficient to protect herbicide applicators or manage clinical intoxication in most cases.[7,26] By contrast, the effects of daily exposure to lower doses over time are evaluated by long-term toxicity assays, including reprotoxicity, multigenerational, carcinogenicity, and delayed neurotoxicity tests.[27] The most sensitive endpoint from different studies is used to estimate a no observed adverse effect level (NOAEL), which is used in the United States as a basis to calculate the chronic reference dose (cRfD). The cRfD reflects chronic dietary exposure scenarios (all populations) and is used to estimate chronic health risks. The European equivalent is the acceptable daily intake (ADI).[28]

A number of limitations have been described in the identification of reference points for the derivation of guidance values for chronic health risks.[28] Table 6.2 is a summary of toxicity endpoints used to derive the cRfD and ADI for common herbicides. Because these two measures are derived from animal experiments typically conducted at three dose levels, there is uncertainty in the determination of the dose-response relationship.[29] Although the NOAEL approach is the most common, benchmark dose methods that derive the toxicity threshold from a statistical approximation are increasingly used, as they are more precise.[28] When human data are sufficient, the comparison to toxicity thresholds can be done by the method of biomonitoring equivalents.[30]

The following sections summarize recent published data on selected major herbicides and, we hope, provide some insights into the possible impact that the new generation of GM crops and their concomitant chemicals could have on public health.

Glyphosate

Glyphosate is the most used herbicide active ingredient globally, with more than 700 GBH commercial formulations.[31] It is also one of the most controversial pesticides in terms of toxicological capability. Debate over its carcinogenicity has been raging

Table 6.2 Toxicity endpoints used to derive the US chronic reference dose (cRfD) and EU acceptable daily intake (ADI) for common herbicides.

Herbicide	*Study type*, toxic effect	US NOAEL	cRfD	ADI
Glyphosate	*Developmental, rabbit* maternal toxicity	175	1.75	0.5
Glufosinate	*Weight of evidence approach from several studies, rat and dog* NOAEL = 6.0 mg/kg/day; brain glutamine synthetase inhibition; alterations in electrocardiogram	6	0.006	0.021
Dicamba	*Reproductive study with metabolite DCSA, rat* offspring LOAEL = 37 mg/kg/day based on decreased pup weights in F1 generation PND 14 and 21 (both sexes) and week 18 (females)	4	0.04	0.3
2,4-D	*2-year, rat* decreased body weight gain and food consumption (females); alterations in hematology and clinical chemistry; decreased T4 (both sexes), glucose (females), cholesterol (both sexes), and triglycerides (females)	5	0.005	0.02
Mesotrione	*Multigeneration, mouse* offspring LOAEL = 2.1 mg/kg/day based on tyrosinemia in F1 and F2a offspring and ocular discharge in F1 pups	–	0.007	0.01
Isoxaflutole	*Combined chronic toxicity/carcinogenicity, rat* LOAEL = 20 mg/kg/day based on liver, thyroid, ocular, and nervous system toxicity (males) and liver toxicity (females)	2	0.02	0.02
Quizalofop-*p*-ethyl	*Combined chronic toxicity/carcinogenicity, rat* LOAEL = 3.7 mg/kg/day based on mild anemia in males and increased number of liver masses and centrilobular enlargement of the liver in both sexes	0.9	0.009	0.009

Table 6.2 Continued

Herbicide	***Study type*, toxic effect**	**US NOAEL**	**cRfD**	**ADI**
Nicosulfuron	*Chronic oral toxicity, dog* LOAEL = 500 mg/kg/day based on increased relative liver and kidney weights in males Supported by *developmental toxicity study, rabbit* NOAEL = 100 mg/kg/day; LOAEL = 500 mg/kg/day	125	1.25	2
Rimsulfuron	*Combined chronic/ carcinogenicity, rat* LOAEL = 121 mg/kg/day in males, 568mg/kg/day in females; NOAEL = 163 mg/kg/ day based on decreased body weight gain and liver effects	11.8	0.118	0.1
Imazamox	*Exempt from the requirement of a tolerance*	–	–	3
Imazethapyr	*Chronic oral toxicity, dog* no toxicity seen at the high dose of 250 mg/kg/day	250	2.5	–
atrazine	*Exposure for lifetime to 70 ppm in diet, rat* decreased body weight gain	3.5	0.035	0.02
Acetochlor	*Chronic oral toxicity, beagle dog* LOAEL = 10 mg/kg/day based on increased salivation and histopathology in testes, kidney, and liver	2	0.02	0.0036
Alachlor	*1-year feeding study, dog* hemosiderosis; hemolytic anemia	1	0.01	–
S-Metolachlor	*1-year chronic toxicity, dog* metolachlor LOAEL = 33 mg/ kg/day based on decreased body weight gain in females	9.7	0.097	0.1
Metolachlor	*2-year feeding study at 150 mg/ kg/day, rat* decreased body weight gain	15	0.15	–
Paraquat	*1-year feeding study, dog* LOAEL = 0.93 mg/kg/day based on increase in severity and extent of chronic pneumonitis	0.45	0.0045	0.004

ADIs were extracted from the EU pesticide database (https://ec.europa.eu/food/plant/pesticides/). NOAEL is the no observed adverse effect level.

since the International Agency for Research on Cancer (IARC) published a monograph classifying it as a Group 2A probable human carcinogen.[31]

Its primary mode of action in plants is to inhibit the binding of phosphoenolpyruvate to the active site of 5-enolpyruvylshikimate-3-phosphate synthase (EPSPS) of the shikimate pathway, causing a shortage in aromatic amino acid biosynthesis. The association of glyphosate to EPSPS is specific, and it is unlikely that this compound can bind to the phosphoenolpyruvate site of other enzymes involved in energy metabolism in mammals.[32] Glyphosate was found not to disrupt estrogen, androgen, and thyroid pathways in the US Environmental Protection Agency (EPA) Endocrine Disruptor Screening Program.[33] Although some independent studies suggested the activation of estrogen receptors at low concentrations,[34] this observation has not been reproducible.[35]

Determination of glyphosate regulatory limits of exposure

Several gaps have been identified in glyphosate health risk assessment. Interested readers can refer to recent consensus statements on concerns over the use of GBHs and risks associated with exposure to their residues.[36,37] The current ADI for glyphosate in the EU is based on developmental effects in rabbits caused by maternal toxic exposure, with a NOAEL of 50 mg/kg body weight/day (mg/kg bw/d).[38] The protective ability of this ADI has been questioned. Some reanalysis of the regulatory data suggested that developmental effects were detected at lower levels.[39] The EPA cRfD is also based on rabbit developmental studies (see Table 6.2).

Previous evaluations of glyphosate toxicity by regulatory agencies used different endpoints to calculate its ADI. The first ADI, in the 1980s, was based on an increased incidence of renal tubular dilation in a rat three-generation reproductive study from a dose of 30 mg/kg bw/d. The NOAEL was 10 mg/kg bw/d and the ADI was 0.1 mg/kg bw/d.[40] From 2002 to 2017, the ADI was based on chronic toxic effects on liver biochemistry in rats exposed to doses greater than 60 mg/kg bw/d, with a NOAEL of 30 mg/kg bw/d. These effects could be attributed to the ability of glyphosate to generate oxidative damage by interfering with mitochondrial activity. Many studies have reported effects on different organs, which can be explained by its oxidative stress-inducing properties.[41] This can explain neurotoxic effects,[42] sex hormone disturbances,[43] and liver function alteration.[44] Glyphosate has been known to negatively affect the function of mitochondria since the late 1970s.[45] Alteration of mitochondrial function is a known mechanism by which reactive oxygen species are generated. Glyphosate modulates the complexes of the mitochondrial respiratory chain.[46–48] Other studies have made a link between mitochondrial dysfunction caused by glyphosate and increased production of reactive oxygen species.[49,50]

Some commentators have also asserted that glyphosate causes toxicity by replacing glycine in protein polypeptide chains, leading to protein misfolding and impaired function.[51,52] However, this notion has been criticized as not being supported by the available evidence at the time[53] and subsequently proven to be incorrect by direct experimentation employing proteomics analysis of mammalian tissue culture cells grown in the presence of a high concentration of glyphosate.[54]

Glyphosate is generally sold and used as a mixture with adjuvants and coformulants. Because of glyphosate's coformulation with ethoxylated tallowamines, a family of compounds with known safety issues, the role of coformulants in the toxicity of glyphosate has become an intense topic of debate.[55] Results from only one long-term toxicity study of a commercial GBH have been published, where Sprague Dawley rats were administered 0.1 ppb (parts per billion) of Roundup Grand Travaux Plus (giving a final glyphosate concentration of 50 ng/L) in their drinking water for 2 years.[56–58] An analysis of urine and blood biochemistry and organ histology suggested structural and functional damage to the liver and kidneys. Follow-up analyses of organs from these rats using molecular profiling techniques confirmed kidney and liver dysfunction at the level of the transcriptome while proteomics and metabolomics revealed nonalcoholic fatty liver disease.[57] Health risks arising from exposure to adjuvants and coformulants are currently not taken into consideration by regulatory agencies. Making such information available would better inform the process of health risk assessment by regulators for improved protection of the public and the environment.[53] Chapter 4 gives an overview of the function, classification, toxicology, and regulation of coformulants and adjuvants.

Glyphosate and carcinogenesis

Whether glyphosate is carcinogenic in humans is still debated, but a large body of evidence suggests that it causes cancer at high doses in laboratory animals.[59] The significance of the animal bioassays is nonetheless doubted because the large number of tests performed implies an increased likelihood of detecting carcinogenic effects by chance.[60] The IARC classified glyphosate as a Group 2A carcinogen, but most regulatory agencies concluded that there was no strong evidence in animals or humans. That is why the IARC did not classify it a Group 1 human carcinogen. The carcinogenic effects in laboratory animals have been linked to the ability of glyphosate to generate oxidative stress. The results of some studies indicated that the oxidative stress is concomitant with DNA damage.[61–64] In addition, antioxidants can mitigate DNA damage, which suggests that it is caused by increased oxidative stress.[65] Overall, 51% of 306 assays (211 from the public literature and 95 from registrant-commissioned regulatory studies) reported genotoxic effects of glyphosate or its formulated products.[66] Whether glyphosate is present at levels sufficiently high to cause cancer is not the only matter of debate. In the EU, pesticides recognized as mutagens, carcinogens, or reproductive toxicants are prohibited unless exposure to humans is considered negligible.[67]

Another pathway through which glyphosate or GBHs may alter patterns of gene expression and contribute to carcinogenesis (and other ailments) is via epigenetic mechanisms, especially changes in DNA methylation. The first study to show changes in DNA methylation involved treating human peripheral blood mononuclear cells (PBMCs) with glyphosate at 0.1–10 mM for 24 hours.[68] Global genomic DNA methylation was decreased at 0.25 mM. In addition, at 0.25 and 0.5 mM there was an increase in *TP53* promoter methylation, suggesting that expression of this tumor suppressor gene would be reduced and thus constitute an oncogenic step.[68] In a follow-up study by the same group, also using human PBMCs, 0.5, 10, and 100 μM not only decreased the overall level of DNA methylation but resulted in localized increases in DNA methylation at the promoters of the *P21* proto-oncogene and *TP53* tumor suppressor gene.[69] Glyphosate also changed the expression of genes involved in the regulation of the cell cycle and apoptosis, specifically decreasing the expression of *P16* and *TP53* tumor suppressors and increasing the expression of *BCL2*, *CCND1*, and *P21* proto-oncogenes.[69] The same group treated PBMCs with glyphosate's main metabolite, aminomethylphosphonic acid (AMPA), at 0.5, 10, and 250 μM for 24 hours.[69] AMPA, like glyphosate, decreased the total levels of genomic DNA methylation at 10 and 250 μM, increased DNA methylation at the *TP53* promoter, and reduced methylation at the *P21* promoter. Of the tumor suppressor and proto-oncogenes investigated, only an increase in expression of *CCND1* was observed; its promoter DNA methylation status, however, was not altered.[69]

Evidence of a more direct link between glyphosate-induced DNA methylation changes and oncogenesis has been described by Duforestel and colleagues, who cultured human nonneoplastic MCF10A mammary epithelial cells in the presence of an extremely low concentration of glyphosate (10 pM) for 21 days.[70] As in the studies with PBMCs,[68,69] the total genomic levels of DNA methylation were decreased, and this seemed to be mediated in part by an increase in TET3 activity. Furthermore, transplantation in mice of glyphosate-treated MCF10A cells that had also been exposed to breast cancer-associated miR182-5p resulted in tumor formation in 50% of the animals, whereas no tumors were observed in the absence of miR182-5p exposure. The authors suggest that the epigenetic changes induced by glyphosate in MCF10A cells prime them for oncogenesis, which can lead to cancer formation when subjected to an insult by a second potent oncogenic agent.[70] A GBH has also been shown to alter DNA methylation at the estrogen receptor gene (*ESR1*) promoter, leading to reduced expression in the mammary gland[71] and increased expression in the uterus[72] of Wistar rats. However, given the very high doses (350 mg/kg bw/d) at which these effects were observed, their relevance to human health is uncertain. Furthermore, as no comparisons with glyphosate alone were undertaken, it is not possible to conclude whether the changes in epigenetic and expression status of *ESR1* were due to the glyphosate, the coformulants, or their combination.

A study demonstrating transgenerational epigenetic effects in rats is also noteworthy.[73] F0 females were administered 25 mg/kg bw/d glyphosate intraperitoneally from day 8 to day 14 of gestation. Offspring were then bred onto the F3 generation. Although the F0s showed no adverse effects, there was a dramatic increase in pathologies in the F2 and, crucially, the F3 transgenerational offspring: prostate disease, obesity, kidney disease, ovarian disease, and higher death rates of late-stage pregnant F2 females or their pups at or immediately after birth. These pathologies were correlated with differential DNA methylation regions in sperm, some of which were associated with genes previously shown to be involved in the observed pathologies.

Glyphosate and the gut microbiome

Glyphosate is not known to be actively metabolized by mammalian cells. In the environment, it undergoes spontaneous degradation to AMPA. It can be metabolized by microorganisms via two major pathways: cleavage of the C-P bond to phosphate and sarcosine or cleavage to AMPA and glyoxylate. Data mining of publicly available gut metagenomes suggested that it can be degraded by some Proteobacteria in the human gut microbiome using the carbon-phosphorus lyase pathway.[74] Its metabolites are not known to have problematic toxicological properties. The degradation of glyphosate into glyoxylate could be toxicologically relevant because it has been reported to increase liver triglyceride and cholesteryl ester levels.[75] However, this was found at doses several orders of magnitude higher than current exposure levels and therefore is likely to be irrelevant for human risk assessment at environmental levels of exposure.

The mechanism of action by which glyphosate acts as a herbicide does not exist in mammalian cells. However, microorganisms inhabiting the digestive tracts of mammals are equipped with the shikimate pathway[76] and thus are potentially sensitive to glyphosate inhibition. The gut microbiome plays important health roles, such as regulating immune function and fermenting indigestible substances. Alterations in gut microbiome composition and function have been linked to the development of human diseases locally in the digestive tract but also in distant organs such as the brain.[77] The number of studies investigating the effects of glyphosate on microbial communities is growing.[78] We have combined high-throughput biology techniques to understand the functional changes caused by glyphosate exposure.[76] We demonstrated that glyphosate inhibits EPSPS in the rat gut microbiome at doses starting from the EU ADI (0.5 mg/kg bw/d). This causes an accumulation of metabolites upstream of the EPSPS catalyzed reaction of the shikimate pathway. Levels of dipeptides involved in the regulation of redox balance were also increased in the gut microbiome. Because of the incompleteness of gut microbiome databases in laboratory rats, we were not able to find which species of microorganisms had their EPSPS inhibited. However, we found that glyphosate caused changes in gut bacteria community abundance. Shotgun metagenomics showed that *Eggerthella* spp. and *Homeothermacea* spp. were increased. Although the translation of these findings to human health remains to be explored, most bacteria found in the human gut are predicted to possess EPSPS enzymes sensitive to inhibition by glyphosate.[74] Further studies using shikimate as a marker could be done in human populations heavily exposed to glyphosate residues in order to understand whether there is likely to be an effect on microbiome function in real-life exposure.

In summary, there is solid evidence that glyphosate and GBH disturb mitochondrial function, leading to redox imbalance and in turn to the production of reactive oxygen species. This can produce organ damage and dysfunction, especially in the liver, where it causes nonalcoholic fatty liver disease. However, information on glyphosate redox effects is available for relatively few cell types. Thus, redox responses need to be investigated in a larger range of cell types and especially in different tissues in animal models. This is important as oxidative stress causes, among other types of cellular damage, genotoxicity. The evidence suggesting that glyphosate alters genomic DNA methylation patterns and brings about changes in gene expression is also restricted to a few cell types. Research on glyphosate and GBH epigenetic effects in vivo are at present limited to just three published studies at doses much higher than in typical human population exposure. More long-term studies in laboratory animals where glyphosate is compared to typical GBH formulations at real-world, environmentally relevant doses are urgently required to clarify epigenetic effects in a wide range of tissues. The importance of understanding these genotoxic and epimutation effects is that the compounds can combine to be major contributors to oncogenesis and other negative health outcomes.

Glufosinate

Glufosinate is a natural amino acid produced by the *Streptomyces* soil bacterial species. Discovered in the 1960s, it is often sold as an ammonium salt for use as a herbicide active ingredient. GM crops tolerant to glufosinate were first commercialized in 1995 with a canola variety, followed by maize in 1997, cotton in 2004, and soybeans in 2011. Tolerance is conferred by the introduction of various genes, including the *bar* gene derived from *Streptomyces hygroscopicus* encoding phosphinothricin *N*-acetyltransferase that acetylates and thus inactivates glufosinate.

Glufosinate is a structural analogue of glutamate, disrupting glutamine synthesis and causing a breakdown in ammonia metabolism in plants.[79] It is a very potent disruptor of ammonia metabolism in plants in which glutamine synthetase (GS) is the main regulator of ammonia concentrations. Mammalian metabolism is less sensitive to glufosinate-mediated increases in ammonia because other pathways control levels of this metabolite, such as carbamoyl phosphate synthetase I (first step of the urea cycle). In the brain, there is an excess capacity of GS, which operates far from its limiting turnover; this explains why brain ammonia concentrations are unchanged by doses of glufosinate that reduced the activity of brain GS by 40% in rats.[80]

Alteration of glutamate metabolism in mammals

Although glufosinate is not likely to alter ammonia homeostasis in mammalian tissues, glutamine has other important roles that glufosinate can disturb, such as excitatory and inhibitory neurotransmission.[80] The EPA estimated the cRfD for human health effects of glufosinate based on rat subchronic and chronic toxicity studies, which revealed an inhibition of brain GS and association with developmental neurotoxic outcomes (Table 6.2). Glufosinate is also classified as a reproductive toxicant.[81] Mechanistic insight was lacking to explain reproductive toxicity, but reduced GS activity may be involved because this enzyme was inhibited in the male rat liver (by 23%–26%) and kidneys (25%–36%) exposed to glufosinate.[82]

Because glufosinate is a structural analogue of glutamate, the disturbance of glutamate metabolism has been one of the more intensely studied toxicological endpoints. Glufosinate binds to NMDA excitatory receptors in assays, which increases the production of nitric oxide.[83] Several studies have reported the disruption of neurogenesis.[84,85] Mice given glufosinate at 2.5–10 mg/kg intraperitoneally three times a week for 10 weeks showed dose-dependent structural changes in the hippocampus and somatosensorial cortex.[85] A recent study showed that mice given subregulatory doses of glufosinate-ammonium (0.2–1 mg/kg bw/d) perinatally via intranasal administration presented behavioral alterations typical of changes observed in animal models of autistic spectrum disorder.[86] A follow-up investigation revealed that such disturbances were associated with impaired neuroblast proliferation and neurogenesis as well as abnormal migration of neural stem cells to the olfactory bulb.[87] Brain transcriptomics data showed that the expression of genes regulating the cytoskeleton, cell proliferation, and cell migration were affected. Another follow-up revealed that such disturbances were associated with autistic symptoms.[87]

Phosphonate herbicides modulate gut microbial activity

A large number of naturally occurring phosphonates, which are structurally related to glufosinate and also to glyphosate and its metabolite aminomethylphosphonate, are potent and useful biocides. Many of these compounds can be synthesized or degraded by microorganisms, including some by mammalian gut microbiota.[88] A recent study in rats suggested that the gut microbiome mediates the toxicity of glufosinate-ammonium on neurodevelopment.[89] Fecal metabolomics showed that the gut microbiome of glufosinate-exposed animals presented an enhancement of pyrimidine metabolism and ubiquinone biosynthesis as well as inhibition of retinol metabolism and fatty acid biosynthesis. The authors also showed that these effects were reversible. However, the gut microbiome is known to participate in the metabolism of retinoic acid,[90] so it can be postulated that the effects of glufosinate were mediated by an alteration in gut microbial metabolism.[90] This is coherent with epidemiological studies showing that children with autistic spectrum disorder had reduced serum retinol levels.[91]

Regulatory limits

Given the potent toxicity in laboratory animals and the suggestion of neurological effects at subregulatory levels, it is still not clear if exposure to glufosinate in humans causes negative health effects. However, developmental defects have been associated with paternal glufosinate-ammonium exposure (adjusted odds ratio [OR] = 2.45, 95% CI 0.78–7.70) in one human case-reference study of 261 matched pairs, although effects were not significant in the adjusted model.[92] Nevertheless, negative neurological outcomes observed at subregulatory levels of exposure in animal model systems suggest that the US cRfD and EU ADI may need to be lowered to ensure public health.

Dicamba

Dicamba is a benzoic acid used as a selective postemergent broad-leaf herbicide. It is often mixed with other active substances in commercial formulations, such as 2,4-D. Dicamba-tolerant GM crops such as Xtend soybeans are already commercialized in the United States.[93] They are engineered to carry genes from the *dmo* family present in soil bacteria, which encode for a mono-oxygenase enzyme that converts dicamba (3,6-dichloro-*o*-anisic acid) into 3,6-dichloro-salicylic acid.[94] Dicamba use initially declined following the introduction of glyphosate-tolerant crops but has been on the rise again since 2010, coinciding with the spread of glyphosate-resistant weeds (Fig. 6.1).

Dicamba acts as a herbicide by mimicking the action of the natural auxin indole-3-acetic acid (IAA), interfering with plant phytohormone responses.[95] This causes an increase in ethylene and abscisic acid levels in plant tissue, which ultimately inhibits plant growth. Although IAA does not have growth-promoting effects in mammalian cells through the auxin signaling machinery, as in plants, a study in mammalian cell lines suggested a role in kidney tissue regeneration.[96] IAA is also a uremic toxin derived from tryptophan metabolism with known pathological roles in humans, as it is a reliable predictor of mortality and cardiovascular events in patients with chronic kidney disease.[97] However, it is not clear if IAA-mimicking herbicides like dicamba have similar effects on mammalian physiology. Chlorophenoxy herbicides such as dicamba are toxic in mammals through different mechanisms, including cell membrane damage, uncoupling of oxidative phosphorylation, and disruption of acetylcoenzyme A metabolism.[98]

Dicamba was not found to be carcinogenic or neurotoxic in regulatory studies, and only produced toxic effects in a mouse long-term study (decreased body weight at 364 mg/kg bw/d).[99] The ADI of 0.3 mg/kg bw/d was calculated from a two-generation reproduction study.[99] In this investigation, reduced weight gain in pups was observed at 105 mg/kg bw/d. The cRfD is lower because it not only takes into account the dicamba technical in the herbicide formulations, but also the plant metabolites of dicamba such as DCSA (3,6-dichlorosalicylic acid) and DCGA (3,6-dichlorogentisic acid).[100] DCSA exposure is an important consideration in the context of dietary exposure because some transgenic food crops (maize, soybeans) were modified to demethylate dicamba and thus produce DCSA. This metabolite may be more toxic than dicamba; a rat reproductive toxicity study showed a decrease in pup weight in both sexes of the F1 generation at 37 mg/kg bw/d. The NOAEL was 4 mg/kg bw/d, which was used to establish a cRfD of 0.04 mg/kg bw/d.[100]

Epidemiological studies looking at the effects of dicamba remain limited. Data from the Agricultural Health Study (AHS) cohort that surveyed 89,000 farmers and their spouses in the United States between 1993 and 1997 associated exposure in pesticide applicators to lung cancer (high exposure, OR = 1.6, 95% CI 0.7–3.4)[101] compared to a low exposure group (but not a no-exposure group). Colon cancer incidence increased in the highest exposure group (relative risk [RR] = 3.29, 95% CI 1.40–7.73), though it was not associated with a significant increase in overall cancer incidence in a meta-analysis.[102] A recent study evaluating whether single-nucleotide polymorphisms involved in hormone homeostasis could alter the effect of pesticide exposure on prostate cancer risk found that men carrying two copies of a variant of the hormone-associated gene steroid 5 alpha-reductase-1 (*SRD5A1*) had a reduced prostate cancer risk associated with low use (OR = 0.62, 95% CI 0.4–0.93) and high use of dicamba (OR = 0.44, 95% CI 0.29–0.68).[103] Dicamba was also associated with the development of hypothyroidism in pesticide applicators in the AHS,[104] as was the case for 2,4-D, another auxin-mimetic herbicide.

2,4-Dichlorophenoxyacetic acid (2,4-D)

Like dicamba, 2,4-D is a synthetic auxin mimicking IAA activity, which selectively kills plants by causing uncontrolled growth. However, it has different mechanisms of toxicity compared to IAA. For instance, 2,4-D but not IAA altered the actin structure in plant cells.[105] The first GM soybean combining tolerance to glyphosate and 2,4-D was commercialized in 2019.[93]

Dioxin contamination

2,4-D was introduced in 1945 by the American Chemical Paint Company. There are now more than 600 products containing 2,4-D derivatives.[106] It has acquired a very controversial status because it was one of the components of defoliants used in the Vietnam War. Agent White was 2,4-D mixed with picloram. Agent Orange was 2,4-D mixed with 2,4,5-trichlorophenoxyacetic acid (2,4,5-T). Agent Orange became infamous for causing serious ill health to exposed individuals and birth defects in their offspring because the 2,4,5-T was contaminated with the highly toxic dioxin TCDD.[107] Despite some controversies,[108] there are currently no data showing that 2,4-D is contaminated with abnormally high levels of TCDD in contrast with historical 2,4,5,-T, which contained on average 3 mg/kg of TCDD (up to 50 mg/kg for some batches).

Thyroid disruption

The cRfD of 2,4-D (0.005 mg/kg bw/d) was calculated using data from a rat chronic toxicity study.[109] Females showed decreased body-weight gain associated with a decrease in food consumption. Some alterations in clinical chemistry (decreased T4, glucose, cholesterol, and triglycerides) occurred in both sexes.

Because 2,4-D disturbs growth hormone activity in plants, its potential endocrine disruptive capability in nontarget organisms has been questioned. Data from studies investigating endocrine disruption and reproductive toxicity are mixed in outcome. Some studies reported thyroid hormone disruption. A bioassay in zebrafish larvae showed decreased thyroxine (T4) immunoreactivity after a 3-day exposure to 24.9 μM.[110] A significant decrease of serum T4 and T3 was observed in a rat toxicity experiment using a sublethal dose (2.4 mL/kg) of a commercial formulation of 2,4-D (Deherban A) via stomach intubation daily for 10 days.[111]

Epidemiological studies indicated potential thyroid disruptive effects, including increased hypothyroidism (OR = 1.35, 95% CI 1.04–1.76) among 22,246 male pesticide applicators in the AHS,[112] confirmed among 35,150 male and female applicators,[104] and a dose-dependent elevation of luteinizing hormone correlating with urine 2,4-D levels in a small sample of 24 agricultural workers and 15 controls.[113]

The presence of urinary 2,4-D in healthy adults has also been linked to increased levels of insulin and thyroid-stimulating hormone in people with low levels of high-density lipoprotein cholesterol.[114]

In 2015, IARC confirmed its 1987 classification of 2,4-D as a Group 2B carcinogen after concluding that there was insufficient human epidemiological evidence to list it as a Group 1.[106] Population-based case-controlled studies indicated mixed results for lymphoma and leukemia and failed to show a robust association with non-Hodgkin lymphoma (NHL). There was some disagreement among the working group, with a minority considering that the evidence was limited. One of the most comprehensive cohorts included production workers from the Dow Chemical Company involved in 2,4-D manufacture from 1945 to 1982. There was no convincing increase in the incidence of all cancers (follow-up until 2007) among employees living in the state of Michigan for the entire period.[115] However, this topic is still controversial. Although one meta-analysis showed that 2,4-D was not associated with NHL (RR = 0.97, 95% CI 0.77–1.22),[116] another recently published meta-analysis taking into account exposure levels revealed an increased risk of NHL for highly exposed subjects (RR = 1.73, 95% CI 1.10–2.72).[117] Mechanistic data provide strong evidence linking 2,4-D to oxidative stress. A 2017 longitudinal study of 30 US maize farmers is the latest example. It found 2,4-D in the urine at higher levels (up to 6 μg/L during the growing season) than nonfarmer controls, and this was associated with elevated 8-hydroxy-2′-deoxyguanosine (8-OHdG), a marker of oxidative damage to DNA, and 8-iso-prostaglandin-F2a (8-isoPGF), a product of lipid peroxidation.[118] This latest survey builds on evidence from animal[119] and in vitro[120] studies, which have demonstrated the potential of 2,4-D to induce oxidative stress.

Effects of auxinic herbicides on the gut microbiome

2,4-D interacts with the gut microbiome, as it mimics IAA, which has been repeatedly shown to have an important role in gut microbial metabolism in humans.[121] However, only one study has investigated the effects of 2,4-D on the mammalian gut microbiome and the subsequent health consequences.[122] It was administered to mice at a dose of occupational exposure relevance (15 mg/kg bw/d). The plasma levels of acylcarnitine were decreased. There were also clear changes in gut microbiome composition and function, in particular a correlation between the perturbations of *Xylanimonas cellulosilytica* by 2,4-D and plasma acylcarnitine levels.

Mesotrione

Mesotrione is relatively new, commercialized in 2001 by Zeneca Agrochemicals (now owned by Syngenta) in the United Kingdom. The mode of toxicity in plants is the alteration of tyrosine metabolism through the inhibition of the enzyme 4-hydroxyphenylpyruvate dioxygenase (HPPD). GM crops developed to tolerate mesotrione have been engineered to carry the avhppd-03 gene from an oat (*Avena sativa*) species that encodes an HPPD variant insensitive to the effects of

HPPD-inhibiting herbicides. Mesotrione-tolerant GM soybean (SYHTØH2) was developed by Bayer CropScience and Syngenta and is also tolerant to glufosinate.[20]

Disruption of tyrosine metabolism in mammals

The inhibition of HPPD is the primary mode of toxicity of mesotrione in mammals.[123,124] The administration of mesotrione to rats elevates plasma tyrosine levels, leading to ocular, liver, and kidney toxicity. HPPD activity was reduced by 70% at a dietary level of 0.5 ppm, and complete inhibition was seen at 10 ppm.[125] The effects are less pronounced in mice due to differences in tyrosine metabolism.[125] Human metabolism is more reflected in mice, so they are considered preferable for human health risk assessment.[125] A study on human volunteers demonstrated the induction of a dose-related peak of tyrosinemia at 12 hours after a single oral dose (0.1, 0.5, or 4 mg/kg bw).[125] Plasma tyrosine concentrations returned to basal levels after 48 hours. The rapporteur EU member state (United Kingdom) in charge of the human health risk assessment of mesotrione concluded that human exposure in a human volunteer study may result in plasma tyrosine concentrations of sufficient magnitude to cause effects on fetal skeletal ossification.[125]

Health risk assessment

Toxicity studies were reviewed by EFSA at the request of the European Commission.[126] Multigenerational studies were the most sensitive to tyrosine metabolism disturbances. A first study in rats revealed a reproductive toxic effect (reduced litter size) at 10 ppm (equivalent to ~1 mg/kg bw/d). Males were generally more sensitive. An increased incidence of hydronephrosis was observed in F2 male pups at all doses tested. The ADI was based on a murine multigenerational study, which revealed decreased organ weight in adults and pups at 10 mg/kg bw/d.[126] Plasma tyrosine concentrations were increased at the lowest dose tested (2 mg/kg bw/d). A safety factor of 200 was added to the ADI calculation to take into account uncertainties related to tyrosinemia observed at 2 mg/kg bw/d, resulting in an ADI of 0.01 mg/kg bw/d (Table 6.2).

The EFSA classified mesotrione as a reproductive toxicant Category 2 for development (delayed ossification in the rat, rabbit, and mouse) and considered that it may be an endocrine disruptor.[126] In male rats, adverse effects were observed in endocrine organs, including increased testes and epididymides weights. An increased incidence of thyroid adenomas occurred in female rats from a dose of 0.57 mg/kg bw/d in a 2-year feeding study, with these incidences lying outside historical ranges at the dose of 190 mg/kg bw/d. Gaps in potential endocrine-disrupting modes of action were considered to be of critical concern.[126] The EPA derived a cRfD of 0.007 mg/kg bw/d using the same studies as EFSA.[127]

Mesotrione is used in substantially lower quantities than high-volume herbicides such as 2,4-D and glyphosate, and no concerns were identified in the health risk assessment of human exposure. However, given the known hazards of this herbicide on human metabolism, changes in patterns of use should be monitored to prevent exposure to levels in the range at which it can produce reprotoxic effects.

Isoxaflutole

Isoxaflutole (5-cyclopropyl-4-(2-methylsulfonyl-4-trifluoromethylbenzoyl)-isoxazole) is an isoxazole herbicide[128] mostly used on maize. It functions similarly to methotrione, working via HPPD enzyme inhibition, despite being of a different structural class. Use is expected to increase with the introduction of isoxaflutole-tolerant GM crops, including the soybean variety FG72 by Bayer CropScience that carries a modified HPPD gene from the *Pseudomonas fluorescens* strain A32.[20]

The cRfD and ADI are both 0.02 mg/kg bw/d (Table 6.2), based on liver, thyroid, ocular, and nervous system toxicity in males, and liver toxicity in females, following 2-year exposure in rats.[129,130] Both the EPA and the EFSA documented carcinogenic effects. A dose-related increase in the incidence of hepatocellular adenomas was observed in male and female CD-1 mice following a 78-week exposure, starting at the lowest dose of 3.6 mg/kg bw/d.[131] A similar increase in liver cancer was observed in a carcinogenicity study in Sprague Dawley rats[131] as well as an increased incidence of thyroid follicular cell adenoma in males. The carcinogenicity NOAEL was 3.2 mg/kg bw/d. A follow-up study investigating the mechanisms underlying these effects found that isoxaflutole affects the pharmacokinetics of thyroxine in a similar fashion to phenobarbital. It was concluded that isoxaflutole induces thyroid tumors through a possible disruption of thyroid-pituitary hormonal feedback mechanisms.[131] Importantly, this proposed mode of action for inducing thyroid tumors can also apply to humans and is thus of public health relevance.

Developmental toxicity effects have prompted EFSA to categorize isoxaflutole as a reproductive Category 2 toxic compound.[129] Pregnant rats exposed to 100–500 mg/kg bw/d showed an increased incidence of fetal and litter anomalies, including growth retardation such as decreased body weight, increased incidence of delayed ossification of sternbrae, and increased incidence of vertebral and rib abnormalities, below the maternal LOAEL of 500 mg/kg bw/d. Studies in rabbits also demonstrated dose-dependent developmental abnormalities below the maternal NOAEL (20 mg/kg bw/d). Pregnant rabbits exposed to 5 mg/kg bw/d gave birth to offspring with increased incidence of 27th presacral vertebrae. At higher doses (20 and 100 mg/kg bw/d), the effects were manifested as an increased amount of postimplantation fetal loss and late resorptions as well as growth retardation in the form of a generalized reduction in skeletal ossification.

Quizalofop (*p*-ethyl and *p*-tefuryl)

Quizalofop, a member of the family of aryloxyphenoxypropionic acids, is used as a herbicide active ingredient in the form of quizalofop-*p*-ethyl and quizalofop *p*-tefuryl.[132] It is increasingly used in combination with glyphosate to combat glyphosate-resistant weeds.[133] This use will be further increased by the cultivation of DAS-40278-9 GM maize carrying the transgene event *aad-1*, introduced to detoxify both 2,4-D and quizalofop.[20]

Quizalofop herbicides show relatively rapid absorption and distribution and rather slow elimination in urine and feces. Only 56%–70% of quizalofop-*p*-ethyl was eliminated after administration to rats. Upon ingestion, it is rapidly metabolized to quizalofop acid, followed by hydroxylation. Only quizalofop acid was detected in the urine of rats administered 10 mg/kg quizalofop-*p*-ethyl. This indicates that quizalofop-*p*-ethyl is rapidly metabolized to quizalofop acid, although bioaccumulation of the former in tissues cannot be excluded based on the current available evidence[134] The different toxicities of the different forms of quizalofop are unclear. However, a recent study investigating the toxicity of quizalofop-*p*-ethyl and its metabolite quizalofop acid on *Eisenia foetida* found that the metabolite displayed much greater toxicity than the parent compound.[135] Also, quizalofop-*p*-ethyl and quizalofop acid showed enantioselective toxicity to earthworms, with racemic mixtures being more toxic than the R(+) forms.[135]

The liver is the main target for quizalofop health effects, as revealed by an analysis of the studies used to establish regulatory guidance values.[136] In a chronic toxicity study of quizalofop-*p*-tefuryl, hypertrophy and hyperplasia of hepatocytes as well as bile stasis and hepatocellular adenomas/carcinomas were observed in male and female rats dosed with 39.5 and 48.7 mg/kg bw/d, respectively. An increased incidence of testicular tumors was also observed, with a degeneration of testicular seminiferous tubules leading to secondary aspermia. The mechanism underlying these effects is related to the mode of action of quizalofop in plants, as peroxisome proliferation was observed. The NOAEL in this study was 1.3 mg/kg bw/d in males and 1.7 mg/kg bw/d in females. Liver toxic effects were also detected in a comparable study in mice. No liver or testicular tumors were detected, though rare renal squamous cell carcinomas were observed. Thus, classification as an R40 toxicant (limited evidence of carcinogenic effects) was proposed. For quizalofop-*p*-ethyl, the NOAEL was 1.6 mg/kg bw/d based on liver and testis toxicities.[137]

In a recent study in which our group compared the potential of glyphosate, 2,4-D, dicamba, mesotrione, isoxaflutole, and quizalofop-*p*-ethyl to induce lipid accumulation in murine 3T3-L1 adipocytes, quizalofop-*p*-ethyl caused a dose-dependent, statistically significant triglyceride accumulation from a concentration of 5 μM. An investigation of gene expression profiles indicated that it exerts its lipid accumulation effects via a peroxisome proliferator-activated receptor gamma (PPARγ)-mediated pathway.[138]

For quizalofop-*p*-ethyl (Table 6.2), the EPA considered that the mild anemia in males and the increased number of liver masses and centribular enlargement of the liver (both sexes) measured after 3.7 mg/kg bw/d in a rat chronic toxicity study (NOAEL = 0.9 mg/kg bw/d) was a reliable point of departure to establish a cRfD of 0.009 mg/kg bw/d.[139] The ADI for quizalofop-*p*-ethyl and quizalofop-*p*-tefuryl were 0.009 mg/kg bw/d and 0.013 mg/kg bw/d. In summary, quizalofop herbicides evidently present numerous hazardous characteristics, including potent liver toxicity and possibly obesogenic effects. A human case report has confirmed that it is a probable inducer of occupational liver injury.[140]

Sulfonylurea and imidazolinones

Different classes of herbicides target acetolactase synthase (ALS), an enzyme critical for the biosynthesis of branched-chain amino acids in plants.[141,142] This includes sulfonylureas (such as nicosulfuron and rimsulfuron), triazolopyrimidines (such as flumetsulam), and imidazolinones (such as imazamox and imazethapyr). They can also have potent antifungal activities because ALS is also present in some microorganisms.[143] These compounds have been used extensively since the 1970s on fruit and cereal crops, with their pattern of use varying considerably depending on the nature of the active principle. For instance, the use of nicosulfuron on maize has severely decreased from 1996 while the use of rimsulfuron has increased during the same period.[144] Although easily biodegradable, this class is a common contaminant of foodstuff and in particular of water.[145]

ALS-inhibiting herbicides are generally considered to be poorly toxic because the target enzyme in plants does not exist in mammals. For instance, in regulatory toxicity experiments (Table 6.2), nicosulfuron and rimsulfuron are only mildly toxic to the liver at high doses and have a NOAEL for chronic effects of 200 mg/kg bw/d (ADI = 2 mg/kg bw/d)[146] and 11.8 mg/kg bw/d (ADI = 0.1 mg/kg bw/d),[147] respectively. However, recent discoveries demonstrating the impact on health stemming from gut microbiome disturbances has raised the question of possible effects of ALS inhibitors on the bacterial communities (ALS-inhibitor sensitive) inhabiting the human and animal gastrointestinal tract.[148]

The other major class of ALS inhibitors is imidazolinones. Imazamox and imazaquin are the two most used since 1992 according to the US Geological Survey pesticide use register.[144] Imazamox is authorized in the EU[149] with an ADI at 3 mg/kg bw/d based on a rabbit developmental study showing agenesis of the intermediate lobe of the lung and cervical hemivertebra at 600 mg/kg bw/d. As a consequence, it was proposed to be classified as a Category 2 reproductive toxin. EFSA also noted that there was insufficient information on the toxicity of its metabolites, which could be responsible for the observed developmental effects. Some other imidazolinones such as imazapyr[150] and imazethapyr[151] are considered minimally toxic. A structural analog, imazapic, caused necrosis of the skeletal muscle in a 1-year study in dogs at 137 mg/kg bw/d; cRfD was 0.137 mg/kg bw/d.[152] Although most imidazolinones are considered poorly toxic, imazalil is classified as a carcinogen by the EPA because it increased the incidence of liver adenomas and combined adenomas/carcinomas in Swiss albino mice and Wistar rats.[153] The ADI is 0.025 mg/kg bw/d, based on the NOAEL in a 1-year dog study showing modification of biochemical parameters, an increase in organ weight, and hepatocyte hypertrophy and histopathological changes.[154]

Although the metabolism of sulfonylurea is well understood, as the compounds are generally cleaved at the sulfonylurea bridge, the metabolism of imidazolinones is more complex and depends on the congener. A biotransformation study in Wistar rats revealed that imazalil is transformed to as many as 25 metabolites, with little of the parent compound excreted.[153]

There are also indications of sulfonylurea or imidazolinone toxicity in human populations in the scientific literature. Exposure to chlorimuron ethyl was associated with an increased risk of rheumatoid arthritis (OR = 1.45, 95% CI = 1.01–2.07) among licensed male applicators in the US AHS.[155] In the same cohort, a high exposure to imazethapyr was associated with increased risk of bladder cancer (RR = 2.73, 95% CI = 1.20–4.68) and colon cancer (RR = 2.73, 95% CI = 1.42–5.25).[156] A more recent study confirmed these findings on a larger population of the same cohort.[157] Although these pesticides were not found to be carcinogenic in long-term toxicity studies, an association with bladder cancer is plausible because aromatic amine compounds have been linked to increased incidences of bladder cancer in other studies.[158] Moreover, DNA methylation of long interspersed nucleotide element 1 (LINE-1) sequences, a hallmark of heightened cancer risk, is increased by imazethapyr in the AHS.[159] The use of sulfonylurea (OR = 2.1, 95% CI = 1.09–4.09) and imidizolinone (OR = 2.6, 95% CI = 1.11–5.87) herbicides was also associated with a higher frequency of first-trimester miscarriages.[160]

Atrazine

Atrazine (2-chloro-4-ethylamino-6-isopropylamino-1,3,5-triazine) is a pre- and post-emergence herbicide that kills plants by interrupting photosynthesis by binding to the plastoquinone-binding protein in photosystem II, resulting in inhibition of the electron transport process.[148] Although atrazine-tolerant crops such as canola were developed,[161] they were abandoned and replaced by varieties with higher yields. Atrazine remains nonetheless one of the most widely used herbicides in the United States, though it was banned in the EU because of its persistence in soil and ground water.

Endocrine-disrupting effects

After several decades of atrazine commercialization, its safety became a major topic of debate in the United States when a study reported an increased incidence of prostate cancer among workers at the Syngenta production plant in St. Gabriel, Louisiana.[162] Ultimately this increase was attributed to intensive prostate specific antigen (PSA) screening resulting in increased prostate cancer diagnosis. Although the results of human epidemiological studies were unclear, results from animal studies demonstrated an endocrine-disrupting potential that can be linked to increased mammary tumorigenesis in Sprague Dawley rats.[163] It was further demonstrated that atrazine at 75–300 mg/kg bw/d by gavage suppressed the estrogen-induced surge of luteinizing hormone (LH) and prolactin in Long-Evans hooded and Sprague Dawley rats.[164] Male reproductive function has also been suggested to be altered.[165] It is now established that atrazine can provoke endocrine-disrupting effects at the level of the hypothalamic-pituitary axis, through a suppression of the hypothalamic release of gonadotropin-releasing hormone, leading to attenuation of the preovulatory LH surge. Atrazine also induces aromatase expression, which increases the conversion of testosterone to estrogen. Aromatase activation was first noted in 1997 in a study on alligators, when researchers tried to understand the cause of endocrine-disrupting effects that caused sex reversal in a historically contaminated lake.[166]

Although these endocrine perturbations are not relevant for cancer assessment because the mechanism by which atrazine causes mammary tumors in Sprague Dawley rats does not exist in humans,[162] they could have reproductive developmental effects (such as a delay in pubertal development) that are relevant to humans. As a consequence, the 6-month LH surge found in rats (NOAEL = 1.8 mg/kg bw/d; LOAEL = 3.65 mg/kg bw/d) was used to establish a cRfD of 0.018 mg/kg bw/d (Table 6.2).[162]

Studies in humans corroborate the endocrine-disrupting effects, particularly on reproductive and developmental endpoints, which are exceptionally vulnerable. Urinary atrazine levels in pregnant women have been associated with fetal growth restriction (OR = 1.5, 95% CI 1.2–2.2) and small head circumference in children (OR = 1.5; 95% CI 1.0–2.2).[167] This follows previous reports linking the consumption of water contaminated with high levels of atrazine during the third trimester and entire pregnancy to the prevalence of small-for-gestational-age.[168] Elevated use of atrazine during spring months in the United States further coincided with a higher risk of birth defects, based on an analysis of 30 million births between 1996 and 2002 and monthly rates of surface water contamination.[169] There have also been several studies linking atrazine to increased cancer incidence. Increased breast cancer risk (OR = 1.2, 95% CI 1.13–1.28) was associated with high levels of triazine exposure in Kentucky counties in 1993–94.[170] The US AHS showed limited associations between atrazine use and thyroid cancer (RR = 4.84; 95% CI 1.31–17.93) and ovarian cancer (RR = 2.91, 95% CI 0.56–13.60), even if chance was not ruled out due to the small number of cases (for ovarian cancers) and because the trend was not monotonic and not always statistically significant (for thyroid cancer).[171] Other, more limited studies on the carcinogenicity of atrazine are available, and the reader can refer to the AHS discussion previously mentioned for more information.

Chloroacetamides (metolachlor, alachlor, and acetochlor)

Chloroacetamide ingredients are commonly used in herbicide formulations. They work by inhibiting gibberellin biosynthesis.[172] Use in the United States was greatly reduced when glyphosate-tolerant GM crops were introduced in 1996.[9] However, the addition of S-metolachlor to glyphosate formulations improved the control of broadleaf weeds. The recent increase in the use of metolachlor could be due to its ability to control glyphosate-resistant weeds.[173] In 2019, the EPA registered a product containing a mixture of dicamba and S-metolachlor for use on dicamba-tolerant GM crops (Tavium, EPA Reg. No. 100-1623).

Chloroacetamides are considered to be slightly toxic for mammals. Major concerns have nonetheless been raised about their carcinogenic potential. The EPA has classified metolachlor as a Group C carcinogen. A NOAEL of 15 mg/kg bw/d was found in a combined chronic toxicity and carcinogenicity study in rats, based on the development of neoplastic nodules/hepatocellular carcinomas. This tumor NOAEL of 15 mg/kg bw/d is comparable to that of 9.7 mg/kg bw/d from a chronic study in dogs showing decreased body weight in females, and which was selected for establishing the cRfD (Table 6.2). Alachlor has been classified as "likely to be carcinogenic to humans at high dose" based on increases in nasal tumors (at 2.5 mg/kg bw/d) and stomach tumors (at 42 mg/kg bw/d) in rats. Evidence of carcinogenicity was also found for acetochlor in rats and mice. In two carcinogenicity studies, rats showed an increase in nasal epithelial tumors and thyroid follicular cell tumors. Cancer caused by chloroacetamides is thought to arise from a nongenotoxic mechanism, the assumption being that there is a threshold at high dose levels, mostly over the maximum tolerable dose. The biotransformation of chloroacetamide herbicides into DNA reactive metabolites (dialkylbenzoquinone imines) by monooxygenase enzymes is believed to constitute the major cancer-causing mechanism.[174] An evaluation of cancer incidence for 49,616 applicators, 53% of whom reported ever using metolachlor, suggested that exposure could increase the risk of developing both liver cancer and follicular cell lymphoma.[175] In the same cohort, there was a strong positive association with the use of alachlor and laryngeal cancer as well as a weaker association with myeloid leukemia.[176] On the other hand, an epidemiology study of workers involved in the manufacture of alachlor did not reveal an increase in either cancer rate or mortality in a group whose exposure greatly exceeded that of pesticide applicators.[177]

Most chloroacetamides have been banned in the EU. Alachlor was banned because of concerns about its fate and behavior in the environment, in particular the formation of degradation products of toxicological concern, as well as because of its possible occupational impacts.[178] Acetochlor was banned because human exposure was found to be over the ADI. Concerns were also raised about the potential for human exposure to t-norchloro acetochlor, a potential surface water genotoxic metabolite.[179] The metolachlor racemate of 50% R- and 50% S-enantiomer is also banned, but an 88% S- and 12% R-mixture is still authorized.[180]

Paraquat

Paraquat (1,1′-dimethyl-4,4′-dipyridinium chloride) is one of the most effective broadleaf weed killers but also one of the most toxic. It is associated with numerous fatalities, mainly caused by accidental or voluntary ingestion. It is banned in the EU but is increasingly used in the United States to mitigate the spread of glyphosate-resistant weeds, which explains a rebound in use in recent years (Fig. 6.1).

It is a bipyridinium quaternary ammonium, like diquat. It acts as a herbicide via an interference with intracellular electron transfer systems. Its potential to induce redox imbalance is also the reason why it is highly toxic to mammals, with a dose as low as 4 mg/kg causing death.[181] Primary effects on the pulmonary system can be explained by its accumulation in lung tissues mediated by the polyamine transport system, which is abundantly expressed in the plasma membrane of alveolar cells.[181] Upon absorption, it interferes with the redox cycling process, resulting in free radical-mediated damage to lipids, proteins, and DNA.

Chronic exposure is associated with the development of Parkinson's disease. This is also due to its interference with redox balance, causing an alteration of mitochondrial function. Impairment in mitochondrial function results in the elevated production of reactive oxygen species, which cause oxidative modification of microtubules and block axonal mitochondrial transport.[182] This exacerbates neuronal dysfunction in Parkinson's patients. A recent review of nine case-control studies concluded that while there was a higher Parkinson's frequency in participants who were exposed to paraquat for longer periods, a causative relationship was not certain, and new studies with a better design are needed.[183]

Concluding remarks and recommendations

The mammalian toxicity of herbicides has been investigated in many studies. It is striking to see that most of the studies are focused on a few usual suspects (glyphosate, 2,4-D, atrazine), and that the toxicology of some major herbicides remains underexplored. There is a snowball effect when some research topics become fashionable, capturing the interest of the scientific community and ultimately hindering interest in exploring new areas. Glyphosate is a good example. Although the toxicology of this ingredient remained out of focus for 30 years, a series of studies suggesting its carcinogenic properties led to public health scandals. This is driving more and more scientists to investigate the toxicity of glyphosate while some of the new ingredients introduced to mitigate glyphosate-resistant weed problems are poorly investigated. In the United Kingdom, the most extensively used herbicides on arable crops in 2018 were glyphosate, diflufenican/flufenacet, and fluroxypyr. A quick search of the PubMed database returned 1130 items for "glyphosate toxicity" but only 10 for "diflufenican toxicity" and 12 for "fluroxypyr toxicity." Another common problem in current herbicide toxicology research is that analytical methods may not be sufficiently sensitive to detect the ability of ingredients to promote chronic diseases. This can be attributed to several factors, including the lack of suitable animal models to detect hormonal or metabolic disturbances,[184] insufficiency in experimental designs in which the relationship to the dose is not always assessed with sufficient resolution,[185] and testing one ingredient at a time, which does not reflect real-life exposure.[186] Although toxicity tests on rodents have always been the first choice to assess the risk of causing human diseases, they can be complemented with new in vitro approaches and biomonitoring from human environmental epidemiology studies.

We cannot ignore the fact that herbicides are used in combination with other classes of pesticides (insecticides, fungicides, nematicides, etc.). For instance, the herbicide Lumax EZ contains a combination of S-metolachlor, atrazine, and mesotrione (EPA Reg. No. 100-1442). Future studies should not only assess the toxicity of mixtures of herbicide active principles but also combinations with other types of pesticides, residues of which are frequently detected in the environment and foodstuffs. This is crucial, as pesticides of different classes acting through different mechanisms can be far more toxic than any one type on its own. A classic example is that the potency of paraquat to induce Parkinson's disease is markedly enhanced by the fungicide maneb.[187]

In perusing the scientific literature for this review, we noticed that there are very few studies taking advantage of cutting-edge molecular profiling methods (transcriptomics, proteomics, metabolomics) to provide deeper insight into toxicity. Omics analyses have at least two advantages in toxicological studies, especially those involving laboratory animals. First, outcomes with health implications can be missed by standard physiological and biochemical measures, particularly in animal studies where the treatment period is subchronic (e.g., 90 days in rats) and with low doses (e.g., at or below the cRfD and ADI). We have found that blood metabolomics in pesticide toxicity studies in rats reveals effects indicative of negative health impacts, which standard blood biochemistry measurements fail to do.[188] Second, the predictive power of transcriptomics and the diagnostic capabilities of proteomics and metabolomics can allow definitive conclusions to be drawn about pesticide toxicity from in vivo studies with fewer animals and shorter exposure periods. Using these analytical approaches would amount to a major saving of resources and much greater benefit to animal welfare. We therefore recommend that multi-omics approaches be employed in future studies investigating the toxicity of both single and combinations of pesticide ingredients.[57,58]

Another major knowledge gap in herbicide and pesticide toxicology in general is epigenetic effects. Environmental factors, including chemical pollutants, can alter gene expression patterns, which can lead to serious disease.[189–191] It is important to include epigenetic effects in the risk assessment of agents both singly and in combination, as there is currently little information on this important topic.[192]

The comparison between the trends in use (Fig. 6.1) and toxicity profiles (Table 6.2) shows that the use of the most toxic herbicides (i.e., chloroacetamides) was reduced when glyphosate-tolerant crops were introduced in the latter 1990s. However, the success of GM crops in terms of weed control was a double-edged sword. The rapid adoption of glyphosate-tolerant crops implied that most farmers relied almost entirely on glyphosate. This heavy dependence resulted in the widespread growth of glyphosate-resistant weeds.[9] As a consequence, glyphosate is now used in combination with other active ingredients on GM crops made to tolerate several herbicides, although the long-term sustainability of such a weed management approach is debatable.[22] Herbicides might appear to be getting safer with respect to human health because toxic ingredients are regularly withdrawn from the market. However, recently introduced compounds are poorly tested, which amplifies the inescapable gap between the introduction of a new herbicide and the detection of its health effects in human populations. Thus, in addition to conducting long-term toxicity studies in laboratory animals with combinations of herbicides and other types of pesticides, it is important to conduct epidemiological surveys of different human population groups to ascertain the distribution and body burden of these food-borne toxicants. A combination of the findings from these two lines of investigation—laboratory toxicity studies and epidemiological surveys—will contribute to more appropriate regulations governing their use being put in place to better protect public health.

Close-Up: A pint of science: Understanding glyphosate health risk assessment with beer

Living on the "Bermondsey beer mile" in London, I (RM) frequently have to answer concerns about the contamination of beer by glyphosate and share a pint of science. Here is an overview of the health risk assessment of dietary glyphosate levels, taking the example of the consumption of 1 L of contaminated beer (Fig. 6.3).

The contamination of foodstuff by glyphosate receives wide media coverage when NGOs promote the results of glyphosate testing campaigns. More and more consumers become distrustful of pesticides. There were repeated stories after 20 μg/L of glyphosate was found in some beers in 2019.[193] In 2016, the safety of glyphosate residues in beer was even addressed by the European Parliament.[194]

To perform a risk assessment, the probability of an outcome (the exposure level) and the severity of this outcome (the toxicity) must be known. Chronic health risks are commonly assessed by comparing environmental levels of exposure to toxicity thresholds calculated from laboratory animal bioassays. This is done in Europe using the EFSA PRIMo (Pesticide Residues Intake Model), which calculates exposure doses based on food contamination data.[195]

In a recent analysis, glyphosate daily intake was estimated for the World Health Organization Global Environment Monitoring System/Food 17 cluster diets,[196] and this included a large range of foodstuffs from different countries. The highest contributor to glyphosate intake was barley in

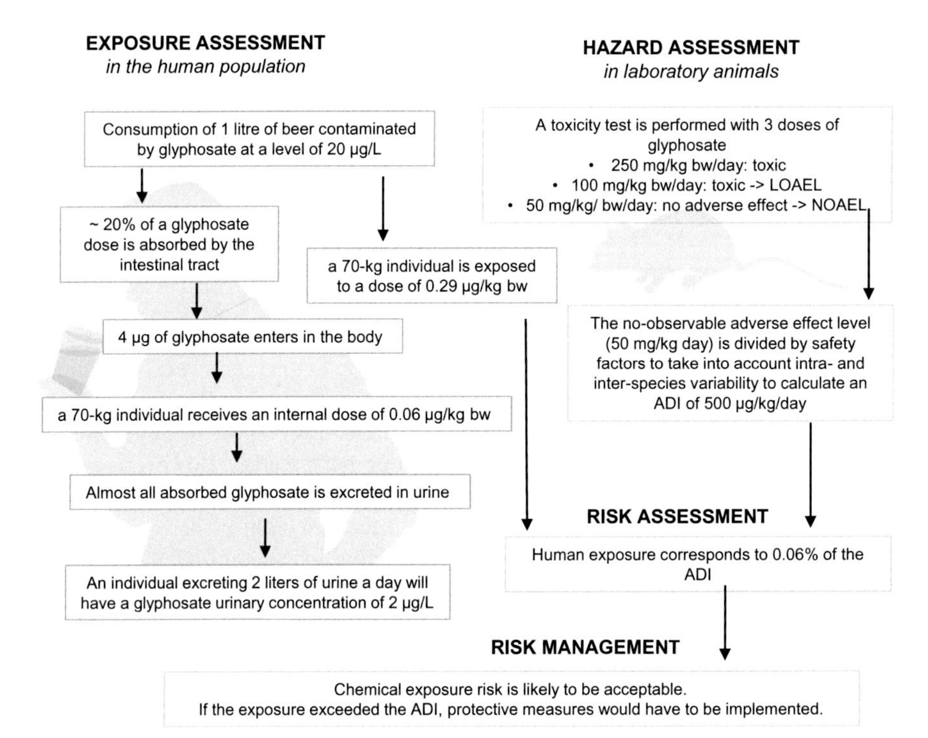

Fig. 6.3 Procedure to evaluate health risks from dietary exposure to a pesticide—From biomonitoring to risk assessment.

Ireland. Although the consumption of stout beer corresponded to an average glyphosate consumption of 1.31 μg/kg bw/day,[196] the highest contributor was actually breakfast cereals.[196] Overall, considering all sources of dietary exposures, glyphosate daily intake for the 17 cluster diets was estimated to range between 1.7 and 4.9 μg/kg bw/day.[196] This calculation fits well with real-life exposure because glyphosate urinary levels are approximately 7 μg/L,[197] suggesting that most exposure is via the diet.

Toxicity tests used by regulators across the world indicate that glyphosate is toxic at levels in the milligram per kilo of body weight range. A battery of tests was used to calculate an acceptable daily intake of 0.5 mg/kg bw/d. However, as extensively described in this chapter, current toxicity tests are insufficient, and it is possible that glyphosate has undetected toxic properties such as epigenetic effects or actions on the gut microbiome.

If I drink 1 L of beer contaminated by 20 μg/L of glyphosate, I will be exposed to a dose of 0.29 μg/kg/bw. This is more than 1000 times lower than the ADI, and more than 100,000 times lower than the NOAEL. The maximum glyphosate daily intake for the 17 WHO cluster diets (4.9 μg/kg bw/day) is approximately 100 times lower than the ADI. This is comparable to the estimation calculated using the EFSA Pesticide Residue Intake Model (EFSA PRIMo), indicating a maximum exposure corresponding to 3% of the ADI. It can be concluded with a reasonable level of certainty that glyphosate is not likely to have health effects in human populations exposed to common environmental levels. Skepticism in science is a healthy attitude, and we will never be fully certain that we are not missing something. For example, glyphosate may have some combined effects with other chemicals. In addition, the dose-dependence of glyphosate absorption in the gut is not well characterized. A recent study showed that urinary excretion of glyphosate in humans is 20 times lower than previously assumed based on animal data.[198]

Decisions in risk management are always a balance between the benefits and the risks. The purpose of herbicide risk management is to ensure a high level of protection while providing weed control strategies for farmers to improve agricultural production. Whether glyphosate is vital in agriculture is another important matter of debate.

Even if we are not sure that the full range of effects is covered by current health guidance values, the discrepancy between actual toxicity and the current risk assessment would have to be massive to reach a toxic level, and I will drink my beer with a clean conscience.

References

1. Kachuri L, Harris MA, MacLeod JS, Tjepkema M, Peters PA, Demers PA. Cancer risks in a population-based study of 70,570 agricultural workers: results from the Canadian census health and Environment cohort (CanCHEC). *BMC Cancer* 2017;**17**(1):343.
2. Juntarawijit C, Juntarawijit Y. Association between diabetes and pesticides: a case-control study among Thai farmers. *Environ Health Prev Med* 2018;**23**(1):3.
3. Bretveld RW, Hooiveld M, Zielhuis GA, Pellegrino A, van Rooij IALM, Roeleveld N. Reproductive disorders among male and female greenhouse workers. *Reprod Toxicol* 2008;**25**(1):107–14.
4. Van Maele-Fabry G, Hoet P, Vilain F, Lison D. Occupational exposure to pesticides and Parkinson's disease: a systematic review and meta-analysis of cohort studies. *Environ Int* 2012;**46**:30–43.
5. Gunier RB, Bradman A, Harley KG, Kogut K, Eskenazi B. Prenatal residential proximity to agricultural pesticide use and IQ in 7-year-old children. *Environ Health Perspect* 2017;**125**(5):057002.

6. Gonzalez-Alzaga B, Hernandez AF, Rodriguez-Barranco M, et al. Pre- and postnatal exposures to pesticides and neurodevelopmental effects in children living in agricultural communities from South-Eastern Spain. *Environ Int* 2015;**85**:229–37.
7. World Health Organization. *Preventing health risks from the use of pesticides in agriculture.* World Health Organization; 2001. 9241590998.
8. Heap I, Duke SO. Overview of glyphosate-resistant weeds worldwide. *Pest Manag Sci* 2018;**74**(5):1040–9.
9. Bonny S. Genetically modified herbicide tolerant crops, weeds, and herbicides: overview and impact. *Environ Manag* 2016;**57**(1):31–48.
10. Ehrlich PR, Harte J. Opinion: to feed the world in 2050 will require a global revolution. *Proc Natl Acad Sci U S A* 2015;**112**(48):14743–4.
11. Van Montagu M. It is a long way to GM agriculture. *Annu Rev Plant Biol* 2011;**62**:1–23.
12. Maiti RK, Satya P. Research advances in major cereal crops for adaptation to abiotic stresses. *GM Crops Food* 2014;**5**(4):259–79.
13. Yeh SD, Kung YJ, Wang HC, Wang SL. Gene-transfer vector comprising helper-component protease gene of papaya ringspot virus for broad-spectrum virus resistance in crops and use thereof. Patent CN102277376B. In: Google Patents; 2012.
14. Giovannoni JJ, Friedman H, Vrebalov J, Elitzur T. Banana MADS-Box Genes for Banana Ripening Control. Patent WO2013022757A2. In: Google Patents; 2013.
15. Kaijalainen SP, Koivu K, Kuvshinov V, Murphy E. Transgenic *Camelina sativa* plant having modified fatty acid contents of seed oil. Patent WO2014009908A3. In: Google Patents; 2014.
16. Young TR, Firoozabady E. Transgenic pineapple plants with modified carotenoid levels and methods of their production. Patent EP1589807A1. In: Google Patents; 2010.
17. Beyer P, Potrykus I. Method for improving the agronomic and nutritional value of plants. In: Google Patents; 2010.
18. Benbrook CM. Trends in glyphosate herbicide use in the United States and globally. *Environ Sci Eur* 2016;**28**(1):3.
19. Coupe RH, Capel PD. Trends in pesticide use on soybean, corn and cotton since the introduction of major genetically modified crops in the United States. *Pest Manag Sci* 2016;**72**(5):1013–22.
20. ISAAA. *GM approval database*; 2019 http://www.isaaa.org/gmapprovaldatabase/.
21. APHIS. Petition for Determination of Nonregulated Status: Bayer/Monsanto; Maize Genetically Engineered for Dicamba, Glufosinate, Quizalofop, and 2,4-Dichlorophenoxyacetic Acid Tolerance with Tissue-specific Glyphosate Tolerance Facilitating the Production of Hybrid Maize Seed. Document ID: APHIS-2020-0021-00012020.
22. Comont D, Lowe C, Hull R, et al. Evolution of generalist resistance to herbicide mixtures reveals a trade-off in resistance management. *Nat Commun* 2020;**11**(1):3086.
23. Wieben CM. *Estimated annual agricultural pesticide use by major crop or crop group for states of the conterminous United States, 1992-2017 (ver. 2.0, May 2020): U.S. Geological Survey data release*; 2019.
24. Solecki R, Davies L, Dellarco V, Dewhurst I, Van Raaij M, Tritscher A. Guidance on setting of acute reference dose (ARfD) for pesticides. *Food Chem Toxicol* 2005;**43**(11):1569–93.
25. EC. *Draft guidance for the setting and application of acceptable operator exposure levels (AOELs)*; 2006.
26. EPA. *Recognition and management of pesticide poisonings*; 2013. Available at https://www.epa.gov/pesticide-worker-safety/recognition-and-management-pesticide-poisonings.

27. World Health Organization. *Principles and methods for the risk assessment of chemicals in food.* World Health Organization; 2009.
28. Chemicals Regulation Directorate, Health & Safety Executive, UK. Investigation of the state of the art on identification of appropriate reference points for the derivation of health-based guidance values (ADI, AOEL and AAOEL) for pesticides and on the derivation of uncertainty factors to be used in human risk assessment. *EFSA Supporting Publications* 2013;**10**(4):413E.
29. Davis JA, Gift JS, Zhao QJ. Introduction to benchmark dose methods and U.S. EPA's benchmark dose software (BMDS) version 2.1.1. *Toxicol Appl Pharmacol* 2011;**254**(2):181–91.
30. Aylward LL, Hays SM. Biomonitoring Equivalents (BE) dossier for 2,4-dichlorophenoxyacetic acid (2,4-D) (CAS No. 94-75-7). *Regul Toxicol Pharmacol* 2008;**51**(3 Suppl):S37–48.
31. Guyton KZ, Loomis D, Grosse Y, et al. Carcinogenicity of tetrachlorvinphos, parathion, malathion, diazinon, and glyphosate. *Lancet Oncol* 2015;**16**(5):490–1.
32. Schonbrunn E, Eschenburg S, Shuttleworth WA, et al. Interaction of the herbicide glyphosate with its target enzyme 5-enolpyruvylshikimate 3-phosphate synthase in atomic detail. *Proc Natl Acad Sci U S A* 2001;**98**(4):1376–80.
33. US EPA. *Endocrine disruptor screening program tier 1 assessments*; 2015.
34. Thongprakaisang S, Thiantanawat A, Rangkadilok N, Suriyo T, Satayavivad J. Glyphosate induces human breast cancer cells growth via estrogen receptors. *Food Chem Toxicol* 2013;**59**:129–36.
35. Mesnage R, Phedonos A, Biserni M, et al. Evaluation of estrogen receptor alpha activation by glyphosate-based herbicide constituents. *Food Chem Toxicol* 2017;**108**(Pt A):30–42.
36. Myers JP, Antoniou MN, Blumberg B, et al. Concerns over use of glyphosate-based herbicides and risks associated with exposures: a consensus statement. *Environ Health* 2016;**15**(1):19.
37. Vandenberg LN, Blumberg B, Antoniou MN, et al. Is it time to reassess current safety standards for glyphosate-based herbicides? *J Epidemiol Community Health* 2017;**71**(6):613.
38. EFSA. Conclusion on the peer review of the pesticide risk assessment of the active substance glyphosate. *EFSA J* 2015;**13**(11):4302.
39. Antoniou M, Habib MEM, Howard CV, Jennings RC, Leifert C, Nodari RO. Teratogenic effects of glyphosate-based herbicides: divergence of regulatory decisions from scientific evidence. *J Environ Anal Toxicol* 2012;**S4**.
40. US EPA. *Pesticide fact sheet. Glyphosate*; 1986. Fact Sheet Number 173. Document 540FS88124. Available at https://nepis.epa.gov/.
41. Mesnage R, Defarge N, Spiroux de Vendomois J, Séralini GE. Potential toxic effects of glyphosate and its commercial formulations below regulatory limits. *Food Chem Toxicol* 2015;**84**:133–53.
42. Astiz M, MJTd A, Marra CA. Effect of pesticides on cell survival in liver and brain rat tissues. *Ecotoxicol Environ Saf* 2009;**72**(7):2025–32.
43. Avdatek F, Birdane YO, Turkmen R, Demirel HH. Ameliorative effect of resveratrol on testicular oxidative stress, spermatological parameters and DNA damage in glyphosate-based herbicide-exposed rats. *Andrologia* 2018;**50**(7):e13036.
44. Pandey A, Dhabade P, Kumarasamy A. Inflammatory effects of subacute exposure of roundup in rat liver and adipose tissue. *Dose-Response* 2019;**17**(2). 1559325819843380.
45. Olorunsogo OO, Bababunmi EA, Bassir O. Effect of glyphosate on rat liver mitochondria in vivo. *Bull Environ Contam Toxicol* 1979;**22**(3):357–64.

46. Pereira AG, Jaramillo ML, Remor AP, et al. Low-concentration exposure to glyphosate-based herbicide modulates the complexes of the mitochondrial respiratory chain and induces mitochondrial hyperpolarization in the Danio rerio brain. *Chemosphere* 2018;**209**:353–62.
47. Wise T, MacDonald GJ, Klindt J, Ford JJ. Characterization of thymic weight and thymic peptide thymosin-beta 4: effects of hypophysectomy, sex, and neonatal sexual differentiation. *Thymus* 1992;**19**(4):235–44.
48. Olorunsogo OO. Modification of the transport of protons and Ca^{2+} ions across mitochondrial coupling membrane by N-(phosphonomethyl)glycine. *Toxicology* 1990;**61**(2):205–9.
49. Bailey DC, Todt CE, Burchfield SL, et al. Chronic exposure to a glyphosate-containing pesticide leads to mitochondrial dysfunction and increased reactive oxygen species production in Caenorhabditis elegans. *Environ Toxicol Pharmacol* 2018;**57**:46–52.
50. Gomes MP, Juneau P. Oxidative stress in duckweed (*Lemna minor* L.) induced by glyphosate: is the mitochondrial electron transport chain a target of this herbicide? *Environ Pollut* 2016;**218**:402–9.
51. Seneff SOL. Glyphosate substitution for glycine during protein synthesis as a causal factor in Mesoamerican nephropathy. *J Environ Anal Toxicol* 2018;**8**:541.
52. Samsel A, Seneff S. Glyphosate, pathways to modern diseases IV: cancer and related pathologies. *J Biol Phys Chem* 2015;**15**:121–59.
53. Mesnage R, Antoniou MN. Facts and fallacies in the debate on glyphosate toxicity. *Front Public Health* 2017;**5**:316.
54. Antoniou MN, Nicolas A, Mesnage R, Biserni M, Rao FV, Martin CV. Glyphosate does not substitute for glycine in proteins of actively dividing mammalian cells. *BMC Res Notes* 2019;**12**(1):1–7.
55. Mesnage R, Benbrook C, Antoniou MN. Insight into the confusion over surfactant co-formulants in glyphosate-based herbicides. *Food Chem Toxicol* 2019;**128**:137–45.
56. Séralini G-E, Clair E, Mesnage R, et al. Republished study: long-term toxicity of a Roundup herbicide and a Roundup-tolerantgenetically modified maize. *Environ Sci Eur* 2014;**26**(1):14.
57. Mesnage R, Renney G, Séralini G-E, Ward M, Antoniou MN. Multiomics reveal non-alcoholic fatty liver disease in rats following chronic exposure to an ultra-low dose of Roundup herbicide. *Sci Rep* 2017;**7**:39328.
58. Mesnage R, Arno M, Costanzo M, Malatesta M, Séralini G-E, Antoniou MN. Transcriptome profile analysis reflects rat liver and kidney damage following chronic ultra-low dose Roundup exposure. *Environ Health* 2015;**14**(1):70.
59. Portier CJ. A comprehensive analysis of the animal carcinogenicity data for glyphosate from chronic exposure rodent carcinogenicity studies. *Environ Health* 2020;**19**(1):18.
60. Crump K, Crouch E, Zelterman D, Crump C, Haseman J. Accounting for multiple comparisons in statistical analysis of the extensive bioassay data on glyphosate. *Toxicol Sci* 2020;**175**(2):156–67.
61. Nwani CD, Nagpure NS, Kumar R, Kushwaha B, Lakra WS. DNA damage and oxidative stress modulatory effects of glyphosate-based herbicide in freshwater fish, *Channa punctatus*. *Environ Toxicol Pharmacol* 2013;**36**(2):539–47.
62. Hong Y, Yang X, Huang Y, Yan G, Cheng Y. Assessment of the oxidative and genotoxic effects of the glyphosate-based herbicide roundup on the freshwater shrimp, *Macrobrachium nipponensis*. *Chemosphere* 2018;**210**:896–906.
63. Hong Y, Yang X, Yan G, et al. Effects of glyphosate on immune responses and haemocyte DNA damage of Chinese mitten crab, *Eriocheir sinensis*. *Fish Shellfish Immunol* 2017;**71**:19–27.

64. Wozniak E, Sicinska P, Michalowicz J, et al. The mechanism of DNA damage induced by Roundup 360 PLUS, glyphosate and AMPA in human peripheral blood mononuclear cells—genotoxic risk assessment. *Food Chem Toxicol* 2018;**120**:510–22.
65. Santo GD, Grotto A, Boligon AA, et al. Protective effect of *Uncaria tomentosa* extract against oxidative stress and genotoxicity induced by glyphosate-Roundup(R) using zebrafish (*Danio rerio*) as a model. *Environ Sci Pollut Res Int* 2018;**25**(12):11703–15.
66. Benbrook CM. How did the US EPA and IARC reach diametrically opposed conclusions on the genotoxicity of glyphosate-based herbicides? *Environ Sci Eur* 2019;**31**(1):2.
67. European Regulation. *No 1107/2009 of the European Parliament and of the Council of 21 October 2009 concerning the placing of plant protection products on the market and repealing Council Directives 79/117/EEC and 91/414/EEC*. Brussels: EU; 2009.
68. Kwiatkowska M, Reszka E, Woźniak K, Jabłońska E, Michałowicz J, Bukowska B. DNA damage and methylation induced by glyphosate in human peripheral blood mononuclear cells (in vitro study). *Food Chem Toxicol* 2017;**105**:93–8.
69. Woźniak E, Reszka E, Jabłońska E, Balcerczyk A, Broncel M, Bukowska B. Glyphosate affects methylation in the promoter regions of selected tumor suppressors as well as expression of major cell cycle and apoptosis drivers in PBMCs (in vitro study). *Toxicol in Vitro* 2020;**63**:104736.
70. Duforestel M, Nadaradjane A, Bougras-Cartron G, et al. Glyphosate primes mammary cells for tumorigenesis by reprogramming the epigenome in a TET3-dependent manner. *Front Genet* 2019;**10**:885.
71. Gomez AL, Altamirano GA, Leturia J, Bosquiazzo VL, Muñoz-de-Toro M, Kass L. Male mammary gland development and methylation status of estrogen receptor alpha in Wistar rats are modified by the developmental exposure to a glyphosate-based herbicide. *Mol Cell Endocrinol* 2019;**481**:14–25.
72. Lorenz V, Milesi MM, Schimpf MG, Luque EH, Varayoud J. Epigenetic disruption of estrogen receptor alpha is induced by a glyphosate-based herbicide in the preimplantation uterus of rats. *Mol Cell Endocrinol* 2019;**480**:133–41.
73. Kubsad D, Nilsson EE, King SE, Sadler-Riggleman I, Beck D, Skinner MK. Assessment of glyphosate induced epigenetic transgenerational inheritance of pathologies and sperm epimutations: generational toxicology. *Sci Rep* 2019;**9**(1):6372.
74. Mesnage R, Antoniou MN. Computational modelling provides insight into the effects of glyphosate on the shikimate pathway in the human gut microbiome. *Curr Res Toxicol* 2020;**1**:25–33.
75. Ford B, Bateman LA, Gutierrez-Palominos L, Park R, Nomura DK. Mapping proteome-wide targets of glyphosate in mice. *Cell Chem Biol* 2017;**24**(2):133–40.
76. Mesnage R, Teixeira M, Mandrioli D, Falcioni L, Ducarmon QR, Zwittink RD, et al. Use of shotgun metagenomics and metabolomics to evaluate the impact of glyphosate or Roundup MON 52276 on the gut microbiota and serum metabolome of Sprague-Dawley rats. *Environ Health Perspect* 2021;**129**(1):17005.
77. Martin CR, Osadchiy V, Kalani A, Mayer EA. The brain-gut-microbiome axis. *Cell Mol Gastroenterol Hepatol* 2018;**6**(2):133–48.
78. Tsiaoussis J, Antoniou MN, Koliarakis I, et al. Effects of single and combined toxic exposures on the gut microbiome: current knowledge and future directions. *Toxicol Lett* 2019;**312**:72–97.
79. Lacuesta M, González-Moro B, González-Murua C, Aparicio-Tejo P, Muñoz-Rueda A. Effect of phosphinothricin (glufosinate) on activities of glutamine synthetase and glutamate dehydrogenase in *Medicago sativa* L. *J Plant Physiol* 1989;**134**(3):304–7.

80. Hack R, Ebert E, Ehling G, Leist KH. Glufosinate ammonium—some aspects of its mode of action in mammals. *Food Chem Toxicol* 1994;**32**(5):461–70.
81. EFSA. Conclusion regarding the peer review of the pesticide risk assessment of the active substance glufosinate. *EFSA J* 2005;**3**(4):27r.
82. Hamann H-JWK, Luetkemeier H, Biedermann K, Werner H-J, Bieler G. *Repeat dose neurotoxicity study in the rat including water maze, functional observation battery and brain-, liver- and kidney-glutamine synthetase enzyme activities glufosinate-ammonium and N-acetyl-L-glufosinate disodium Code. AE F039866 00 TK50 A133.* Generated by: RCC Umweltchemie AG, Itingen, CH; Toxicology Document No: C008991. (unpublished report); 2000.
83. Lantz SR, Mack CM, Wallace K, Key EF, Shafer TJ, Casida JE. Glufosinate binds N-methyl-D-aspartate receptors and increases neuronal network activity in vitro. *Neurotoxicology* 2014;**45**:38–47.
84. Calas AG, Richard O, Meme S, et al. Chronic exposure to glufosinate-ammonium induces spatial memory impairments, hippocampal MRI modifications and glutamine synthetase activation in mice. *Neurotoxicology* 2008;**29**(4):740–7.
85. Meme S, Calas AG, Montecot C, et al. MRI characterization of structural mouse brain changes in response to chronic exposure to the glufosinate ammonium herbicide. *Toxicol Sci* 2009;**111**(2):321–30.
86. Laugeray A, Herzine A, Perche O, et al. Pre- and postnatal exposure to low dose glufosinate ammonium induces autism-like phenotypes in mice. *Front Behav Neurosci* 2014;**8**:390.
87. Herzine A, Laugeray A, Feat J, et al. Perinatal exposure to glufosinate ammonium herbicide impairs neurogenesis and neuroblast migration through cytoskeleton destabilization. *Front Cell Neurosci* 2016;**10**:191.
88. Kafarski P. Phosphonates: their natural occurrence and physiological role. In: *Biological role of phosphorus*. IntechOpen; 2019.
89. Dong T, Guan Q, Hu W, et al. Prenatal exposure to glufosinate ammonium disturbs gut microbiome and induces behavioral abnormalities in mice. *J Hazard Mater* 2020;**389**:122152.
90. Iyer N, Vaishnava S. Vitamin A at the interface of host-commensal-pathogen interactions. *PLoS Pathog* 2019;**15**(6):e1007750.
91. Guo M, Zhu J, Yang T, et al. Vitamin A and vitamin D deficiencies exacerbate symptoms in children with autism spectrum disorders. *Nutr Neurosci* 2019;**22**(9):637–47.
92. Garcia AM, Benavides FG, Fletcher T, Orts E. Paternal exposure to pesticides and congenital malformations. *Scand J Work Environ Health* 1998;**24**(6):473–80.
93. Nandula VK. Herbicide resistance traits in maize and soybean: current status and future outlook. *Plants (Basel, Switzerland)* 2019;**8**(9):337.
94. Green JM, Owen MD. Herbicide-resistant crops: utilities and limitations for herbicide-resistant weed management. *J Agric Food Chem* 2011;**59**(11):5819–29.
95. Gleason C, Foley RC, Singh KB. Mutant analysis in Arabidopsis provides insight into the molecular mode of action of the auxinic herbicide dicamba. *PLoS One* 2011;**6**(3):e17245.
96. Cernaro V, Medici MA, Leonello G, et al. Auxin induces cell proliferation in an experimental model of mammalian renal tubular epithelial cells. *Ren Fail* 2015;**37**(5):911–3.
97. Dou L, Sallée M, Cerini C, et al. The cardiovascular effect of the uremic solute indole-3 acetic acid. *J Am Soc Nephrol* 2015;**26**(4):876.
98. Bradberry SM, Proudfoot AT, Vale JA. Poisoning due to chlorophenoxy herbicides. *Toxicol Rev* 2004;**23**(2):65–73.
99. EFSA. Reasoned opinion on the modification of the MRL for dicamba in genetically modified soybean. *EFSA J* 2013;**11**(10):3440.

100. US EPA. *Dicamba; Pesticide tolerances. Federal Register*. vol. 81, no. 236; 2016. /Thursday, December 8, 2016/Rules and Regulations. [EPA–HQ–OPP–2010–0496, EPA–HQ–OPP– 2012–0841; FRL–9954–37].
101. Alavanja MC, Dosemeci M, Samanic C, et al. Pesticides and lung cancer risk in the agricultural health study cohort. *Am J Epidemiol* 2004;**160**(9):876–85.
102. Samanic C, Rusiecki J, Dosemeci M, et al. Cancer incidence among pesticide applicators exposed to dicamba in the agricultural health study. *Environ Health Perspect* 2006;**114**(10):1521–6.
103. Christensen CH, Barry KH, Andreotti G, et al. Sex steroid hormone single-nucleotide polymorphisms, pesticide use, and the risk of prostate cancer: a nested case–control study within the agricultural health study. *Front Oncol* 2016;**6**:237.
104. Shrestha S, Parks CG, Goldner WS, et al. Pesticide use and incident hypothyroidism in pesticide applicators in the agricultural health study. *Environ Health Perspect* 2018;**126**(9):97008.
105. Takahashi M, Umetsu K, Oono Y, Higaki T, Blancaflor EB, Rahman A. Small acidic protein 1 and SCFTIR1 ubiquitin proteasome pathway act in concert to induce 2,4-dichlorophenoxyacetic acid-mediated alteration of actin in Arabidopsis roots. *Plant J* 2017;**89**(5):940–56.
106. Loomis D, Guyton K, Grosse Y, et al. Carcinogenicity of lindane, DDT, and 2,4-dichlorophenoxyacetic acid. *Lancet Oncol* 2015;**16**(8):891–2.
107. Sciences UNAo. Committee to Review the Health Effects in Vietnam Veterans of Exposure to Herbicides (Ninth Biennial Update); Board on the Health of Select Populations; Institute of Medicine. *Veterans and Agent Orange: Update 2012*. Washington, DC: National Academies Press; 2014. 2014 Mar 6. 3, Exposure to the Herbicides Used in Vietnam.
108. Holt E, Weber R, Stevenson G, Gaus C. Polychlorinated dibenzo-p-dioxins and dibenzofurans (PCDD/Fs) impurities in pesticides: a neglected source of contemporary relevance. *Environ Sci Technol* 2010;**44**(14):5409–15.
109. US EPA. *Reregistration eligibility decision for 2,4-D*; 2005. List A. Case 0073. EPA-HQ-OPP-2004-0167. Available at https://archive.epa.gov/pesticides/reregistration/web/pdf/24d_red.pdf.
110. Raldua D, Babin PJ. Simple, rapid zebrafish larva bioassay for assessing the potential of chemical pollutants and drugs to disrupt thyroid gland function. *Environ Sci Technol* 2009;**43**(17):6844–50.
111. Kobal S, Cebulj-Kadunc N, Cestnik V. Serum T3 and T4 concentrations in the adult rats treated with herbicide 2,4-dichlorophenoxyacetic acid. *Pflugers Arch* 2000;**440**(5 Suppl): R171–2.
112. Goldner WS, Sandler DP, Yu F, et al. Hypothyroidism and pesticide use among male private pesticide applicators in the agricultural health study. *J Occup Environ Med* 2013;**55**(10):1171–8.
113. Garry VF, Tarone RE, Kirsch IR, et al. Biomarker correlations of urinary 2,4-D levels in foresters: genomic instability and endocrine disruption. *Environ Health Perspect* 2001;**109**(5):495–500.
114. Schreinemachers DM. Perturbation of lipids and glucose metabolism associated with previous 2,4-D exposure: a cross-sectional study of NHANES III data, 1988-1994. *Environ Health* 2010;**9**:11.
115. Burns C, Bodner K, Swaen G, Collins J, Beard K, Lee M. Cancer incidence of 2,4-D production workers. *Int J Environ Res Public Health* 2011;**8**(9):3579–90.

116. Goodman JE, Loftus CT, Zu K. 2,4-Dichlorophenoxyacetic acid and non-Hodgkin's lymphoma, gastric cancer, and prostate cancer: meta-analyses of the published literature. *Ann Epidemiol* 2015;**25**(8):626–36. e624.
117. Smith AM, Smith MT, La Merrill MA, Liaw J, Steinmaus C. 2,4-Dichlorophenoxyacetic acid (2,4-D) and risk of non-Hodgkin lymphoma: a meta-analysis accounting for exposure levels. *Ann Epidemiol* 2017;**27**(4):281–9. e284.
118. Lerro CC, Beane Freeman LE, Portengen L, et al. A longitudinal study of atrazine and 2,4-D exposure and oxidative stress markers among iowa corn farmers. *Environ Mol Mutagen* 2017;**58**(1):30–8.
119. Pochettino AA, Bongiovanni B, Duffard RO, Evangelista de Duffard AM. Oxidative stress in ventral prostate, ovary, and breast by 2,4-dichlorophenoxyacetic acid in pre- and postnatal exposed rats. *Environ Toxicol* 2013;**28**(1):1–10.
120. Bongiovanni B, Ferri A, Brusco A, et al. Adverse effects of 2,4-dichlorophenoxyacetic acid on rat cerebellar granule cell cultures were attenuated by amphetamine. *Neurotox Res* 2011;**19**(4):544–55.
121. Chanclud E, Lacombe B. Plant hormones: key players in gut microbiota and human diseases? *Trends Plant Sci* 2017;**22**(9):754–8.
122. Tu P, Gao B, Chi L, et al. Subchronic low-dose 2,4-D exposure changed plasma acylcarnitine levels and induced gut microbiome perturbations in mice. *Sci Rep* 2019;**9**(1):4363.
123. European Commission. *Opinion on the evaluation of mesotrione in the context of council directive 91/414/eec concerning the placing of plant protection products on the market.* SCP/MESOTRI/002-Final; 2002.
124. Lewis RW, Botham JW. A review of the mode of toxicity and relevance to humans of the triketone herbicide 2-(4-methylsulfonyl-2-nitrobenzoyl)-1,3-cyclohexanedione. *Crit Rev Toxicol* 2013;**43**(3):185–99.
125. Mesotrione D. *Renewal assessment report prepared according to the Commission Regulation (EU) N° 1107/2009*; 2015. Mesotrione Volume 3—B.6 (AS). Available at http://dar.efsa.europa.eu/dar-web/provision.
126. European Food Safety Authority. Peer review of the pesticide risk assessment of the active substance mesotrione. *EFSA J* 2016;**14**(3):4419.
127. US EPA. *Federal Register*. vol. 66, no. 120; 2001. Thursday, June 21, 2001/Rules and Regulations. Mesotrione; Pesticide Tolerance. [OPP–301138; FRL–6787–7].
128. Pallett KE, Cramp SM, Little JP, Veerasekaran P, Crudace AJ, Slater AE. Isoxaflutole: the background to its discovery and the basis of its herbicidal properties. *Pest Manag Sci* 2001;**57**(2):133–42.
129. EFSA. Peer review of the pesticide risk assessment of the active substance isoxaflutole. *EFSA J* 2016;**14**(2):4416.
130. US EPA. *Federal Register*. vol. 76, no. 235; 2011. Wednesday, December 7, 2011/Rules and Regulations. Isoxaflutole; Pesticide Tolerances. [EPA–HQ–OPP–2010–0845; FRL–8885–8].
131. US EPA. *Memorandum. Carcinogenicity peer review of isoxaflutole*. Office of Prevention, Pesticides, and Toxic Susbtances; 1997. Available at https://www3.epa.gov/pesticides/chem_search/cleared_reviews/csr_PC-123000_6-Aug-97_075.pdf.
132. Price LJ, Herbert D, Moss SR, Cole DJ, Harwood JL. Graminicide insensitivity correlates with herbicide-binding co-operativity on acetyl-CoA carboxylase isoforms. *Biochem J* 2003;**375**(Pt 2):415–23.
133. Bo T, Jingkai Z, Calvin GM, Nalewaja JD. Efficacy of glyphosate plus bentazon or quizalofop on glyphosate-resistant canola or corn. *Weed Technol* 2007;**21**(1):97–101.

134. Liang Y, Wang P, Liu D, et al. Enantioselective metabolism of quizalofop-ethyl in rat. *PLoS One* 2014;**9**(6):e101052.
135. Ma L, Liu H, Qu H, et al. Chiral quizalofop-ethyl and its metabolite quizalofop-acid in soils: enantioselective degradation, enzymes interaction and toxicity to *Eisenia foetida*. *Chemosphere* 2016;**152**:173–80.
136. European Food Safety Authority. Conclusion regarding the peer review of the pesticide risk assessment of the active substance quizalofop-P. *EFSA J* 2009;**7**(7):205r.
137. Mesnage R, Biserni M, Wozniak E, Xenakis T, Mein CA, Antoniou MN. Comparison of transcriptome responses to glyphosate, isoxaflutole, quizalofop-p-ethyl and mesotrione in the HepaRG cell line. *Toxicol Rep* 2018;**5**:819–26.
138. Biserni M, Mesnage R, Ferro R, et al. Quizalofop-p-ethyl induces adipogenesis in 3T3-L1 adipocytes. *Toxicol Sci* 2019;**170**(2):452–61.
139. US EPA. *Quizalofop ethyl; Pesticide tolerances. Federal Register*. vol. 81, no. 231; 2016. Thursday, December 1, 2016/Rules and Regulations. [EPA–HQ–OPP–2015–0412; FRL–9950–89].
140. Elefsiniotis IS, Liatsos GD, Stamelakis D, Moulakakis A. Case report: mixed cholestatic/hepatocellular liver injury induced by the herbicide quizalofop-p-ethyl. *Environ Health Perspect* 2007;**115**(10):1479–81.
141. Tan S, Evans RR, Dahmer ML, Singh BK, Shaner DL. Imidazolinone-tolerant crops: history, current status and future. *Pest Manag Sci* 2005;**61**(3):246–57.
142. Whitcomb CE. An introduction to ALS-inhibiting herbicides. *Toxicol Ind Health* 1999;**15**(1–2):231–9.
143. Lee YT, Cui CJ, Chow EW, et al. Sulfonylureas have antifungal activity and are potent inhibitors of *Candida albicans* acetohydroxyacid synthase. *J Med Chem* 2013;**56**(1):210–9.
144. USGS. *Estimated annual agricultural pesticide use*; 2017. Available at https://water.usgs.gov/nawqa/pnsp/usage/maps/compound_listing.php.
145. USDA. *USDA pesticide data program*; 2017. Available at https://www.ams.usda.gov/sites/default/files/media/2013%20PDP%20Anuual%20Summary.pdf.
146. EFSA. Conclusion regarding the peer review of the pesticide risk assessment of the active substance nicosulfuron. *EFSA J* 2008;**6**(1):120r.
147. EFSA. Conclusion regarding the peer review of the pesticide risk assessment of the active substance Rimsulfuron. *EFSA J* 2005;**3**(8):45r.
148. Richie DL, Thompson KV, Studer C, et al. Identification and evaluation of novel acetolactate synthase inhibitors as antifungal agents. *Antimicrob Agents Chemother* 2013;**57**(5):2272–80.
149. EFSA. Peer review of the pesticide risk assessment of the active substance imazamox. *EFSA J* 2016;**14**(4):4432.
150. US EPA. *Reregistration eligibility decision for imazapyr*; 2006. Available at https://archive.epa.gov/pesticides/reregistration/web/html/status.html.
151. US EPA. *Imazethapyr; Pesticide tolerance. Federal Register*. vol. 67, no. 168; 2002. Thursday, August 29, 2002/Rules and Regulations. [OPP–2002–0189; FRL–7193–4].
152. US EPA. *Imazapic; Pesticide tolerances. Federal Register*. vol. 78, no. 159; 2013. Friday, August 16, 2013/Rules and Regulations. [EPA–HQ–OPP–2012–0384; FRL–9394–8].
153. US EPA. *Reregistration eligibility decision for imazalil*; 2003. Chemical list b. Case no. 2325. Available at https://archive.epa.gov/pesticides/reregistration/web/html/status.html.
154. EFSA. Conclusion on the peer review of the pesticide risk assessment of the active substance imazalil. *EFSA J* 2010;**8**(3):1526.

155. Meyer A, Sandler DP, Beane Freeman LE, Hofmann JN, Parks CG. Pesticide exposure and risk of rheumatoid arthritis among licensed male pesticide applicators in the agricultural health study. *Environ Health Perspect* 2017;**125**(7):077010.
156. Koutros S, Lynch CF, Ma X, et al. Heterocyclic aromatic amine pesticide use and human cancer risk: results from the U.S. Agricultural Health Study. *Int J Cancer* 2009;**124**(5):1206–12.
157. Koutros S, Silverman DT, Alavanja MC, et al. Occupational exposure to pesticides and bladder cancer risk. *Int J Epidemiol* 2015;**45**(3):792–805.
158. Vineis P, Pirastu R. Aromatic amines and cancer. *Cancer Causes Control* 1997;**8**(3):346–55.
159. Alexander M, Koutros S, Bonner MR, et al. Pesticide use and LINE-1 methylation among male private pesticide applicators in the Agricultural Health Study. *Environ Epigenetics* 2017;**3**(2).
160. Garry VF, Harkins M, Lyubimov A, Erickson L, Long L. Reproductive outcomes in the women of the Red River Valley of the north. I. The spouses of pesticide applicators: pregnancy loss, age at menarche, and exposures to pesticides. *J Toxicol Environ Health A* 2002;**65**(11):769–86.
161. Miller TW, Callihan RH. Interference between triazine-resistant *Brassica napus* and *Panicum miliaceum*. *Weed Res* 1995;**35**(6):453–60.
162. US EPA. *Interim reregistration eligibility decision for atrazine case no. 0062*; 2006. EPA-HQ-OPP-2003-0367. Available at https://archive.epa.gov/pesticides/reregistration/web/html/status.html.
163. Stevens JT, Breckenridge CB, Wetzel LT, Gillis JH, Luempert 3rd LG, Eldridge JC. Hypothesis for mammary tumorigenesis in Sprague-Dawley rats exposed to certain triazine herbicides. *J Toxicol Environ Health* 1994;**43**(2):139–53.
164. Cooper RL, Stoker TE, Tyrey L, Goldman JM, McElroy WK. Atrazine disrupts the hypothalamic control of pituitary-ovarian function. *Toxicol Sci* 2000;**53**(2):297–307.
165. Stoker TE, Laws SC, Guidici DL, Cooper RL. The effect of atrazine on puberty in male wistar rats: an evaluation in the protocol for the assessment of pubertal development and thyroid function. *Toxicol Sci* 2000;**58**(1):50–9.
166. Crain DA, Guillette Jr LJ, Rooney AA, Pickford DB. Alterations in steroidogenesis in alligators (*Alligator mississippiensis*) exposed naturally and experimentally to environmental contaminants. *Environ Health Perspect* 1997;**105**(5):528–33.
167. Chevrier C, Limon G, Monfort C, et al. Urinary biomarkers of prenatal atrazine exposure and adverse birth outcomes in the PELAGIE birth cohort. *Environ Health Perspect* 2011;**119**(7):1034–41.
168. Ochoa-Acuna H, Frankenberger J, Hahn L, Carbajo C. Drinking-water herbicide exposure in Indiana and prevalence of small-for-gestational-age and preterm delivery. *Environ Health Perspect* 2009;**117**(10):1619–24.
169. Winchester PD, Huskins J, Ying J. Agrichemicals in surface water and birth defects in the United States. *Acta Paediatr* 2009;**98**(4):664–9.
170. Kettles MK, Browning SR, Prince TS, Horstman SW. Triazine herbicide exposure and breast cancer incidence: an ecologic study of Kentucky counties. *Environ Health Perspect* 1997;**105**(11):1222–7.
171. Freeman LE, Rusiecki JA, Hoppin JA, et al. Atrazine and cancer incidence among pesticide applicators in the agricultural health study (1994-2007). *Environ Health Perspect* 2011;**119**(9):1253–9.
172. Fuerst EP. Understanding the mode of action of the chloroacetamide and thiocarbamate herbicides. *Weed Technol* 1987;**1**(4):270–7.

173. Clewis SB, Wilcut JW, Porterfield D. Weed management with S-metolachlor and glyphosate mixtures in glyphosate-resistant strip- and conventional-tillage cotton (*Gossypium hirsutum* L.). *Weed Technol* 2006;**20**(1):232–41.
174. Coleman S, Linderman R, Hodgson E, Rose RL. Comparative metabolism of chloroacetamide herbicides and selected metabolites in human and rat liver microsomes. *Environ Health Perspect* 2000;**108**(12):1151–7.
175. Silver SR, Bertke SJ, Hines CJ, et al. Cancer incidence and metolachlor use in the Agricultural Health Study: an update. *Int J Cancer* 2015;**137**(11):2630–43.
176. Lerro CC, Andreotti G, Koutros S, et al. Alachlor use and cancer incidence in the agricultural health study: an updated analysis. *J Natl Cancer Inst* 2018;**110**(9):950–8.
177. Acquavella JF, Delzell E, Cheng H, Lynch CF, Johnson G. Mortality and cancer incidence among alachlor manufacturing workers 1968-99. *Occup Environ Med* 2004;**61**(8):680–5.
178. EU Commission. *Review report for the active substance alachlor finalised in the Standing Committee on the Food Chain and Animal Health at its meeting on 4 April 2006*; 2007. Review report. Document SANCO/11703/2011 rev 1. Available at https://ec.europa.eu/food/plant/pesticides/.
179. EU Commission. *Review report for the active substance acetochlor finalised in the Standing Committee on the Food Chain and Animal Health at its meeting on 11 October 2011*; 2011. Document SANCO/11703/2011 rev 1. Available at https://ec.europa.eu/food/plant/pesticides/.
180. EU Commission. *Review report for the active substance S-Metolachlor*; 2001. Document SANCO/1426/2001—rev. 3. Available at https://ec.europa.eu/food/plant/pesticides/.
181. Dinis-Oliveira RJ, Duarte JA, Sanchez-Navarro A, Remiao F, Bastos ML, Carvalho F. Paraquat poisonings: mechanisms of lung toxicity, clinical features, and treatment. *Crit Rev Toxicol* 2008;**38**(1):13–71.
182. Stykel MG, Humphries K, Kirby MP, et al. Nitration of microtubules blocks axonal mitochondrial transport in a human pluripotent stem cell model of Parkinson's disease. *FASEB J* 2018;**32**(10):5350–64.
183. Vaccari C, El Dib R, Gomaa H, Lopes LC, de Camargo JL. Paraquat and Parkinson's disease: a systematic review and meta-analysis of observational studies. *J Toxicol Environ Health B Crit Rev* 2019;**22**(5–6):172–202.
184. Patisaul HB, Fenton SE, Aylor D. Animal models of endocrine disruption. *Best Pract Res Clin Endocrinol Metab* 2018;**32**(3):283–97.
185. Wignall Jessica A, Shapiro Andrew J, Wright Fred A, et al. Standardizing benchmark dose calculations to improve science-based decisions in human health assessments. *Environ Health Perspect* 2014;**122**(5):499–505.
186. Tsatsakis AM, Kouretas D, Tzatzarakis MN, et al. Simulating real-life exposures to uncover possible risks to human health: a proposed consensus for a novel methodological approach. *Hum Exp Toxicol* 2017;**36**(6):554–64.
187. Thiruchelvam M, McCormack A, Richfield EK, et al. Age-related irreversible progressive nigrostriatal dopaminergic neurotoxicity in the paraquat and maneb model of the Parkinson's disease phenotype. *Eur J Neurosci* 2003;**18**(3):589–600.
188. Mesnage R, Teixeira M, Mandrioli D, et al. Multi-omics phenotyping of the gut-liver axis allows health risk predictability from in vivo subchronic toxicity tests of a low-dose pesticide mixture. *bioRxiv* 2020. 2020.2008.2025.266528.
189. Cortessis VK, Thomas DC, Levine AJ, et al. Environmental epigenetics: prospects for studying epigenetic mediation of exposure-response relationships. *Hum Genet* 2012;**131**(10):1565–89.

190. Ladd-Acosta C, Fallin MD. The role of epigenetics in genetic and environmental epidemiology. *Epigenomics* 2015;**8**(2):271–83.
191. Tiffon C. The impact of nutrition and environmental epigenetics on human health and disease. *Int J Mol Sci* 2018;**19**(11):3425.
192. Collotta M, Bertazzi PA, Bollati V. Epigenetics and pesticides. *Toxicology* 2013;**307**:35–41.
193. Cook K. *Fund UPE. Bottoms up: Glyphosate pesticide in beer and wine*; 2019.
194. EU Parliament. *Question for written answer E-001668-16 to the Commission. Subject: German beer contaminated with glyphosate*; 2016. Available at https://www.europarl.europa.eu/doceo/document/E-8-2016-001668_EN.html.
195. EFSA, Brancato A, Brocca D, et al. Use of EFSA pesticide residue intake model (EFSA PRIMo revision 3). *EFSA J* 2018;**16**(1):e05147.
196. Stephenson CL, Harris CA. An assessment of dietary exposure to glyphosate using refined deterministic and probabilistic methods. *Food Chem Toxicol* 2016;**95**:28–41.
197. Gillezeau C, van Gerwen M, Shaffer RM, et al. The evidence of human exposure to glyphosate: a review. *Environ Health* 2019;**18**(1):2.
198. Zoller O, Rhyn P, Zarn JA, Dudler V. Urine glyphosate level as a quantitative biomarker of oral exposure. *Int J Hyg Environ Health* 2020;**228**:113526.

Direct herbicide effects on terrestrial nontarget organisms belowground and aboveground

Johann G. Zaller[a] *and Carsten A. Brühl*[b]

[a]Institute of Zoology, University of Natural Resources and Life Sciences (BOKU), Vienna, Austria, [b]iES Landau, Institute for Environmental Sciences, University of Koblenz-Landau, Landau, Germany

Chapter outline

Introduction

Evidence is mounting that commercial herbicide formulations affect not only target plant pests such as wildflowers and grasses in agricultural fields, forests, and bodies of water but also a variety of nontarget organisms. Most information on the hazards to wild plants and wildlife is based on the knowledge of herbicide mobility and persistence in the environment, application rate, and toxicity.

In Europe, active substances used in pesticide products are approved for marketing (Regulation EC No. 1107/2009) under the requirement that they do not have any

Herbicides. https://doi.org/10.1016/B978-0-12-823674-1.00004-3

harmful effects on human or animal health or any unacceptable effects on the natural world. The environmental aspect particularly includes the consideration of the impact on nontarget species, including the behavior of those species, and on biodiversity and the ecosystem.[1] European Commission Regulation (EU) No. 283/2013 states that the potential impact on biodiversity and the ecosystem, including potential indirect effects via alteration of the food web, shall be considered.

This chapter first provides an overview of published research describing herbicide effects on various organisms, including plants, soil microorganisms, insects, and vertebrates within and outside agricultural fields. The focus is on organisms inhabiting terrestrial ecosystems, although herbicides are also often used to control aquatic weeds and algae, and they are leached from fields into water bodies. Particularly in aquatic biota, a plethora of studies have revealed that a broad range of herbicides representing a variety of chemical classes induce embryotoxicity and teratogenicity in nontarget fish, amphibia, and invertebrates: organ malformation, delayed hatching, growth suppression, and embryonic mortality.[2–4] Dealing fully with the effects on aquatic environments would go beyond the scope of this chapter; comprehensive reviews are cited for the interested reader.[5, 6] However, a short overview on herbicide effects in aquatic food webs is provided in Chapter 8. The focus here is on the direct effects at the species and population levels. A comprehensive review on acute and chronic toxicological effects of glyphosate and metabolites on organisms of the kingdom animalia is provided by Gill et al.[7] The toxicological effects of glyphosate have been traced from lower invertebrates to higher vertebrates and include genotoxicity, cytotoxicity, nuclear aberration, hormonal disruption, chromosomal aberration, and DNA damage. The consequences for ecosystem functions and properties, including potentially higher susceptibility of crops to various diseases,[8] are addressed in Chapter 8.

Around 70,000 publications can be found in the scientific database Scopus with the search term "herbicide*" between 1927 and 2019. In terms of sheer number of publications, most research on herbicide effects on nontarget organisms deals with mammals, insects, fish, and, to a lesser extent, birds (Fig. 7.1). This reflects the fact that environmental risk assessments (ERAs) are required to include certain surrogate mammals (e.g., mice, rats), insects (e.g., honeybees), fish (e.g., rainbow trout), or birds (e.g., Japanese quail). The majority of mammal studies are for obligatory ecotoxicological testing in rats for human risk evaluation; they do not consider wildlife mammals such as mice, shrews, moles, and bats exposed in agricultural landscapes.

However, not all of these publications address nontarget effects per se, and with this coarse literature search it is not always possible to distinguish whether terrestrial and aquatic organisms have been investigated, or whether investigations were conducted in the laboratory, mesocosm, or field. The best-studied herbicide classes regarding the number of publications were: 2,4-D 13,459; paraquat 12,232; atrazine 11,818; glyphosate 10,310; diuron 2426; sulfonylureas 1770; simazine 1769; pendimethalin 1741; dicamba 1689; and glufosinate 1399. The number of publications gives some indication of an agent's agricultural importance and reflects the public debate on environmental or ecotoxicological impact. Moreover, several of these publications also include sociological and political science matters.

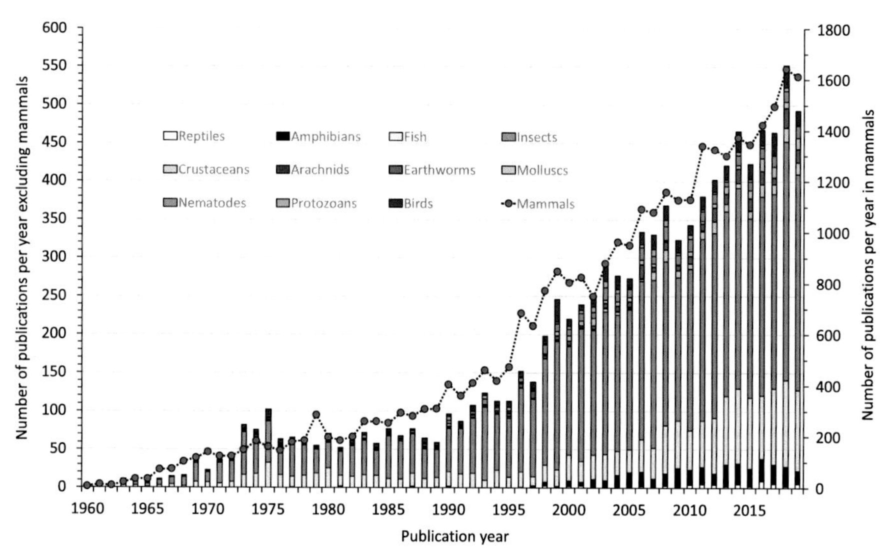

Fig. 7.1 Number of publications considering herbicides and nontarget organisms, 1927–2019. Search was done in June 2020 in the Scopus database using the term herbicide* combined with the shown categories (e.g., herbicide* AND reptile*) for each year. Categories with few publications are not always depicted.

One might think that 70,000 published studies would cover all aspects of ecotoxicology. However, this chapter will reveal that most of the studies consider either active ingredients or formulated herbicides assessed on single species and thus do not reflect field-realistic situations. Moreover, ecological relationships in ecosystems, including competitive interactions, population trends, life history traits, biodiversity interactions, and long-term effects, are rarely investigated.

Most research on herbicides remains a scattered assemblage of data recorded at the molecular, cellular, physiological, or individual level for different species and rarely considers parameters at the population or community level. This reflects the requirements of ERAs aiming to evaluate the effects on a single species of selected test organisms and trying to extrapolate the obtained no-effect concentrations to safe levels for populations and communities. In the ERA of chemicals, such safe levels are then compared with predicted or measured exposure levels to estimate the risk and set standards to protect the structure and functioning of ecosystems.[9]

The reader can consult several comprehensive reviews of the effects of herbicides on nontarget organisms.[10, 11] The main objective here is to provide an overview of existing knowledge by presenting examples of particularly innovative approaches. Because of the many modes of action of herbicides, establishing their common effects is unlikely. Factors that confound or moderate understanding are emphasized.

Table 7.1 gives an overview of the acute effects of popular herbicides on surrogate organisms. The measured endpoint is mortality. Most herbicides do not appear to be acutely toxic to nonplants, despite linuron's and pinoxade's effects on predatory mites. However, most of the data are from short-term studies addressing the effect

Table 7.1 Ecotoxicology of selected herbicides on nontarget organisms.

Compound	Mammals acute oral LD_{50} (mg kg^{-1} body)	Birds acute LD_{50} (mg kg^{-1} body)	Insects contact acute LD_{50} (μg bee^{-1})	Earthworms acute LC_{50} (mg kg^{-1} soil)	Other arthropods LR_{50} or effect in %
Atrazine	1869 Rat	4237 Japanese quail	>100 Honeybee	79 No species info	No data. Lacewings
Diuron	>2000 Rat	1104 Bobwite quail	>102 Honeybee	>798 Compost worm	Dose: 5 kg ha^{-1} Predatory mites
Ethofumesate	>2000 Rat	>2000 Mallard	>50 Honeybee	134 Compost worm	LR_{50} : 1 kg ha-1 Predatory mites
Flazasulfuron	>5000 Rat	>2000 Bobwhite quail	>100 Honeybee	>16 Compost worm	Dose: 0.02 kg ha^{-1} Predatory mites
Fluometuron	>5000 Rat	2974 Mallard	>102 Honeybee	>500 Compost worm	Dose: 2 kg ha^{-1} Predatory mites
Fluroxypyr	>2000 Rat	>2000 Mallard	>102 Honeybee	>500 Compost worm	Dose: 0.6 kg ha^{-1} Predatory mites
Glufosinate-ammonium	416 Mouse	>2000 Japanese Quail	>345 Honeybee	>1000 Compost worm	Dose: 0.6 kg ha^{-1} Aphid parasitoid
Glyphosate	>2000 Rat	>2250 Bobwite quail	>100 Honeybee	>5600 Compost worm	Dose: 3.6 kg ha^{-1} Predatory mites
Isoproturon	1826 Rat	1401 Unknown spp.	200 Honeybee	>1000 Compost worm	Dose: 25 kg ha^{-1} Predatory mites
Linuron	1146 Rat	314 Bobwite quail	98 Honeybee	>500 Compost worm	Dose: 1 kg ha^{-1} Predatory mites
Metamitron	1183 Rat	1302 Japanese quail	>100 Honeybee	914 Compost worm	LR_{50} 14 kg ha^{-1} Predatory mites
Phenmedipham	>5000 Rat	>2500 Mallard	>100 Honeybee	36 Compost worm	LR_{50} 0.5 kg ha^{-1} Aphid parasitoid
Pinoxaden	>5000 Rat	>2250 Mallard	>100 Honeybee	>500 Compost worm	LR_{50} 1.8 g ha^{-1} Predatory mites
Simazine	>5000 Rat	4640 Mallard	97 Honeybee	1000 Compost worm	Dose: 1.1 kg ha^{-1} Predatory mites
Terbuthylazine	>1000 Rat	1236 Bobwite quail	>32 Honeybee	>142 Compost worm	LR_{50} 0.8 kg ha^{-1} Predatory mites

Color denotes toxicity for the species: *green* = low, *orange* = moderate, *red* = high. LD_{50} = median lethal dose required to kill half the test population, LC_{50} = median lethal concentration required to kill half the test population, LR_{50} = median lethal residue.
Data derived from PPDB. *Pesticide properties database*. University of Hertfordshire; 2020. Available from: https://sitem.herts.ac.uk/aeru/ppdb/en/Reports/2631.htm [Accessed 9 June 2020].

of one agent on one species while in the field, numerous pesticides are applied during the cropping periods. They can affect species that are interacting with others, and exposed species may additionally be stressed by environmental or climatic factors. Sublethal endpoints such as effects on behavior and reproduction that reduce the population are scarcely addressed.

Soil contamination

Soil contamination has become an issue of increasing concern due to the wide use of herbicides in conventional agriculture and the high persistence of some of them. Soil consists of a solid phase of minerals and organic matter, a porous phase that holds gases, and a liquid phase of water and dissolved nutrients. Soil characteristics are important in the fate of herbicides, and intrinsic characteristics make them more or less susceptible to leaching, degradation, and accumulation, thereby influencing the potential soil contamination. Table 7.2 lists herbicides found in the soil matrix[12] and their properties regarding environmental fate.[13]

The degradation of an active ingredient is the process when the substance is no longer stable and begins to break down. During this process, metabolites can be formed by means of enzymes in the target plant or may be degraded in the environment. The most important metabolite of glyphosate is aminomethylphosphonic acid (AMPA), which itself is a phytotoxin causing injuries in, for instance, glyphosate-resistant soybeans.[14] Degradation time (DT) is a measure of the rate of breakdown. The DT_{50}, or half-life, is the time required for the breakdown of one-half of the volume of a substance, in this case a herbicide in soil. A herbicide is considered nonpersistent if its DT_{50} is less than 30 days, and moderately persistent if the DT_{50} is between 30 and 100 days. No commonly used herbicide is highly persistent (see Table 7.2). Solubility in water and thus susceptibility to water erosion or leaching is high for atrazine, fluroxypyr, glyphosate, and metamitron, and moderate or low for the other herbicides.

The bioaccumulation potential (measured as the octanol-water partition coefficient) also varies considerably and is high only for pinoxaden and terbuthylazine.

Drift potential is commonly assessed via the substance's vapor pressure (the pressure at which a liquid is in equilibrium with its vapor at 20°C). It is a measure of the tendency of a material to vaporize; the higher the vapor pressure, the greater the potential for vaporization. According to guidelines of the European Food Safety Authority (EFSA), a herbicide is highly volatile if the vapor pressure >10 mPa, which is the case only for an atrazine metabolite.[15] However, these values must be interpreted with caution; for instance, many spray drift incidents have been reported for dicamba, yet according to its vapor pressure, the drift potential would be considered low.

Leaching potential is assessed via the GUS index, the groundwater ubiquity score.[16] A GUS > 2.8 indicates high leachability. The herbicides in Table 7.2 vary greatly, with atrazine and its metabolites fluometuron and terbuthylazine showing high leaching potential and most others being less prone to leaching.

Soil adsorption and mobility are assessed by a sorption coefficient that measures the tendency to bind to soils, corrected for soil organic carbon content. Values vary substantially depending on the soil type, the soil pH, the acid-base properties of the compound, and the type of organic matter. Among compounds in Table 7.2, mobility is high, especially for dicamba, flazasulfuron, and fluroxypyr. Consequently, herbicides are frequently found in off-target soil and water bodies, indicating that a complex suite of factors determines the environmental fate.

Table 7.2 Soil degradation in field (DT_{50}), solubility in water (S_w), bioaccumulation, volatility (V_p), leaching potential (GUS), and soil mobility (K_{oc}) of selected herbicidal active ingredients.

Compound	DT_{50} (days)	Sw (mg/L)	Bioaccum. Log *P*	V_p (mPa)	GUS index	K_{oc} (mL/g)
Atrazine	29 NP	35 L	2.7 M	0.039 L	3.2 H	100 MM
Atrazine deisopropyl[a]	n.d.	980 H	1.15 L	n.d.	n.d.	130 MM
Atrazine-desethyl[b]	45 MP	2700 H	1.51 L	12.44 H	4.37 H	110 MM
Dicamba	3.9 NP	250,000 H	-1.88 L	1.67 L	1.72 L	12.36 VM
Diuron	229 P	35.6 L	2.87 M	0.00115 L	1.83 T	813 SM
Ethofumesate	37.8 MP	50 M	2.7 M	0.36 L	3.04 H	n.d.
Flazasulfuron	10 NP	2100 H	-0.06 L	0.013 L	1.55 L	46.2 M
Fluometuron	89.8 MP	111 M	2.28 L	0.125 L	3.92 H	80 MM
Fluroxypyr	51 MP	6500 H	0.04 L	38×10^{-7} L	3.7 H	68 M
Glufosinate ammonium	7 NP	500,000 H	-4.01 L	0.031 L	1.03 L	600 SM
Glyphosate	23.8 NP	10,500 H	-3.2 L	0.0131 L	-0.29 L	1424 SM
Isoproturon	23 NP	70.2 M	2.5 L	0.0055 L	2.61 T	122 MM
Linuron	48 MP	63.8 M	3.0 M	0.051 L	2.11 T	843 SM
Metamitron	11.1 NP	1770 H	0.85 L	74×10^{-5} L	2.16 T	78 MM
Phenmedipham	16.7 NP	1.8 L	2.7 M	7×10^{-7} L	1.17 L	1775 SM
Pinoxaden	0.6 NP	200 M	3.2 H	0.0002 L	-0.32 L	349 MM
Simazine	90 MP	5 L	2.3 L	81×10^{-5} L	2.3 L	130 MM
Terbuthylazine	21.8 NP	6.6 L	3.4 H	0.151 L	2.19 T	231 MM

Characterization based on University of Hertfordshire Pesticide Properties DataBase.[13]
Color denotes evaluation of specific parameter: *green* = low, *orange* = moderate, *red* = high.
Soil degradation in the field based on DT_{50}: NP-nonpersistent (DT_{50} < 30 days), MP-moderately persistent (DT_{50} 30–100 days), P-persistent (DT_{50}: 100–365 days, VP-very persistent (DT_{50} > 365 days).
Solubility in water at 20°C (S_w): L-low (S_w < 50 mg/L), M-moderate (S_w 50–500 mg/L), H-high (S_w > 500 mg/L).
Bioaccumulation based on log *P*—octanol-water partition coefficient: L-low (log *P* < 2.7), M-moderate (log *P* 2.7–3), H-high (log *P* > 3.0).
Volatility measured as vapor pressure at 25°C (V_p): L-low (V_p < 5.0 mPa), M-moderate (V_p 5.0–10.0 mPa), H-high (V_p > 10.0 mPa).
Leaching potential index (GUS): L-low (GUS < 1.8), T-transition state (GUS 1.8–2.8), H-high (GUS > 2.8).
Soil adsorption and mobility (K_{oc}): NM-nonmobile (K_{oc} > 4000 mL/g), SM-slightly mobile (K_{oc} 500–4000 mL/g), MM-moderately mobile (K_{oc} 75–500 mL/g), M-mobile (K_{oc} 15–75 mL/g), VM-very mobile (K_{oc} < 15 mL/g).
[a] Metabolite of Simazine;
[b] metabolite of Atrazine.; n.d., no data available.

In contrast to water bodies, no systematic environmental monitoring exists to regularly assess pesticide residue levels in soil. One of the first comprehensive studies analyzed 75 arable soils in the Czech Republic several months after the last pesticide application. Triazine was the most frequently occurring herbicide, present in 89% of soils.[17] Another study assessed 317 agricultural topsoil samples from various agricultural fields across Europe and found that more than 80% contained pesticide residues, with 166 different combinations.[12] Glyphosate and its metabolite AMPA showed the highest concentrations besides the long-banned insecticide DDT and its metabolites and broad-spectrum fungicides (boscalid, epoxiconazole, tebuconazole). These compounds occasionally exceeded their predicted environmental concentrations but were generally below the respective toxic endpoints for standard in-soil organisms. However, studies show that chronically low-level herbicide stress can favor herbicide resistance in weed species.[18] Fifteen herbicides or their metabolites were found in another study,[19] but the levels were weakly correlated with soil or herbicide properties. Glyphosate was found in 21% of samples, and 42% contained AMPA. The maximum concentrations found in soils were 2.1 mg/kg for glyphosate and 1.9 mg/kg for AMPA. The researchers stress that some contaminated soils are in areas highly susceptible to water and wind erosion, and that particulate transport can contribute to human and environmental exposure to glyphosate and AMPA residues.

Glyphosate and AMPA have been found in soils of the Pampas in Argentina, with intensive use of glyphosate for weed control associated with glyphosate-tolerant crops.[20] They appeared in almost all investigated solid matrices, with maximum concentrations among the highest reported in the world: 8105 μg/kg glyphosate and 38,939 μg/kg AMPA in soil, 3294 μg/kg glyphosate and 7219 μg/kg AMPA in sediment, and 584 μg/kg glyphosate and 475 μg/kg AMPA in suspended particulate matter. Surface water had a lower detection frequency, with maximum concentrations in whole water of 1.80 μg/L of glyphosate and 1.90 μg/L of AMPA. Based on these findings, the authors consider glyphosate and AMPA as "pseudopersistent" pollutants.

Few long-term studies on the fate of herbicides in soil have been done. In sugar beet fields in Finland, the order of persistence over a 2-year period was ethofumesate > glyphosate > phenmedipham > metamitron > glufosinate-ammonium.[21] Seasonal variation was very high, and dissipation almost ceased during winter. Ethofumesate and glyphosate persisted for several months, until the following spring.

Besides a herbicide's chemical properties, its fate in soil depends on several parameters. It can interact strongly with soil components by forming complexes with metal ions in solution, and by being adsorbed on soil particles, including clay minerals.[22] An early laboratory study found that the rate of degradation for three dinitroanilines was directly correlated with soil temperature and moisture content.[23] A compound may adsorb onto clay particles and soil organic matter and remain unchanged for varying lengths of time (see Table 7.2; diuron is the longest, with a DT_{50} of 229 days). Glyphosate adsorption is strongly influenced by cations associated with the soil, for instance, phosphate.[24]

The uptake, translocation, degradation, and complexation behaviors of glyphosate were reviewed by Singh et al.,[25] who concluded that the long half-life period of glyphosate and its metabolites under different environmental conditions is a major concern. Degradation and leaching dynamics are also greatly dependent on the biological activity in the soil.[26] Matters are even more complex, as herbicides themselves affect the activity and survival of soil biota[27] and interfere with soil types and abiotic parameters such as pH, temperature, and moisture.[28]

Nontarget plants

Nontarget plant species include monocots (e.g., grasses, orchids, lilies) and dicots (most others) and are defined as those that grow outside the cropping area. All plants inside the cropping area other than the crop are recognized as targets and called weeds. Nontarget plants in the agricultural landscape grow in seminatural habitats such as field margins, hedges, and copses.[29]

Controlling or eradicating weeds is, of course, the aim of using herbicides. Thus, herbicide use has also contributed to a decline in the weed flora in Europe.[30] There is also strong evidence for a decline in the size of arable weed seedbanks, with only a few species dominating and being able to adapt to herbicide use in agriculture.[31]

Nontarget plants, including native plants in field margins and crop plants in adjacent fields, are affected in their habitat by spray drift from agricultural fields, and by legacy or carryover effects of herbicide residues in soil from previous treatments. Spray drift is the part of the pesticide application deflected away from the target area during or following applications.[32] Narrow field margins can receive direct herbicide overspray that is up to 50% of the field rate.[33]

Plant growth in the area surrounding agricultural fields that have had herbicide application is likely to be affected. For herbicide approval, laboratory studies are conducted to assess the drift potential onto crop plants following specific guidelines of the Organization for Economic Co-operation and Development (OECD).[34] However, this does not address the herbicide sensitivity of wild plants and how they can be protected. The risk of spray drift to nontarget terrestrial plants is assessed by using plant characteristics (or endpoints) sensitive to the mode of action. In the official guidelines of the OECD, biomass is used as a vegetative endpoint to assess the risks. However, reproductive endpoints such as seed and fruit yield should also be assessed.[35] For example, in a field study, exposure of hawthorn (*Crataegus monogyna*) to metsulfuron-methyl resulted in 100% loss of berries at an application rate of 0.05 g AI/ha (AI means active ingredient while the vegetative endpoints [leaves] remained unaffected).[36] Others also showed hormesis—a favorable biological response to low exposure to toxins[37]—occurring in some plant species at low dosages of metsulfuron-methyl and in both vegetative and reproductive endpoints.[35]

Two extensive databases from the US Environmental Protection Agency (US EPA) and the Canadian regulatory authorities were used to examine plant species sensitivity patterns to herbicides.[38] Crop species were not consistently more or less sensitive than noncrop species. However, noncrop species of grasses had higher sensitivity in 15 cases out of 23, and only in 5 cases were crops more sensitive. The authors concluded that the number of plant species tested was insufficient. In a comparison of the direct toxicity of herbicides toward plants using European Union (EU) and German Federal Environment Agency data, only glyphosate and dicamba provided enough endpoints for a comparison of wild and crop plant sensitivity.[39] For regulatory purposes, only crop plants were tested. Wild plants were tested only in research studies, and they showed a hundred-fold higher sensitivity. Some plant families occurring in natural field margins (e.g., Lamiaceae) are known to have a very high herbicide sensitivity. Some crop species often assessed in nontarget plant studies for regulatory purposes showed generally low sensitivity; among them were oats, onions, and soybeans. The authors recommend that these species be excluded from regulatory testing and replaced by more sensitive wild plants. To get a more precise estimate, several herbicides need to be assessed in studies containing several wild and crop species and designed with exactly matching conditions. It is also remarkable that only higher plants are subjected to ERAs, and the sensitivity of terrestrial algae, mosses, ferns, and lichens is assessed in only a very few scientific studies and is never used for regulatory purposes. A recent evaluation funded by the European pesticide producers' European Crop Protection Association (ECPA) concluded that there were no consistent differences in sensitivity between wild plant species and crop species. However, the analysis was influenced by very heterogeneous data.[40]

Besides direct herbicide effects on plant growth, sublethal effects on the reproduction of nontarget plants are likely as well. In field margins, with a sulfonylurea product (Atlantis) applied at a realistic overspray rate of 30% of the field dosage, the flower intensity in the common buttercup *Ranunculus acris* was suppressed and there was an 85% reduction in flower density.[33] In a parallel monitoring along 11 km of field margins, *R. acris* was recorded at only 2% frequency when cereal crops adjoined field margins but 85% frequency when meadows without herbicide use were adjoining. A reduction of flowers leads to indirect food web-related effects on pollinators (see Chapter 8) and also affects the seed set. In a follow-up study, the plant frequency and reproductive capacity of four wild plant species (*R. acris, Lathyrus pratensis, Vicia sepium, Rumex acetosa*) were investigated; plant frequencies of all four species were significantly reduced by the herbicide and by fertilizer treatment. The combination of herbicide and fertilizer resulted in additive effects on the frequency of two species. Furthermore, herbicide suppressed the formation of flowers and, hence, led to reduced seed production of three species. In some herbicide-treated plots, no fruits were formed. The seed weight of *R. acris* was reduced in all herbicide plots by more than 70%, but this did not affect germination. Reproduction is a highly sensitive endpoint. For example, herbicide effects on flower formation were observed up to a distance of 10 m from the field edge.[41]

Studies indicate that ERAs focusing only on the effects on the biomass of nontarget plants underestimate the full herbicide effect. Together with the described differences in sensitivity, this suggests the necessity of an increase in uncertainty factors and leads to the conclusion that ERA schemes do not protect wild plants in their habitats. Absent a corrective, the long-run effect will be a decrease in plant biodiversity.[42] Because plants are important components of the food web in agricultural landscapes by providing pollen, nectar, or biomass, higher trophic levels are affected indirectly and therefore biodiversity as a whole declines (see Chapter 8). To reverse this trend, it is central to protect plant diversity and to address herbicide nontarget effects with a sound ERA approach.

Herbicide resistance of weeds

Current weed management practices tend to focus on property-level decisions that usually neglect landscape-scale interactions. Glyphosate, 2,4-D, and dicamba spray drift have been reported to cause severe injury to sensitive vegetation and crops, especially when best practices are not used.[43–46] Repeated drift exposure causes natural selection for herbicide resistance. The spray drift of glyphosate, 2,4-D, and dicamba on *Amaranthus* spp., specifically Palmer amaranth and waterhemp, in field margins and ditches has been shown to rapidly select for *Amaranthus* ecotypes with reduced herbicide sensitivity over just two generations.[47] Many weed species have evolved herbicide resistance following recurrent low-concentration application or drift.[18, 48–50] This level of exposure gradually selects for metabolism alleles present in the standing genetic variation of the population.[51, 52] Some researchers suggest that low herbicide rates can induce new stress-related mutations and epigenetic alterations, leading to reduced herbicide sensitivity and higher resistance.[53]

Herbicides target specific enzymes, and mutations are selected that confer resistance-endowing amino acid substitutions, decreasing herbicide binding.[54] Where herbicides bind in an enzyme catalytic site, very few mutations give resistance while conserving enzyme functionality. Where they bind away from a catalytic site, many resistance-endowing mutations may evolve. Increasingly, resistance evolves due to mechanisms limiting the herbicide reaching target sites. Especially effective are herbicide-degrading cytochrome P450 enzymes able to detoxify existing herbicides.

Resistance to herbicides that target 5-enolpyruvylshikimate-3-phosphate synthase (EPSPS; e.g., glyphosate), 4-hydroxyphenylpyruvate dioxygenase (HPPD; e.g., mesotrione, benzofenap), photosystem II, protoporphyrinogen oxidase (PPO; e.g., diphenylether, thiadiazole), auxin receptors, microtubule assembly, and acetolacte synthase (ALS; e.g., sulfonylureas, imidazolinones) have been reported to create some of the most troublesome weed species in the United States.[55–57] Moreover, pollen-mediated gene flow was identified as a major contributor to resistance dissemination.[58] In dose-response experiments, multiple herbicide resistances in a waterhemp (*Amaranthus tuberculatus*) population showed threefold resistance to 2,4-D, sevenfold to atrazine and fomesafen, 22-fold to glyphosate, and 14-fold to mesotrione.[57] Dicamba and glufosinate were the only agents that provided effective control of the resistant population. These are among the first results confirming the existence of weed populations (*Amaranthus*) that are resistant to six different herbicidal modes of action.

A broad-scale analysis of data from 150 winter wheat fields in France showed that herbicides (16 were used by the investigated farms: 2,4-MCPA, amidosulfuron, bromoxynil phenol, carfentrazone-ethylene, clodinafop-propargyl, clopyralid, cloquintocet-mexyl, diclofop-methyl, diflufenicanile, florasulam, fluroxypyr, glyphosate, iodosulfuron-methyl-sodium, isoproturon, mecoprop-P, and mesosulfuron-methyl) were more effective at controlling rare rather than abundant weed species.[59] Interestingly, no relationship was found between crop yields and herbicide use, which led the authors to suggest that reducing the use of herbicides by up to 50% could maintain crop production while encouraging the diversity of weed communities.

Nontarget crop plants

The effects of herbicides on crops are less often considered, but it can be an issue in arable and perennial crops. Carryover effects leading to physiological disorders, decreased tuber quality, and lower yield were reported for the potato (*Solanum tuberosum*), for example when tembotrione and atrazine were previously applied at postemergence in corn (*Zea mays*).[60] Tembotrione and atrazine residues in soil also reduced crop dry mass and yields in sugar beet even 8 months after herbicide application.[61]

Flazasulfuron application to weeds growing under grapevines in vineyards in the Champagne region in France evoked alterations in the nontreated grapevines regarding their photosynthetic activity (85% decrease in leaf gas exchange, 88% decrease in photosynthetic pigment concentration) and a marked disorganization of the leaf plastids, altering growth and yellowing the leaves throughout the growing season.[62] The herbicide also caused a decrease in the levels of leaf starch and soluble carbohydrates. The toxicity appeared to be transient and was overcome the following year. Flazasulfuron, glufosinate, and glyphosate altered nutrient composition in grapevine roots, leaves, grape juice, and xylem sap that was collected 11 months after herbicide application.[63] To what extent such effects ultimately affect the sturdiness of grapevines against diseases or affect fermentation processes in wine production is unclear.

Invasive plants

Invasive species can hinder the establishment and growth of native species and alter several ecosystem properties, including nutrient cycling, fire regimes, erosion processes, and hydrology. Controlling invasive plants is then a necessary and usually expensive step toward restoration. The great challenge is to maintain or enhance native biodiversity while reducing exotics. Common techniques are mowing, tillage, grazing, manual removal, and herbicide application. These methods are often used in combination across varying temporal and spatial scales. In many US ecosystems, the use of selective herbicides has become the primary method.[64] In prairies, this can be partly accomplished by grass-specific agents.[65] However, the long-term effects of invasive species control by herbicides on the composition of plant communities are not well documented.[66]

In one study, native and exotic plant community diversity and composition were measured across areas that differed in burning history and grass-specific herbicide application.[67] Areas treated with a grass-specific herbicide generally had lower exotic (*Cytisus scoparius*—Scotch broom) and higher native cover. The authors concluded that simply eradicating exotics will not be sufficient to restore prairies if native species do not also expand.[68] They found that native species are often seed-limited, and without seed addition, herbicides and burning resulted in increases in nontarget exotics.

In a review of 372 articles published from 2000 to 2019 on the control of undesirable plants (both invasive and overabundant native species), grasses and forbs were found to be the most-studied invasives in restoration sites.[69] Nonchemical interventions (mostly mowing and prescribed fire) were used in more than half the studies globally. Chemical methods (mainly glyphosate spraying, used in 40% of projects using herbicides) were most common in rich countries with nonchemical methods most common in less-developed ones.

Effects on belowground fauna

Soil fauna is exposed to herbicides via direct overspray or oral uptake from the contaminated soil. The in-soil communities of invertebrates and microorganisms are the most diverse part inhabiting agricultural landscapes. Yet the current risk assessment, at the first tier, examines only a selection of invertebrate model species (e.g., compost worm species *Eisenia fetida/andrei*, collembola species *Folsomia candida/fimetaria*, predatory mite species *Hypoaspis aculeifer*), and two microorganism-mediated processes (nitrogen and carbon transformation).[70] Proposals have been made for specific protection goals for in-soil organisms being key drivers for relevant ecosystem services in agricultural landscapes such as nutrient cycling, soil structure, pest control, and biodiversity.[1]

Still, fairly little is known about the impact of increased herbicide use on soil biota and the ecosystem services they provide. This in part reflects the diversity of applied chemicals and the diversity and function of soil ecological communities, which renders a full systematic assessment almost impossible. One review of the effects of herbicides on soil biology[71] states that the database of knowledge is simply too small to draw sound conclusions. Another emphasizes the lack of a suitable framework for the routine evaluation of pesticide effects on soil microbial communities and functions.[72] Since then, comprehensive reviews of such effects have been published by Rose et al.[11] and Van Bruggen et al.[73] Those reviews give useful background, especially on specific herbicides. This chapter has a more ecological orientation.

Soil invertebrates provide services that contribute to plant growth and overall ecosystem functioning. Soil fauna is classified according to body dimensions: macrofauna >1 cm length, >2 mm width, and visible to the naked eye; mesofauna <1 cm length.[74] Macrofauna include annelids (earthworms), insects (adults and larvae), Myriapoda (millipedes, centipedes), Isopoda (woodlice), Araneae (spiders, scorpions), Gastropoda (slugs, snails), and a few vertebrates such as moles. In terms of abundance, biomass, and impact on the soil environment, earthworms and ants are considered the most important macrofauna components of temperate soils. Mesofauna comprise nematodes, mites, and collembolans; all play a strong role in redistributing and breaking down organic matter, recycling nutrients, regulating microbial communities, and interacting with plant roots and other soil biota. To the best of our knowledge, no study has yet compared the susceptibility of differently sized soil taxa to herbicides.

Nematodes (roundworms)

These taxa make up a very diverse group of soil animals, with tens of thousands of species and millions of individuals per square meter (30 cm deep) of soil.[75] They show very diverse lifecycles, from predatory to parasitic, and may be beneficial or detrimental to crop plants or other species. Predatory nematodes can control agricultural pests (cutworms, corn earworm moths, slugs), regulate bacterial populations, and affect nitrogen cycles.[76] Plant-parasitic species such as the root-knot nematode attack crops and are vectors for plant viruses. So nontarget effects of herbicides on nematodes can have broad consequences for agroecosystem functions.

More than 300 publications between1967 and 2019 considered the effects of herbicides on nematodes. It appears that nematode communities are rather unaffected by glyphosate (range 0.6–4.5 kg/ha) and other formulations (cloransulam plus *S*-metolachlor plus sulfentrazone for soybeans, a commercially available mixture of acetochlor and atrazine) in a controlled growth chamber experiment.[77] Simazine at 2.7 g/ha and diuron at 1.8 kg/ha had no significant effects on nematode or food-web structure.[78] As with all soil biota, the direct effects of herbicides on nematode communities are always mediated by plant, rhizosphere, and overall soil conditions.

Weed control can affect nematode communities in rice paddies. Treatment with paraquat, chlormethoxynil, a mixture of thiobencarb and simetryne, and oxadiazon favored plant-parasitic nematodes in the soil compared to hand-weeded control plots.[79] In contrast, predaceous nematodes, which live mostly in the surface stratum, were drastically decreased under thiobencarb and chlormethoxynil, which could make rice plants more susceptible to disease because ecological interactions in the soil are disrupted.

A meta-analysis that quantified the impacts of herbicides on the abundance of nematodes of five trophic groups (bacterivores, fungivores, plant parasites, omnivores, predators) showed a decrease in fungivores and predators but an increase in bacterivores, plant parasites, and omnivores.[80] The decrease in predators suggests that herbicides can disturb soil food webs. Bacterivores seem to be less sensitive than fungivores and predators. The increase in omnivores may be due to the increased amount of food for them after weed control. Results also suggest that herbicides may increase the risk of plant-parasitic nematode abundance in a field. However, multiple herbicide treatments in a 15-year field experiment did not significantly affect nematode communities under turfgrass compared with equivalent controls without herbicide.[81]

Collembola (springtails)

This diverse group of small arthropods with thousands of species worldwide is omnivorous, free-living, and prefer moist conditions. They are believed not to directly engage in the decomposition of organic matter but contribute indirectly by fragmentating it, thereby also controlling soil microbial communities.[82] They are considered bioindicators of soil conditions. Only about 70 publications on the effects of herbicides on collembola appeared between 1978 and 2019. The following are notable examples

A study investigated the response of collembolans to four types of herbicides (glyphosate, atrazine, 2,4-D, nicosulfuron) in a no-till maize-turnip rotation in Brazil and found that atrazine and 2,4-D caused the highest reduction of the springtail population.[83] Another study in maize found that glufosinate or terbuthylazine applied at conventional rates reduced protozoa, mites, and springtails in one soil but not in another.[84] Transgenic herbicide-tolerant soybean varieties and their associated weed management systems affected the abundance of 21 surface-active springtail species during three successive soybean growing seasons.[85] Effects on springtails appeared to be due to a combination of direct herbicide effects and associated changes in weed cover and soil disturbance.

Investigations in vineyards in Romania showed that springtail activity density in areas neighboring rows treated with flazasulfuron, glufosinate, or glyphosate was positively correlated with interrow tillage and herbicide application.[86] The authors interpreted this as the result of a herbicide-induced decrease in potential competitors and predators, stimulation of microorganisms, and increased input of nutrients contained in herbicides.

Very few studies have compared the effects of herbicide formulation with those of the pure active ingredient. A study compared commercial glyphosate-based herbicide (GBH) formulations (Roundup LB Plus, Touchdown Quattro, Roundup PowerFlex) with their respective glyphosate salt forms (isopropylammonium, diammonium, potassium) on the collembola species *Sminthurinus niger*, which is ubiquitous in many agricultural soils.[28] Results showed that both GBHs and active ingredients applied at recommended dosages increased the surface activity of springtails compared to control plots (Fig. 7.2). Moreover, springtail activity was higher under GBHs than under corresponding active ingredients, suggesting additional effects by the coformulants. The stimulation was much higher in soil with high organic matter content, showing that different soil properties can influence herbicide effects.

Worms

Enchytraeidae (potworms)

These creatures resemble very small earthworms, comprise a couple of hundred species worldwide, and are found in various habitats, including forests and agricultural land. Like earthworms, they influence soil structure and organic matter dynamics by affecting microbial communities. Römbke, Schmelz, and Pélosi[87] recently published an overview of herbicide effects on them. Only four herbicides have been tested so far: phenmedipham, bromoxynil, atrazine, and 2,4-5-T. Responses varied from the low toxicity of 2,4,5-T ($LC_{50} = 14{,}150$ mg AI/kg soil[88]) to the high toxicity of atrazine on the reproduction of *Enchytraeus albidus* at very low concentrations (1–2 mg AI/kg soil).[89] Bromoxynil at low concentrations caused high mortality of *Fridericia bulbosa*.[90] Reproduction and avoidance behavior are more sensitive than mortality as endpoints. Differences in acute sensitivity between species seem to be low (within a factor of 2 in the same soils). However, there is an influence of soil properties on toxicity, leading to differences of EC_{50} for reproduction by almost a factor of

Fig. 7.2 Experimental units with model weeds (*Amaranthus*) to test the direct effects of herbicides on the activity of collembolans (top) and earthworms (middle). Bottom: system to test the avoidance behavior of earthworms to soils contaminated with herbicide residues. (Images: J. Zaller.)

30, and differences in avoidance due to soil properties varying between <1 and 252 mg AI/kg soil dry weight.[87] Enchytraeid reproduction was severely affected in all samples after application of mesotrione, which is used in forest restoration in Southern Brazil.[91] Atrazine caused a dose-dependent increase in avoidance, in which more than 50% of enchytraeids avoided a soil concentration of 38 mg/kg.[92] Sensitivity may vary even within subpopulations of the same species, further complicated by the finding that soil characteristics strongly influence the variability in responses.[92]

For instance, terbuthylazine had no toxic effects on enchytraeids, soil mites, nematodes, and soil microbes when applied at 1–53 kg/ha AI.[93] But when applied as a herbicide formulation (Gardoprim), it was acutely toxic to enchytraeids (24 kg/ha; no observable effect level of 10 kg/ha) and to gamasid soil mites (50 kg/ha). The formulation was more toxic to gamasid mites when applied on the soil surface than when mixed into the humus.

Earthworms

Earthworms comprise about 6000 species and represent a large fraction of soil biomass in many temperate ecosystems. They play an important role in soil functioning; as ecosystem engineers, they influence organic matter dynamics, soil structure, and the microbial community.[94–96] They actively participate in soil aeration, water infiltration, and the mixture of soil horizons, and also represent an important source of food for many other organisms such as birds and moles. Earthworms have also been used as bioindicators for soil quality and the effects of agrochemicals.[97, 98] The compost worm (*E. fetida*) is a sentinel species in the ERAs of pesticides.[1]

Among the soil fauna, earthworms are the best-studied organisms to assess herbicide effects, with more than 250 publications between 1975 and 2019. Contrasting effects of herbicides on earthworms have been described in the literature, with differences arising from earthworm species, ecotypes, and the nature of the study. A review by Pelosi et al.[99] found that herbicides appear to be less toxic on survival and reproduction compared to insecticides and fungicides. Nevertheless, herbicides show effects at different organizational levels such as infraindividual (gene expression and physiology), individual, population (life-history traits, population density, and behavior), and community (biomass and density). Herbicides can disrupt enzymatic activity, increase individual mortality, decrease fecundity and growth, change individual behavior such as the feeding rate, and decrease the overall community biomass and density.

Using available databases that provide information on almost 400 pesticides, a review found that fewer than 5% of pesticides (among them four herbicides) have an LC_{50} below or equal to 10 mg/kg soil, which is considered moderately to highly toxic for *E. fetida*.[13] Another review focused on earthworm reproduction and found a no observed effect concentration (NOEC) <10 mg/kg for 12 herbicides,[99] of which sulfonylureas were the most toxic. Especially in the older literature, herbicides are considered not toxic to earthworms.[95, 100]

Earthworms are commonly distinguished in three ecological groups according to their feeding habits: endogeics or horizontal burrowers, anecics or vertical burrowers (nightcrawlers), and epigeics or litter dwellers. Most herbicide studies were done on epigeics, of which the compost worm *E. fetida* or the closely related *E. andrei* accounted for about a third, due to the fact that they are also approved surrogate species in official ERAs such as those required in the EU. Standardized laboratory test protocols with earthworms as surrogates for ERAs are available from the OECD. Among these, acute mortality as the main endpoint is required for the registration of pesticides. The contact filter paper and artificial soil test exposure protocols with LC_{50} in *E. fetida* as the metric of interest have received the most attention,[101] with the latter also being adopted by the European Union[102] and the EPA.[103] In the paper test, the worms are exposed by contact with the pesticide. In the artificial soil test, the chemicals get into contact with and are additionally ingested by worms.

It appears that distinctive response parameters of worms are differentially sensitive to herbicides. The survival, growth, and reproduction of *E. andrei* responded differently to various chemicals.[104] Cocoon production was the most-sensitive parameter for the herbicides paraquat and phenmedipham while cocoon hatchability was most sensitive for insecticides and fungicides.

Wang et al. assessed 45 pesticides against *E. fetida.*[105] In soil tests, the 23 herbicides from 10 chemical groups induced variable toxicity responses that partly differ from the assessment shown in Table 7.1. However, because the assessment of multiple herbicides within one experimental setting is more homogenous, as exactly the same facility and earthworm populations are used, some results are mentioned here. Among the most toxic herbicides to *E. fetida* were glufosinate-ammonium, terbutryn, and *S*-metolachlor, with LC_{50} values ranging from 167 mg/kg (glufosinate) to 188 mg/kg (*S*-metolachlor). Paraquat, fluoroglycofen, and pyraflufen-ethyl were among the least toxic, with $LC_{50} > 1000$ mg/kg.[105] Interestingly, different herbicides within the same chemical group have different toxicities to *E. fetida*: the toxicity of acetanilide herbicide *S*-metolachlor was significantly higher than that of pretilachlor, and the toxicity of triazine terbutryn was also much higher than that of atrazine. This study shows that nontarget effects of certain herbicide classes can vary considerably, even with just one earthworm species. Hence, it is not possible to estimate the effects on earthworms by knowing only the herbicide class.

The extent to which contact tests in ERAs reflect the situation in agricultural fields is open to debate. Several herbicides (acetochlor, anilofos, flutamone, pretilachlor, *S*-metolachlor, and terbutryn) that registered as very toxic in contact testing showed low toxicity in soil testing.[106–108] According to the required tests for pesticide regulation, most herbicides in use are not directly toxic to earthworms, although they may exert considerable indirect effects through their influence on weeds as a source of organic matter on which worms feed in soil.

The effects of glyphosate on earthworms were addressed in more than 50 publications between 1992 and 2019, 19 of which were conducted using *E. fetida* or *E. andrei*. It seemed to have little effect on endogeics (topsoil feeders) while epigeics (surface litter feeders) were more sensitive.[109] *Eisenia fetida* lost approximately half its body mass after 28 days when exposed to 8 mg kg^{-1}.[110] In another study with *E. fetida*, 10, 100, 500, and 1000 mg/kg produced a significant gradual reduction in mean weight at all concentrations.[111] In the same study, 2,4-D caused 100% mortality at 500 and 1000 mg/kg after 1 week, and 30% mortality at 1 mg/kg after 14 days.

Eisenia fetida also showed strong avoidance behavior in field soil treated with glyphosate at 1.4 kg/ha (Fig. 7.2).[112] Avoidance also occurred when *E. fetida* was exposed to a formulation containing 5% glyphosate by mass (as isopropylamine salt).[113] Glyphosate can also alter ecological interactions among earthworms, mycorrhizal fungi, and plants, leading to reduced mycorrhizal plant colonization and modified feeding behavior.[114] Roundup had no direct impact on the density of the dominant earthworm *Pheretima carnosus* in a field experiment,[115] and there was also no direct effect on the growth, mortality, and behavior in a related lab experiment. However, the authors concluded that Roundup affected earthworm casting activity through variation of environmental factors, such as litter amount, that are related to herbicide activity on weeds.

Two anecic (*Lumbricus terrestris, Octodrilus complanatus*) and one endogeic species (*Aporrectodea caliginosa*) were taken from vineyards treated or not treated with a

GBH (Roundup 360) and transferred into terraria with uncontaminated soil, and glyphosate was applied again.[116] Earthworms with a herbicide experience had higher mortality after the new application than those from untreated fields. Glyphosate application reduced cocoon production for *L. terrestris* by 70% when worms came from an untreated field but only 50% for worms with previous glyphosate experience. These results indicate some resistance mechanisms in earthworms; nevertheless, the glyphosate impact is still serious with up to 26% mortality, especially in the deep-burrowing species (*L. terrestris, O. complanatus*).

More relevant for the situation in the field with several earthworm species are reports of species-specific responses of earthworms to herbicides. In a greenhouse experiment, a model weed community consisting of a grass, a nonleguminous herb, and a legume species was established and sprayed with two Roundup products often used in private gardens (Roundup Alphée and Roundup Speed, each with 7.2 g/L glyphosate).[27] The surface casting activity of the vertically burrowing species *L. terrestris* almost ceased 3 weeks after herbicide application while the activity of the soil-dwelling species *A. caliginosa* was little affected. However, the reproduction (number of hatchlings) of the soil dwellers was reduced by 56% within 3 months.

Very few studies have compared the toxicity of glyphosate-based herbicides (GBHs) to the toxicity of the AI glyphosate for earthworms. A literature review revealed negative, positive, and nonsignificant impacts on a suite of outcomes.[117] Responses varied with initial body mass and environmental characteristics such as soil temperature. Only initially heavy worms (*E. fetida*) grown in heated soil responded to contamination by growing heavier than their uncontaminated counterparts. Lighter worms and those grown in a cooler temperature were unchanged. Those living in unheated soil survived fewer minutes during a stress test, with herbicide-exposed ones surviving for the shortest duration overall; the initial body mass did not affect the outcome of the stress test. These findings provide important clues to explain variations in the literature, where in most cases neither initial body mass nor soil temperature is considered.

In another experiment, the effects of two GBHs (Roundup Ready-to-Use III, Roundup Super Concentrate) were compared with their AI isopropylamine salt in *E. fetida*.[118] A relatively high dosage of 26.3 mg/kg soil was applied. Worms living in soil with pure glyphosate lost 26% of their biomass and survived a stress test 66% of the time relative to the control worms living in uncontaminated soil. Worms with the GBH contaminations did not lose body mass and survived the stress test as well as the controls. The authors suggest that nutrients and other nondeclared coformulants in the GBHs offset the toxic effects of glyphosate by spurring microbial growth and speeding glyphosate degradation.

Studies done under real agricultural conditions have the problem that the effects of management practices and myriad other factors can be misattributed to the effects of the herbicide per se. For instance, earthworms interact with plants and might differ in their susceptibility to herbicides depending on which plants are present in the experimental fields. A field experiment in Poland showed earthworm density and biomass (*A. caliginosa, Aporrectodea rosea,* and *L. terrestris*) being unaffected by herbicide on fava beans (Corum 502.4 SL; with bentazon and imazamox at 1.25 L/ha) and spring barley (Mustang 306 SE; with florasulam and 2,4-D EHE at 0.5 L/ha).[119] The additional application of several fungicides in the barley fields had no direct or interactive effect on earthworms. In the bean fields, by contrast, herbicides and fungicides decreased worm density and biomass. Both effects were stronger with higher soil organic matter content.

Atrazine applied at recommended rates had no significant impact on the number of earthworms but reduced reproduction at 21 mg/kg in one soil.[120] Another study found an LC_{50} of atrazine on *E. fetida* of 15 mg/kg and concluded that chronic reproduction should be preferred over mortality as a more-sensitive endpoint.[121]

Under laboratory conditions, no effect on the mortality of isoproturon on mature *L. terrestris* was seen at 1.4 g/kg soil after 60 days of exposure,[122] but residues reduced the earthworm growth rate by 28%. Another study found that soils treated with the formulated herbicides penoxsulam (Mikado) or sulcotrione (Vipera) were more frequently avoided by *E. andrei* than the corresponding rates of the active ingredients only.[123]

A multiyear survey on 15 conventional farms in France found that an increase in the use of a variety of herbicides (among them glyphosate, isoproturon, pendimethalin) correlated with decreased earthworm density.[124] The three studied species differed in their sensitivity to recommended herbicide treatment: 25% reduction in the endogeic species (*Aporrectodea chlorotica*) and about 40% in the epigeic (*Lumbricus castaneus*) and anecic species (*L. terrestris*). Regardless of whether the effects were direct or indirect, the authors suggest that a reduction in herbicide and other pesticide application would increase earthworm populations in agricultural fields.[99]

Very little is known about how the herbicide exposure of earthworms might affect the impact of other toxic substances. For instance, glyphosate and its functional groups such as amine, carboxylate, and phosphonate can react with metal ions to form metal complexes. In a laboratory experiment, there was no observable toxicity of glyphosate to *E. fetida*, but the presence of glyphosate reduced the acute toxicity of copper.[125]

Taken collectively, it becomes clear that the commonly used indicator organisms—earthworms (*E. fetida, E. andrei*), collembolans (*F. candida, F. fumatoria*), and soil mites (*H. aculeifer*)—represent a very small fraction of the millions of species living in agricultural soils. When the sensitivity of current standard species to several pesticides was compared to that of other species from the same taxonomic group, the conclusion was that standards are not always the most sensitive, thus underestimating the toxicity in the official ERA.[1] This uncertainty should be more widely acknowledged by authorities approving pesticides. Furthermore, shifts in soil faunal communities are not addressed, although changes in community structure are known to be the most significant effects of some pesticides.[126] Other fundamental shortcomings of ERAs result in a weaker protection of soil biodiversity.[42] In conclusion, herbicide effects on soil invertebrates are variable depending on the herbicide and the test species. There are indications that size triggers different responses, and that environmental and soil parameters influence effects.

Microorganisms

Microorganisms include bacteria, archaea, protists, algae, and fungi and are extremely diverse, with millions of species. The relationship of herbicides and microorganisms was the subject of more than 3200 publications between 1950 and 2019, covering microorganisms in soil, on leaf surfaces, and on or within the bodies of vertebrates and invertebrates. Herbicide effects thereon are important because of their eminent role in nutrient and carbon cycling (for which regulatory agencies require data) and metabolic activity, and their interaction with other biota affecting crop health and production.

Soil microbial activity in carbon cycling can be measured in many ways. The most common measures to assess the effects of herbicides on soil microbes are heterotrophic respiration, the activity of enzymes involved in soil carbon cycling, and organic matter decomposition and mineralization. From the few review papers available, the emerging picture is one of compound-specific effects on particular soil functions. For assessing a herbicide's potential impact on soil health as related to plant nutrition and disease, adverse effects on phosphatase activity by glyphosate,[127] inhibition of nitrification by simazine,[128] and adverse effects on pathogen-antagonistic *Pseudomonas* bacteria by acetochlor and chlorimuron-ethyl[129, 130] are examples.

As with other soil taxa, most studies of herbicide effects on soil microorganisms have been short-term. This chapter considers measures of activity, diversity, and abundance. Parameters of ecosystem functioning such as decomposition are dealt with in Chapter 8. For an in-depth discussion of the effects on enzyme activity and nutrient cycles, the reader may refer to Rose et al.[11]

Herbicides can have both positive and negative impacts on microorganisms. Direct effects include reduction of activity and biomass but also stimulation of growth of herbicide-degrading microorganisms and the activity of resistant heterotrophic microbial groups promoted by dead biomass from sensitive organisms.[131, 132] Keeping in mind the crucial importance of microorganisms for all living organisms via their microbiomes, it is important to not only assess the impact of herbicides on microbial communities but also on microbiomes and the consequences on the holobionts they form with plants and animals.

Glyphosate residues in nectar have been shown to disturb the gut microbiota of honeybees[133] and may also affect plant microbiome composition.[134] On the other hand, rats with deficient gut microbiota are more sensitive to *s*-triazine, highlighting the fact that gut microbiota can mitigate herbicide toxicity.[135] In a vineyard, herbicides containing glyphosate, flazasulfuron, or glufosinate were applied beneath grapevines, and their effects on soil microbial communities were assessed.[132] More colony-forming units of bacteria, yeasts, and molds were found for flazasulfuron than for glyphosate, glufosinate, or mechanical weeding, which had similar unit numbers. In the same experiment, the abundance of the fungus genus *Mucor* was higher under flazasulfuron than under glufosinate or mechanical weeding; *Mucor* was completely absent under glyphosate. Several other fungi taxa were found exclusively with a specific treatment. Total soil bacterial counts for the three herbicides were on average 260% higher than under mechanical weeding (though because of high variability this was not statistically significant). The authors suggest that these alterations of soil microorganisms could have knock-on effects on other parts of the grapevine system. Indeed, in a companion study, glyphosate, glufosinate, or flazasulfuron altered the nutrient composition in grapevine roots, leaves, grape juice, and xylem sap.[63] Xylem sap collected 11 months after herbicide application contained on average 70% more bacteria than under mechanical weeding. Such legacy effects for the following season have largely been ignored in research and risk assessment.

Many academic studies have been done with various cultivation-dependent and cultivation-independent methods to assess the effect of herbicides on the abundance, diversity, and activity of soil microorganisms.[72] Most of these studies found both negative or positive effects leading to the adaptation of microorganisms to herbicide biodegradation. For microorganisms with biodegradation capacity, biological systems involved in the herbicide response are often highly specific and characterized by enzymatic reactions that closely depend on the chemical structure of the herbicide.[136] For instance, in order to mitigate the environmental effects of atrazine, atrazine-degrading bacteria have been used.[120] It is noteworthy that catabolic gene systems used here are often regulated by and located on plasmids, allowing their dispersion in the soil bacterial community by horizontal gene transfer under favorable selection pressure exerted by repeated exposure to a herbicide.[73]

Commercial herbicide formulations may elicit toxic effects that differ from those of the pure active ingredients. Glyphosate-based agents have been shown to reduce microbial biomass at the field-recommended dose.[137] In contrast, in the laboratory soil microbial and fungal biomass were unaffected by a relatively high dose of 26 mg/kg soil of glyphosate (isopropylamine salt) or GBHs (Roundup Ready-to-Use, Roundup Super Concentrate).[118] A study of the effects of increasing doses (0 79 mg/kg) of glyphosate and the formulation Roundup CT on substrate-induced respiration and enzyme activity in three contrasting agricultural soils showed that both variables—formulation and soil characteristics—were important factors controlling potential effects on soil functionality.[138] A recent review revealed much controversy about the effects of glyphosate on microbial communities and activities in the soil and rhizosphere.[73] In studies comparing soil treated with glyphosate to untreated control soil, microbial communities seemed to recover from short-term glyphosate treatment[139, 140] with minor or no effect on global microbial structure, biomass, or activity,[141, 142] probably due to the great diversity and compensatory ability of soil microorganisms.

Despite studies showing no herbicide effect or even a stimulation of microorganisms, other studies suggest negative effects of GBHs on microbial count and activity. The herbicide Glyset I.P.A. (48% glyphosate) was added to soil at 0, 50, 100, and 200 mg/kg and incubated in the laboratory at 30°C[143] for several weeks. Bacteria, fungi, and actinomycetes activity and counts decreased dose-dependently while soil respiration was unchanged. In a field study, by contrast, glyphosate at 0.84 kg/ha increased soil respiration and enzyme activity in one year but not in the other.[144] The authors suggest that higher soil moisture and glyphosate acted together to increase microbial activity. In a lab study using field soil, glyphosate at 2.2 mg/kg showed an increase in CO_2 respiration and in enzyme activity in microbes.[145] Here, indications of a legacy effect of previous glyphosate application were found because soil that had been exposed for several years showed the strongest response.

A few experiments have been done in a realistic setting including plants that are sprayed with the herbicide. In a greenhouse experiment where clover (*Trifolium repens*) was treated with Roundup Speed, a 30% increase of soil microbial respiration was found.[146] Multivariate analysis revealed small but significant differences between the microbial communities of treated and untreated pots. Bacterial phospholipid fatty acids decreased following herbicide application. Mycorrhizal and fungal phospholipid fatty acids were not affected, and no difference in fatty acid markers of Gram-negative and Gram-positive bacteria was found. Phospholipids are the primary type composing cellular membranes and are widely used in microbial ecology as chemotaxonomic markers of bacteria and other organisms. Effects were especially pronounced in the clover rhizosphere and were likely due to herbicide-induced changes in root exudate composition. Adverse effects of glyphosate on crop-beneficial soil bacteria, including several *Pseudomonas* species, have been reported[147] and will be addressed in a wider ecosystem context in Chapter 8.

Besides glyphosate, dose-response relationships were generated in a laboratory experiment for diuron (1.5, 7.5, and 150 mg AI/kg soil) and thiram (3.5, 17.5, and 350 mg AI/kg soil) on soil respiration, bacterial count, and changes in the culturable fraction of bacteria.[148] Test parameters were generally unaffected by the lowest dose of each, which corresponded to the recommended field rates. The highest doses of thiram suppressed substrate-induced respiration while diuron increased the respiration rate. Overall, the total numbers of bacteria increased in herbicide-treated soil. In pretilachlor-treated soil, microbial biomass carbon increased through the first 15 days and thereafter decreased to the initial level.[149]

Glufosinate, a herbicide that is chemically similar to glyphosate, appears to have a greater impact on microbial communities.[150] At 3 kg/ha, it caused transient changes in the eubacterial and *Pseudomonas* population structure[151] and altered the active bacterial communities in the rhizosphere of canola crops (*Brassica napus*, *B. rapa*, and *B. juncea*), with generally higher active populations of key groups. However, there were no significant changes in bacterial community structure in either conventional or transgenic herbicide-tolerant maize.[152] Likewise, there was no impact on bacterial community structure by glufosinate 0.6 kg/ha on glufosinate-resistant oilseed rape or maize.[153]

Most sulfonylureas applied at conventional rates have shown no impact on microbial biomass after a few weeks or months.[11] The long-term (5 or 10 year) application of chlorimuron-ethyl to soybean fields significantly reduced culturable bacteria and actinomycetes but increased fungal counts. Molecular profiling of bacterial and fungal DNA also indicated significant shifts in the community structures of both microbial groups, resulting in reduced diversity.[154]

Only a few studies have compared the effects of formulations to their active ingredients. Pure mesotrine and its formulation (Callisto) at recommended doses showed no change in microbial communities, but there were effects at doses exceeding recommended levels.[155] In another study, Callisto had about a 30% greater effect than mesotrione in reducing chlorophyll concentrations and decreasing the diversity of cyanobacterial populations in soil.[156]

Very rarely are interactions of different pesticide classes investigated. A greenhouse experiment showed that wheat seed dressing with insecticides (neonicotinoid) and fungicides (strobilurin and triazolinthione) and subsequent glyphosate application (Roundup LB Plus, 360 g/L, isopropylamine salt) had no effect on litter decomposition, soil basal respiration, microbial biomass, or specific respiration.[157]

Reviews of direct effects of herbicides on soil microorganisms suggest that they are minor or transient when the agents are applied at recommended doses.[11] However, in many cases the corresponding processes and mechanisms are still poorly understood. Even though herbicide effects on soil microbes appear to be negligible, this does not rule out an effect on finer-scale population dynamics.[132, 146, 158] Furthermore, assessing toxicity to soil ecosystems is complex because contamination is low-level and diffuse when it originates from the continuous use of poorly degradable herbicides. It should be emphasized that only a few studies have addressed the effects of low-level, chronic exposure. Monitoring the appearance of herbicide-tolerant bacteria could be a valuable tool for assessing the soil microflora response, for example, by using specific biomarkers.[159]

Several classes of herbicides showed direct antifungal effects and were able to inhibit the growth of fungal pathogens belonging to different taxonomical groups.[8] Thus, their activity against phytopathogens or symbiotic fungi does not depend on their specific mode of action. Glyphosate, for instance, has also been observed to alter mammalian pathogenic fungi[160] and was active against apicomplexan parasites that cause diseases such as malaria and toxoplasmosis.[161] Indirect effects of herbicides on crop diseases will be discussed in more detail in Chapter 8.

In conclusion, contrasting results on the effects of herbicides on microorganisms highlight the difficulties in comparing or summarizing data from studies that use different methods, different herbicide modes of action, and a variety of settings. Until a standard procedure to assess these effects is established, the comparison of multiple methods for assessment is prudent to cover methodological biases.

Mycorrhizal fungi

Arbuscular mycorrhizal fungi (AMF) colonize the roots of more than 80% of terrestrial plants, improving their growth and survival and thereby having a key role in ecosystem structure and function. About 175 publications considered the effects of herbicides on them between 1970 and 2019. Effects vary broadly and can be negative, neutral, and positive.[162] Clearly the occurrence of direct and indirect pathways will depend on the plant cover at the time of herbicide application. The consequences of this practice on the plant community structure will vary with the mycorrhizal dependence of the species composition regardless of the pathway involved.

For example, glyphosate at 0.8–3.0 mg/kg applied directly to the soil reduced the spore count even at the lowest dose[163, 164] but had no effect when applied to foliage.[164]

Total root colonization for both plant species tested (*Paspalum dilatatum, Lotus tennis*) was similarly decreased by herbicide application, with no difference between the applied dosage. The number of arbuscules for foliage application was 20% lower than for soil application. This may suggest that the reduced colonization was due to a reduction in photosynthate supply to roots rather than a direct inhibitory effect on the AMF.

In a pot experiment, a GBH (Roundup Speed) significantly decreased root mycorrhization, soil AMF spore biomass, vesicles, and propagules but increased AMF hyphal biomass in the soil.[114] In an experimental vineyard, within-row herbicide treatment with flazasulfuron, glufosinate, or glyphosate reduced mycorrhization by an average of 53% compared to mechanical weeding.[63] Glyphosate application in autumn reduced the mycorrhizal colonization and growth of both the target and nontarget grasses in the following spring in a field experiment in Finland.[165] Moreover, glyphosate residues were found in weeds and crop plants in the ensuing growing season.

The mechanisms by which mycorrhizal fungi respond to herbicides are not well understood, but it seems obvious that effects can be modulated by the identity and dose of the herbicide, the timing of application, and the soil type. Microbial communities display microevolutionary responses within a short time. Transient effects of herbicides on populations and communities and their function in ecosystems are regularly seen, but studies suggest a high capacity to recover, perhaps by tolerant microorganisms that benefit from herbicide application.

Effects on aboveground fauna

The half-life of herbicides in the environment ranges from <1 month to >1 year (Table 7.1). Wildlife within fields is most likely to be exposed to herbicides, particularly when fields are planted with crops (e.g., corn, soybean, wheat, cotton) that are routinely sprayed. Wildlife is also likely to be exposed in noncrop habitats adjoining croplands, primarily from direct overspray (especially during aerial application) and drift during volatilization and after application.

Arthropods

Insects make up the bulk of terrestrial diversity, and reports of insect declines suggest that 40% of insect species in temperate countries may face extinction over the next few decades.[166] The main factors are habitat losses, invasive species, climate change, and pollution, including agrochemicals.[167] About 1800 publications considered herbicides and insects between 1955 and 2019. The density of insect pests and nontarget insects may be affected by herbicides commonly used for weed management via three main routes: direct mortality to insects present in the field during and immediately following application, induced plant defenses that increase resistance to insect herbivores, and alteration of the quantity and composition of weed populations, which in turn changes the structure of insect communities found subsequently in the crop. In this chapter, only direct effects are considered. Indirect effects via food source contamination will be addressed in Chapter 8.

Direct effects are frequently found. A study showed effects on the mortality of rice pests by several herbicides commonly used in US rice production: 2,4-D (Weed Rhap A-4D), clomazone (Command 3ME), thiobencarb (Bolero 8EC), imazethapyr (Newpath), 3′,4′-dichloropropionanilide (Propanil 4SC), and propanil and thiobencarb (Ricebeaux).[168] Herbicides can also cause negative direct impacts on predatory Arachnida and Coleoptera behavior and survival that itself might affect pest insects.[169] Glufosinate-ammonium at concentrations used for weed control in orchards has been shown to affect nymphs and adults of several beneficial and predatory mite and insect species.[170] Herbicides produced sublethal, concentration-dependent declines in the population of the aphid *Metopolophium dirhodum.*[171]

Atrazine is known to affect male reproduction in vertebrates and invertebrates, but less is known about its effects on other life history traits. A study assessed the effects of five chronic exposure levels on a variety of fitness traits in the common fruit fly (*Drosophila melanogaster*).[172] Atrazine exposure decreased the proportion pupated, the proportion emerged, and adult survival. Development time was also affected, and exposed flies pupated and emerged earlier than controls. Although development time was accelerated, body size was actually larger in some of the exposed flies. Many of the traits showed nonmonotonic dose-response curves, where the intermediate concentrations produced the largest effects.

Bees

Bees may come in contact with herbicides by being sprayed directly, by touching a sprayed plant part, or orally via foraging for nectar and pollen.[173, 174] The effects on bees were addressed in 195 studies between 1968 and 2019, of which 120 consider honeybees. Bees are exposed to a diversity of agricultural chemicals, as demonstrated by Mullin et al.[175] who found 121 pesticides and metabolites in beehives and associated hive materials. A small group of commercially bred *Apis* and non-*Apis* species comprises the principal agents of agricultural crop pollination. However, the plight of solitary wild bees is crucially important. Diverse pollinator communities are integral for the proper functioning of ecological systems.

Honeybees are social insects and among the most important pollinators in agricultural ecosystems. Increased mortality of colonies has been attributed to several factors but is still not fully understood. Very early experiments tested herbicide formulations by direct overspray.[176] Formulations and combinations of herbicides and adjuvants were sprayed on small cages containing *Apis mellifera*. Daily counts of dead bees were made for 14 days and this showed that the herbicides MSMA, paraquat, and cacodylic acid were highly toxic. Sublethal doses have also been reported to alter sleep behavior, which has an essential role in neural and energetic homeostasis. When honeybees were fed a sugar solution containing 50 ng/L of glyphosate, sleeping patterns were disrupted due to metabolic stress.[177] The glyphosate concentrations in this study were chosen according to reports from agricultural landscapes and expected environmental concentration.[178]

Sublethal concentrations of Roundup were applied to the hypopharyngeal glands of nursing honeybee workers.[179] The results were changes in the cellular ultrastructure of the glands, early degeneration of the rough endoplasmic reticulum, and morphological and structural changes in the mitochondria. No changes were noted in the amount of royal jelly produced; qualitative changes were not assessed. This is the first study to evaluate the effect of Roundup that suggests damage to the development and survival of bee colonies.

Most studies of herbicide effects on bees have been done in Europe and the United States, mainly on honeybees,[180] which comprise just a few species (*Apis mellifera, A. cerana, A. dorsate*, and others). There are approximately 20,000 known wild bee species worldwide (bumble, stingless, mason, leafcutter, etc.). Generally, herbicide effects on mortality are well represented in the literature in comparison to sublethal effects. Most studies have been in the lab rather than the field,[180] and in just one species. Only nine investigated effects on multiple species in the same study. The most widely investigated agent is glyphosate (15 studies), followed by atrazine (6), 2,4-D (5), paraquat (5), and simazine (4); other herbicides were investigated in only a single study. Active ingredients were studied more frequently than formulations. Most studies used oral exposure; topical exposure has been less well investigated. This may be due to difficulties in determining or mimicking field-realistic topical exposure in an experimental setup, or due to the assumption that oral is the most likely route in nature.

Glyphosate has long been considered innocuous to animals, including bees, because it targets an enzyme found only in plants and microorganisms. However, bees rely on a specialized gut microbiota that benefits growth and provides defense against pathogens. Glyphosate at documented environmental concentrations has been shown to reduce gut microbiota species, increasing the bee's susceptibility to infection by opportunistic pathogens.[133] Glyphosate exposure of young worker bees increased their mortality after exposure to the opportunistic pathogen *Serratia marcescens*. Sublethal concentrations might impair bee health and their effectiveness as pollinators.

Field realistic doses of glyphosate in *A. mellifera* chronically and acutely reduced their sensitivity to sucrose and learning behavior, especially chronic exposure,[181] compared to untreated controls. Acute exposure decreased elemental learning and short-term memory retention. These results imply that field-realistic glyphosate concentrations can reduce the sensitivity to nectar reward and impair associative learning. However, no effect on foraging-related behavior was found. This study also shows that forager bees could become a source of constant inflow of nectar with glyphosate residues that could then be distributed among nestmates, stored in the hive, and have long-term negative consequences on colony performance.

Studies of herbicide effects on wild bees and other flower-visiting insects are scarce (only 18 between 1996 and 2019). Bees in wheat fields and grasslands are exposed to many pesticides. In one study, 19 pesticides and metabolites were found: atrazine in 19% and metolachlor in 9% of the bees sampled.[182] Agricultural pesticides other than insecticides need to be explored to assess the contribution of different pesticide classes to the decline in *Apis* and non-*Apis* species in recent decades.[183]

Most studies were done in western industrialized countries. Contributions from other parts of the world are especially interesting. In sub-Saharan Africa, for example

Ghana, many rural farmers apply higher than recommended dosages of herbicides because they have not been properly trained. In a laboratory study, *A. mellifera* and the wild species *Hypotrigona ruspolii* were subjected to contact for 24 h with the recommended and twice the recommended concentrations of the GBH Sunphosate 360 SL, with distilled water as the control.[184] More bees died after contact with plants freshly sprayed with the herbicide than on herbicide-treated filter paper. In both cases, more bees died after contact with the higher concentration. The findings suggest that both these beneficial insects may be killed if they are sprayed or come into contact with plants that have been freshly sprayed with GBHs.

Spiders

Spiders (Araneae) are a very diverse animal order with more than 48,000 species worldwide. The majority prey on insects and other smaller animals. Very few are herbivorous. Given their important role in agroecosystems,[185] it is surprising that only 126 studies investigated herbicide effects on them between 1982 and 2019.

Contact exposure of the wolf spider *Pardosa milvina* to the GBH Roundup Klasik Pro over 2 months resulted in lower survival and affected behavior (locomotion time and distance) compared to topically exposed and control groups.[169] Comparing the herbicide and the surfactant Wetcit separately and in combination showed that both pure surfactant and GBH + surfactant tank mixes decreased predatory activity.[186] The authors concluded that testing pesticide tank mixes is highly important because they are what is actually applied. Atrazine has been shown to affect various life history traits of *P. milvina*.[187] The main effects were delayed maturation, increased molting errors, decreased production of an egg sac after mating, and reduced adult lifespan. Interestingly, the total number of eggs produced from a single mating was increased in the presence of atrazine through the production of multiple egg sacs.

Mites are small arthropods of the class Arachnida and are related to spiders. Predatory mites (phytoseiidae) feed on thrips and other small herbivores and are surrogate species in ERAs. Therefore, they are among the most-studied natural enemy groups in the field of pesticide nontarget effects. Many natural enemies, including phytoseiids, are associated with weeds due to the presence there of alternative prey, shelter, and floral resources. Nontarget effects on the predatory mite *Phytoseiulus persimilis*, and the primary pest that it controls, the red spider mite (*Tetranychus urticae*), were studied for several herbicides.[188] *S*-metolachlor was highly toxic to the biocontrol mite *P. persimilis* (80%–90% mortality) but had minimal effect on the pest *T. urticae*. Dicamba, oxyfluorfen, and napropamide also caused moderate mortality levels of *P. persimilis* (21%–74%). These results show that herbicides affect the biocontrol potential in agricultural systems. Some herbicide effects on predatory mites are also presented in Table 7.1.

Covering more invertebrate species living aboveground would be beyond the scope of this chapter. Of course, they can also be directly affected by herbicides. Just as an example, newly hatched land snails (*Helix aspersa*) were exposed for more than 5 months to soil and/or food contaminated with the GBH Bypass or the glufosinate-based herbicide Basta at the recommended and 10 fold field doses.[189]

No effects on survival and growth were found, but both herbicides decreased the maturation of the genital tract. Glyphosate was also found in the snail body. As these snails are also consumed by people, there is a potential risk of transferring herbicides to the food chain.

Amphibians, reptiles, birds, and small mammals

To estimate the risk of pesticides for terrestrial vertebrate species, only birds and rodent mammals are obligatory in ERAs. Amphibians, reptiles, and bats are not, at least in the EU.[190] The following account focuses on the direct effects in terrestrial ecosystems. A very comprehensive review of pesticide effects on the biodiversity of birds and mammals is provided by Jahn et al.[191] We will refer to some of these findings and add recently published data.

Amphibians

Amphibians are among the most threatened vertebrate species worldwide. With a life cycle that encompasses aquatic and terrestrial phases as well as migration to and from spawning waters, they are exposed to pesticides in the water and on land.[192, 193] In the aquatic and terrestrial life stages, they are assumed to be covered by the risk assessment for aquatic invertebrates and fish or mammals and birds. Detrimental effects on terrestrial amphibians are likely because their skin is highly permeable to allow gas, water, and electrolyte exchange with the environment. Moreover, the dermal uptake processes of chemicals are two orders of magnitude faster than in mammals,[194] suggesting that, for an amphibian in a crop field, dermal uptake is a likely exposure route.[195]

Generally, the herbicide exposure of amphibians depends on in-field presence, herbicide application, and crop interception. Several studies show that amphibians and herbicides overlap regularly in fields. The extent of overlap and interception varies between years, crops, and amphibian species. From 1977 to 2019, 423 publications addressing herbicides and amphibians appeared, most of them limited to aquatic developmental stages.

Pesticide absorption and toxicity studies for terrestrial life stages, focusing on dermal exposure through contact with treated soil and direct overspray, have been reviewed.[195] As stated above, the cutaneous absorption of chemicals is significant, and percutaneous passage is higher in amphibians than in mammals. Rapid, substantial uptake of atrazine from treated soil by toads (*Bufo americanus*) has been described.[196] Herbicide effects on terrestrial amphibians were mostly studied in contact test designs. In a US study, the juveniles of two southern high plains species, Great Plains toads (*B. cognatus*) and New Mexico spadefoots (*Spea multiplicata*), were exposed to environmentally relevant concentrations of several widely used herbicide formulations.[197] Besides filter paper, natural soil was included as a substrate to increase environmental realism. A GBH (Roundup Weed and Grass Killer Ready-to-Use Plus) intended for lawns and gardens caused significant mortality in both species. An agricultural GBH (Roundup WeatherMAX) and a glufosinate-based herbicide (Ignite 280 SL) did not affect the short-term survival of either species, suggesting that glyphosate formulations can differ considerably in their effects on amphibians. Gray treefrogs (*Hyla versicolor, H. chrysoscelis*) avoided oviposition in pools contaminated with Roundup.[198] Juvenile western toads (*Bufo boreas*) and Cascades frogs (*Rana cascadae*) avoided urea-soaked paper towels but not urea in soil,[199] even though exposure to urea-treated soil resulted in significant mortality of both species.[200] Juvenile *B. americanus* also did not avoid soils contaminated with atrazine.[196] These results suggest that some amphibians are not able to avoid contaminated terrestrial substrates.

In the first study of exposure of juvenile European common frogs (*Rana temporaria*) following an agricultural overspray scenario with seven pesticides, the herbicides CurolB (bromoxynil-octanoate) and Dicomil (fenoxaprop-*P*-ethyl) resulted in 60% and 40% mortality at the recommended label rate. Dicomil also caused 40% mortality at 10% of the label rate.[201] This degree of toxicity is alarming, and large-scale negative effects of terrestrial pesticide exposure on amphibian populations seem likely.

The direct toxicity of herbicides to juvenile amphibians migrating through fields can affect populations by reducing their gene flow in agricultural landscapes.[202, 203] A 2-year field study in eastern Germany showed that in maize, up to 17% of the reproducing population of the fire-bellied toad *Bombina bombina* encountered a single herbicide application on bare soil without weeds present.[204] Such single events might, depending on the toxicity of the formulation, be detrimental for local population persistence.

The weight of the evidence leads to the conclusion that terrestrial herbicide exposure might be underestimated as a driver of the decline in frog and other amphibian populations. The risk assessment procedures in place do not protect this animal group. Exposure and toxicity studies for terrestrial amphibian life stages are scarce, and the reported data indicate the need for further research.

Pesticide uptake is highest when individuals are directly oversprayed.[205] In a study of the migration of four temperate amphibian species, it was suggested that up to 100% of the population could come into contact with glyphosate.[206] Temporal coincidence between amphibian migration and herbicide application to crops is documented.[204] In winter cereals, on average, applications were temporally coincident with 13% of common spadefoot (*Pelobates fuscus*) and 37% of the northern crested newt (*Triturus cristatus*) with 50%–70% interception of the crop canopy. In maize, up to 17% of the reproducing population of fire-bellied toads encountered a single application out of a total of 14 herbicides used in the study systems during bare soil/emergence without interception. The highest relative population share affected by a single herbicide (71%) was determined for *T. cristatus* during stem elongation (80% interception) of winter rape.

As exposure of amphibians to herbicides seems inevitable, a high risk for populations can be anticipated. One stark result of laboratory overspray experiments was a 60% mortality within 7 days for European common frogs at recommended field application rates of CurolB (active ingredient bromoxyniloctanoate) and 40% for overspray with Dicomil.[201]

Reptiles

Reptiles comprise turtles, crocodilians, snakes, and lizards, with more than 10,000 species globally. Yet very little is known about herbicide effects on them, with only 68 publications between 1981 and 2019.

Dermal exposure to two commercial GBHs (Agpro Glyphosate 360, Yates Roundup Weedkiller, both at the label-specified concentrations of 144 mg AI/L) on the New Zealand common skink (*Oligosoma polychroma*) was studied.[207] Agpro Glyphosate 360 contained the surfactant ethoxylated tallow amine while the surfactant in Yates Roundup Weedkiller was unknown. Neither product had a significant impact on body mass. However, skinks treated with Yates Roundup Weedkiller selected significantly higher temperatures across 3 weeks following exposure. The authors suggest that this heat-seeking behavior could be a fever response, an increase in metabolism to counteract physiological stress.

The toxicity of glyphosate, clopyralid, triclopyr, metsulfuron-methyl, and haloxyfop-methyl was investigated for a screening-level risk assessment using the New Zealand western fence lizard (*Sceloporus occidentalis)* as a test organism in an oral test design.[208] Only triclopyr was acutely toxic (LD_{50} 550 μg/g). No sublethal endpoint was tested.

Reptiles that lay eggs in the soil of agricultural landscapes can be subject to herbicide uptake, as presumably could the eggs and the hatchlings. To the best of our knowledge, this has never been investigated. The data gap for herbicide effects on reptiles acutely and chronically is large and concerning.

Birds

These warm-blooded vertebrates inhabit all ecosystems, with about 10,000 species. In contrast to reptiles, they are among the best-studied wild animals, with 524 publications on herbicides between 1963 and 2019. Unfortunately, a coarse search based on key words does not allow a clear differentiation between publications reporting direct and indirect effects.

Birds are an integral part of ecotoxicology and ERA. Avian toxicity data for risk assessments are produced in feeding studies for two required bird species: Japanese quail (*Coturnix japonica*) or bobwhite quail (*Colinus virginianus*) and Mallard duck (*Anas platyrhynchos*) (for acute mortality values, see Table 7.1). To cover the uncertainty in the sensitivity of other species, a factor of 10 is included in the calculation. Toxicity tests of other bird species are rarely performed. One study orally dosed American kestrel (*Falco sparverius*) nestlings with three different diphenyl ether herbicides. All three showed effects, especially nitrofen.[209] Five hundred mg/kg produced 100% mortality; 250 mg/kg reduced growth as reflected by decreased body weight, crown-rump length, and bone length, and resulted in increased liver weight, indicating hepatotoxicity. Kestrel nestlings depend on parental care, like all passerine species. However, both standard test species of quail and the Mallard duck, belong to species whose nestlings are very mobile, feed themselves very early in life, and do not depend on their parents feeding them. So, the question remains how well these ERAs can describe the toxicological situation for the majority of bird species.

In an ERA, the avian toxicity data are linked to other bird species using contaminated food uptake estimations in a daily dietary dose approach. The focal species concept for ERA in the EU includes the grey partridge or pigeon for medium herbivores, the goose for large herbivores, the wren for insectivores, and the linnet for granivores. Direct effects through oral poisoning are thought to be not very relevant.[210] In an ERA comparison, it became evident that estimated doses of dicamba and glyphosate exceeded toxicity thresholds at maximum application rates in forest use,[211] and triclopyr and 2,4-D exceeded toxicity thresholds at typical application rates in two and eight scenarios. Vegetation-eating and insect-eating birds and mammals were at most risk. Technical-grade glyphosate is also considered slightly toxic to birds.[212]

In an ERA using birds, only the oral uptake of contaminated food is addressed. However, dermal uptake by the feet, especially when occurring in crop fields, might need to be incorporated.[213] Indirect effects via changes in the availability and quality of insect and plant food following herbicide treatment are more important.

Recently, a study analyzed changing patterns of pesticide use in agriculture in Great Britain over the 1990–2016 period with respect to the risk posed to corn buntings (*Emberiza calandra*).[214] The analysis was based on annual toxic loads measured as the total number of LD_{50} doses. The herbicide pendimethalin was ranked fifth. Its use increased by 297%, from 1357 tonnes applied in 2016 against about 341 tonnes in 1990. While the authors are well aware of the limitations of their approach (e.g., in reality only a tiny proportion of herbicides applied will be ingested by birds), their analysis suggests that possible effects on birds should be given more attention.

Sublethal effects of GBHs on reproductive health and the risk that residues pose to livestock birds (hens, geese, turkeys) have been reviewed.[215] This topic will not be elaborated here because we are focusing on wildlife species.

Small terrestrial mammals and bats

Herbicides may have direct effects on herbivorous small mammal species via the consumption of toxic plant material, but direct effects are rarely studied. Small mammals are a diverse group with hundreds of species, including mice, shrews, gophers, badgers, and weasels, often living in agricultural land and therefore subject to contamination. However, they have been seldom used as indicators of biodiversity responses to environmental changes. Only 64 publications were found for "herbicide*" AND "small mammals" between 1971 and 2019.

Mammal toxicity data for ERAs are generated with the rat (*Rattus norvegicus*; for values, see Table 7.1). The same endpoint is used in human risk assessment. Most modern herbicides have low mammalian toxicity.[216] However, DNA damage caused by paraquat was highlighted more than 40 years ago,[217] and a variety of stress responses led to its recognition as a pneumotoxicant. It is also known to induce neurodegeneration.[218] Terbuthylazine produced DNA damage in human and mouse cells,[219] and simazine caused immunotoxicity in mice.[220]

As for birds, a daily dietary dose for oral exposure is calculated in ERAs using the available toxicity data and a selected focal species such as the shrew, wood mouse, (*Apodemus sylvaticus*), and European hare (*Lepus europaeus*). In addition to oral exposure, dermal uptake and even overspray need to be considered for mammals, as was recently documented for the hare.[221]

A review of records on vertebrate wildlife incidents with pesticides in Europe registered few poisonings by approved herbicide application between 1990 and 1994 in the United Kingdom.[222] A hedgehog (*Erinaceus* spp.) was poisoned after spraying of paraquat in grassland and one hare (*Lepus europaeus*) died after spraying. In Europe, reasons for the decline of brown hare populations, a herbivorous small mammal species, have been reviewed.[223] An analysis of wildlife incident data confirmed the exposure of hares to herbicides in stubbles, grassland, potatoes, and dormant alfalfa crops. Though paraquat residues were detected in the tissues and stomach contents of hares, whether this exposure was the cause of death was uncertain. In the United Kingdom and in France, only 2% and 0.06% of hare incidents, respectively, were confirmed to be due to paraquat over a period of 23 and 11 years, respectively.[223] However, hares were likely to be deterred from consuming sprayed vegetation and showed reduced foraging on affected stubble.[223]

Of the incidents with small mammals recorded over the period 1990–94 in France, the United Kingdom, and the Netherlands, 53%–66% were due to deliberate pesticide abuse.[222] Most of the studies were in the United Kingdom.[191] Information on direct effects on mammal populations in agricultural landscapes is scarce.[10] Analyzing both direct and indirect effects is difficult because studies are often done within the wide scope of agricultural intensification and a broad variety of negative impacts.

Bats are the only mammals capable of active flight. There are more than 1400 bat species worldwide, about 50 in the EU and the United States. They differ dramatically from other small insectivorous mammals in temperate regions in that they hibernate and have only one offspring at a time. The necessity of including bats in mammalian ERAs was recently addressed by EFSA.[224] Exposure to herbicides is likely via the oral route, but no studies are available for herbicide residues on flying insects and therefore the risk cannot be calculated. It is also unclear whether the applied uncertainty factors used in ERAs are protective.[225]

Endocrine disruption

Sublethal effects of herbicides affecting the hormone system of animals–endocrine disruption—is a heavily researched topic, as shown by more than 800 publications between 1978 and 2019. For decades, studies of endocrine-disrupting chemicals (EDCs) have challenged traditional concepts in toxicology, particularly because EDCs can have effects at very low doses that are contradictory to the classical dose-response relations.[226] Epidemiological studies also show that environmental exposure to EDCs is associated with human diseases and disabilities. A detailed discussion would go beyond the scope of this chapter, but the interested reader is directed to current reviews.[227, 228]

Among the most prominent herbicides related to endocrine disruption is atrazine. Water-borne atrazine contamination of wild leopard frogs (*Rana pipiens*) in different regions of the United States resulted in up to 92% of males showing gonadal abnormalities, retarded development, and hermaphroditism.[229] They were both demasculinized (chemically castrated) and completely feminized as adults.[230] Detrimental effects were also observed in other vertebrates such as species of salmon,[231] a caiman,[232] and rodents.[233–235]

When female wallabies (*Macropus eugenii*) were exposed to atrazine-contaminated drinking water throughout pregnancy, parturition, and lactation, their male offspring showed altered testis function and reduced penis length.[236] It can therefore be expected that exposure to herbicides acting as endocrine disruptors contributes to the dramatic population decline of wallabies in Australia because their distribution overlaps agricultural areas with heavy herbicide applications.

Endocrine-disrupting effects are also reported for GBHs (Roundup Flex, 480 g/L glyphosate). Subtoxic exposure resulted in lower testosterone levels and altered gut microbiome in Japanese quail females and males (*C. japonica*).[237] In another study with *C. japonica*, GBH (Roundup Flex) was shown to transfer directly from mothers to eggs, and embryonic development tended to be poorer in the eggs of GBH-exposed parents compared to control parents.[238]

Exposure of Friesian ewe lambs to GBH (Roundup Full II, 540 g/L glyphosate) altered their adult fertility by decreasing uterine differentiation and functionality.[239] The authors suggest that the GBH may be responsible for uterine subfertility by acting as an EDC. Early postnatal exposure by injection of the GBH Roundup Full II induced long-term alterations in the mammary gland morphology of aging female Wistar rats.[240]

Ingaramo et al. summarized the endocrine-disrupting effects of exposure to glyphosate and other GBHs at environmentally relevant doses for various vertebrate species and showed alterations of the development and differentiation of ovarian follicles and uterus, affecting fertility when exposure occurred before puberty.[241]

Herbicides act not only as environmental estrogens, antiestrogens, and/or antiandrogenic chemicals, but they have also been shown to exert adverse effects on hippocampal neurogenesis.[242] All these sublethal effects on the endocrine and neurological systems have barely been studied in wild animal species.

Conclusions

Herbicides, although designed to specifically eradicate plants, have numerous direct effects on nonplant organisms. This might sound banal but it is important to note, as there is still a broad assumption among laypersons and scientists that pesticides are specific and affect only target pests.[243] Most effects are documented from a few selected surrogate species that are regularly used in environmental risk assessment. Wildlife organisms vary considerably in susceptibility to herbicides, but little is known about individual variations or differences within species. The variety of modes of action of herbicides makes it difficult to discern general response patterns among nontarget organisms. Thus, our understanding of effects thereon is scattered and mainly limited to single-species studies while investigations of populations, communities, and their ecological interactions are rare.

Studies are conducted at different levels of complexity ranging from controlled lab experiments through mesocosm and field studies to modeling. New approaches using population dynamics and food web modeling, including herbicide effects on population growth rate or age structure and different sensitivities of life history traits, are in development.[244] Despite recent promising achievements, population modeling is not yet developed enough to fully assess herbicide impacts on, for instance, endangered species or the decline in biodiversity.[215,246] A global threat to the sustainability of agricultural sectors with high herbicide use is the multiple resistance developed by weed populations after generations of exposure.

There are many knowledge gaps in herbicide effects on nontarget organisms in terrestrial ecosystems:

- Multispecies interactions within agricultural fields and landscapes.
- Long-term effects of high- and low-level chronic herbicide exposure in communities of animals and plants in species with short and long generation times.
- Sequences and interactions of herbicides with other pesticides applied during the cropping year. The sheer number of possible combinations makes such an assessment extremely difficult.
- Comparisons between effects on active ingredients that are mainly considered in ERAs versus formulated herbicide products that are actually applied in the field.
- Legacy and chronic effects of previous herbicides on subsequent crop and weed populations.
- Herbicide drift and its impact on the overall biodiversity and health of ecosystems and people.
- Endocrine-disruptive activity of herbicides that goes beyond acute and chronic frank toxicity. To date, more than 120 endocrine-disruptive pesticides are known, including several herbicides.[247]
- A lack of spatial and temporal precision in data on real herbicide and pesticide load and exposure for ecosystems.

Most of these gaps can be addressed effectively by interdisciplinary approaches taking into account complex real-world ecological interactions.

References

1. Ockleford C, Adriaanse P, Berny P, et al. Scientific opinion addressing the state of the science on risk assessment of plant protection products for in-soil organisms. *EFSA J* 2017;**15**(2). https://doi.org/10.2903/j.efsa.2017.4690, e04690.
2. Faust M, Altenburger R, Boedeker W, Grimme LH. Additive effects of herbicide combinations on aquatic non-target organisms. *Sci Total Environ* 1993;**134**:941–52. https://doi.org/10.1016/S0048-9697(05)80101-9.
3. Hasenbein S, Lawler SP, Connon RE. An assessment of direct and indirect effects of two herbicides on aquatic communities. *Environ Toxicol Chem* 2017;**36**(8):2234–44. https://doi.org/10.1002/etc.3740.
4. Breckels RD, Kilgour BW. Aquatic herbicide applications for the control of aquatic plants in Canada: effects to nontarget aquatic organisms. *Environ Rev* 2018;**26**(3):333–8. https://doi.org/10.1139/er-2018-0002.

5. Van den Brink PJ, Blake N, Brock TCM, Maltby L. Predictive value of species sensitivity distributions for effects of herbicides in freshwater ecosystems. *Hum Ecol Risk Assess* 2006;**12**:645–74.
6. Malaj E, von der Ohe PC, Grote M, et al. Organic chemicals jeopardize the health of freshwater ecosystems on the continental scale. *Proc Natl Acad Sci* 2014;**111** (26):9549–54. https://doi.org/10.1073/pnas.1321082111.
7. Gill JPK, Sethi N, Mohan A, Datta S, Girdhar M. Glyphosate toxicity for animals. *Environ Chem Lett* 2018;**16**(2):401–26. https://doi.org/10.1007/s10311-017-0689-0.
8. Kortekamp A. Unexpected side effects of herbicides: modulation of plant-pathogen interactions. In: Kortekamp A, editor. *Herbicides and environment.* InTech; 2011. p. 85–104.
9. van Gestel CAM. Soil ecotoxicology: state of the art and future directions. *ZooKeys* 2012;**176**:275–96.
10. Freemark K, Boutin C. Impacts of agricultural herbicide use on terrestrial wildlife in temperate landscapes: a review with special reference to North America. *Agric Ecosyst Environ* 1995;**52**(2-3):67–91. https://doi.org/10.1016/0167-8809(94)00534-L.
11. Rose MT, Cavagnaro TR, Scanlan CA, et al. Impact of herbicides on soil biology and function. In: Donald LS, editor. *Advances in agronomy.* Academic Press; 2016. p. 133–220.
12. Silva V, Mol HGJ, Zomer P, Tienstra M, Ritsema CJ, Geissen V. Pesticide residues in European agricultural soils—a hidden reality unfolded. *Sci Total Environ* 2019; **653**:1532–45. https://doi.org/10.1016/j.scitotenv.2018.10.441.
13. PPDB. *Pesticide properties database*. University of Hertfordshire; 2020. Available from: https://sitem.herts.ac.uk/aeru/ppdb/en/Reports/2631.htm [Accessed 9 June 2020].
14. Reddy KN, Rimando AM, Duke SO. Aminomethylphosphonic acid, a metabolite of glyphosate, causes injury in glyphosate-treated, glyphosate-resistant soybean. *J Agric Food Chem* 2004;**52**(16):5139–43. https://doi.org/10.1021/jf049605v.
15. EFSA. Guidance on the assessment of exposure of operators, workers, residents and bystanders in risk assessment for plant protection products. *EFSA J* 2014;**12**(10):3874. 55 pp https://doi.org/10.2903/j.efsa.2014.3874.
16. Gustafson DI. Groundwater ubiquity score: a simple method for assessing pesticide leachability. *Environ Toxicol Chem* 1989;**8**:339–57.
17. Hvězdová M, Kosubová P, Košíková M, et al. Currently and recently used pesticides in Central European arable soils. *Sci Total Environ* 2018;**613-614**:361–70. https://doi.org/10.1016/j.scitotenv.2017.09.049.
18. Ashworth MB, Walsh MJ, Flower KC, Powles SB. Recurrent selection with reduced 2,4-D amine doses results in the rapid evolution of 2,4-D herbicide resistance in wild radish (Raphanus raphanistrum L.). *Pest Manag Sci* 2016;**72**(11):2091–8. https://doi.org/10.1002/ps.4364.
19. Silva V, Montanarella L, Jones A, et al. Distribution of glyphosate and aminomethylphosphonic acid (AMPA) in agricultural topsoils of the European Union. *Sci Total Environ* 2018;**621**:1352–9. https://doi.org/10.1016/j.scitotenv.2017.10.093.
20. Primost JE, Marino DJG, Aparicio VC, Costa JL, Carriquiriborde P. Glyphosate and AMPA, "pseudo-persistent" pollutants under real-world agricultural management practices in the Mesopotamic Pampas agroecosystem, Argentina. *Environ Pollut* 2017;**229**:771–9. https://doi.org/10.1016/j.envpol.2017.06.006.
21. Laitinen P, Siimes K, Eronen L, et al. Fate of the herbicides glyphosate, glufosinate-ammonium, phenmedipham, ethofumesate and metamitron in two Finnish arable soils. *Pest Manag Sci* 2006;**62**(6):473–91. https://doi.org/10.1002/ps.1186.

22. Székács A, Darvas B. Forty years with glyphosate. In: Hasaneen MN, editor. *Herbicides—properties, synthesis and control of weeds.* InTech; 2012. p. 247–84.
23. Zimdahl RL, Gwynn SM. Soil Degradation of Three Dinitroanilines. *Weed Sci* 1977;**25**(3):247–51. https://doi.org/10.1017/S0043174500033397.
24. Gimsing AL, dos Santos AM. Glyphosate. In: Nowack B, VanBriesen J, editors. *Biogeochemistry of chelating agents. ACS Symp Ser*, American Chemical Society; 2005. p. 263–77.
25. Singh S, Kumar V, Datta S, et al. Glyphosate uptake, translocation, resistance emergence in crops, analytical monitoring, toxicity and degradation: a review. *Environ Chem Lett* 2020;**18**(3):663–702. https://doi.org/10.1007/s10311-020-00969-z.
26. Borggaard OK, Gimsing AL. Fate of glyphosate in soil and the possibility of leaching to ground and surface waters: a review. *Pest Manag Sci* 2008;**64**:441–56.
27. Gaupp-Berghausen M, Hofer M, Rewald B, Zaller JG. Glyphosate-based herbicides reduce the activity and reproduction of earthworms and lead to increased soil nutrient concentrations. *Sci Rep* 2015;**5**. https://doi.org/10.1038/srep12886, 12886.
28. Maderthaner M, Weber M, Takács E, et al. Commercial glyphosate-based herbicides effects on springtails (Collembola) differ from those of their respective active ingredients and vary with soil organic matter content. *Environ Sci Pollut Res* 2020;**27**(14):17280–9. https://doi.org/10.1007/s11356-020-08213-5.
29. Brühl CA, Alscher A, Hahn M, et al. *Protection of biodiversity in the risk assessment and risk management of pesticides (plant protection products & biocides) with a focus on arthropods, soil organisms and amphibians. Federal Environment Agency (Germany)*; 2015. 259 pp.
30. Marshall EJP, Brown VK, Boatman ND, Lutman PJW, Squire GR, Ward LK. The role of weeds in supporting biological diversity within crop fields. *Weed Res* 2003;**43**(2):77–89. https://doi.org/10.1046/j.1365-3180.2003.00326.x.
31. Andreasen C, Jensen HA, Jensen SM. Decreasing diversity in the soil seed bank after 50 years in Danish arable fields. *Agric Ecosyst Environ* 2018;**259**:61–71. https://doi.org/10.1016/j.agee.2018.02.034.
32. Matthews G, Bateman R, Miller PR. *Pesticide application methods.* 4th ed. Wiley-Blackwell; 2014.
33. Schmitz J, Schäfer K, Brühl CA. Agrochemicals in field margins - assessing the impacts of herbicides, insecticides, and fertilizer on the common buttercup (*Ranunculus acris*). *Environ Toxicol Chem* 2013;**32**(5):1124–31. https://doi.org/10.1002/etc.2138.
34. OECD. *Test no. 208: terrestrial plant test: seedling emergence and seedling growth test.* OECD; 2006. 21 pp.
35. Nelemans JB, van Wijngaarden RPA, Roessink I, Arts GHP. Effects of the herbicide metsulfuron-methyl on a plant community, including seed germination success in the F1 generation. *Front Environ Sci* 2017;**5**:10. https://doi.org/10.3389/fenvs.2017.00010.
36. Kjaer C, Strandberg M, Erlandsen M. Metsulfuron spray drift reduces fruit yield of hawthorn (*Crataegus monogyna* L.). *Sci Total Environ* 2006;**356**:228–34.
37. Belz RG, Duke SO. Herbicides and plant hormesis. *Pest Manag Sci* 2014;**70**(5):698–707. https://doi.org/10.1002/ps.3726.
38. Boutin C, Rogers CA. Pattern of sensitivity of plant species to various herbicides—an analysis with two databases. *Ecotoxicology* 2000;**9**(4):255–72.
39. Schmitz J, Stahlschmidt P, Brühl CA. *Protection of terrestrial non-target plant species in the regulation of environmental risks of pesticides.* Federal Environment Agency (Germany), TEXTE; 2015. 148 pp.

40. Christl H, Morilla J, Hoen T, Zumkier U. Comparative assessment of the intrinsic sensitivity of crop species and wild plant species to plant protection products and their active substances and potential implications for the risk assessment: a literature review. *Integr Environ Assess Manag* 2019;**15**(2):176–89.
41. Marrs RH, Williams CT, Frost AJ, Plant RA. Assessment of the effects of herbicide spray drift on a range of plant species of conservation interest. *Environ Pollut* 1989;**59**:71–86.
42. Brühl CA, Zaller JG. Biodiversity decline as a consequence of an inappropriate environmental risk assessment of pesticides. *Front Environ Sci* 2019;**7**(177). https://doi.org/10.3389/fenvs.2019.00177.
43. Buehring NW, Massey JH, Reynolds DB. Shikimic acid accumulation in field-grown corn (*Zea mays*) following simulated glyphosate drift. *J Agric Food Chem* 2007;**55**:819–24.
44. Reddy KN, Ding W, Zablotowicz RM, Thomson SJ, Huang Y, Krutz LJ. Biological responses to glyphosate drift from aerial application in non-glyphosate-resistant corn. *Pest Manag Sci* 2010;**66**(10):1148–54. https://doi.org/10.1002/ps.1996.
45. Egan JF, Barlow KM, Mortensen DA. A meta-analysis on the effects of 2,4-D and dicamba drift on soybean and cotton. *Weed Sci* 2014;**62**(1):193–206. https://doi.org/10.1614/WS-D-13-00025.1.
46. Jones GT, Norsworthy JK, Barber T, Gbur E, Kruger GR. Off-target movement of DGA and BAPMA dicamba to sensitive soybean. *Weed Technol* 2019;**33**(1):51–65. https://doi.org/10.1017/wet.2018.121.
47. Vieira BC, Luck JD, Amundsen KL, Werle R, Gaines TA, Kruger GR. Herbicide drift exposure leads to reduced herbicide sensitivity in Amaranthus spp. *Sci Rep* 2020;**10**(1):2146. https://doi.org/10.1038/s41598-020-59126-9.
48. Busi R, Gaines TA, Walsh MJ, Powles SB. Understanding the potential for resistance evolution to the new herbicide pyroxasulfone: field selection at high doses versus recurrent selection at low doses. *Weed Res* 2012;**52**(6):489–99. https://doi.org/10.1111/j.1365-3180.2012.00948.x.
49. Busi R, Girotto M, Powles SB. Response to low-dose herbicide selection in self-pollinated Avena fatua. *Pest Manag Sci* 2016;**72**(3):603–8. https://doi.org/10.1002/ps.4032.
50. Busi R, Powles SB. Evolution of glyphosate resistance in a Lolium rigidum population by glyphosate selection at sublethal doses. *Heredity* 2009;**103**(4):318–25. https://doi.org/10.1038/hdy.2009.64.
51. Busi R, Neve P, Powles S. Evolved polygenic herbicide resistance in Lolium rigidum by low-dose herbicide selection within standing genetic variation. *Evol Appl* 2013;**6**(2):231–42. https://doi.org/10.1111/j.1752-4571.2012.00282.x.
52. Neve P, Busi R, Renton M, Vila-Aiub MM. Expanding the eco-evolutionary context of herbicide resistance research. *Pest Manag Sci* 2014;**70**(9):1385–93. https://doi.org/10.1002/ps.3757.
53. Gressel J. Low pesticide rates may hasten the evolution of resistance by increasing mutation frequencies. *Pest Manag Sci* 2011;**67**(3):253–7. https://doi.org/10.1002/ps.2071.
54. Powles SB, Yu Q. Evolution in action: plants resistant to herbicides. In: Merchant S, Briggs WR, Ort D, editors. *Annual review of plant biology. Annual reviews, vol. 61. Annual Review of Plant Biology*; 2010. p. 317–47.
55. Bagavathiannan MV, Norsworthy JK. Multiple-herbicide resistance is widespread in roadside Palmer amaranth populations. *PLoS One* 2016;**11**(4). https://doi.org/10.1371/journal.pone.0148748, e0148748.
56. Schulz B, Segobye K. 2,4-D transport and herbicide resistance in weeds. *J Exp Bot* 2016;**67**(11):3177–9. https://doi.org/10.1093/jxb/erw192.

57. Shergill LS, Barlow BR, Bish MD, Bradley KW. Investigations of 2,4-D and multiple herbicide resistance in a Missouri waterhemp (Amaranthus tuberculatus) population. *Weed Sci* 2018;**66**(3):386–94. https://doi.org/10.1017/wsc.2017.82.
58. Sarangi D, Tyre AJ, Patterson EL, et al. Pollen-mediated gene flow from glyphosate-resistant common waterhemp (Amaranthus rudis Sauer): consequences for the dispersal of resistance genes. *Sci Rep* 2017;**7**. https://doi.org/10.1038/srep44913, 44913.
59. Gaba S, Gabriel E, Chadœuf J, Bonneu F, Bretagnolle V. Herbicides do not ensure for higher wheat yield, but eliminate rare plant species. *Sci Rep* 2016;**6**:30112. https://doi.org/10.1038/srep30112. https://www.nature.com/articles/srep30112#supplementary-information.
60. Reis MR, Aquino LÂ, Melo CAD, Silva DV, Dias RC. Carryover of tembotrione and atrazine affects yield and quality of potato tubers. *Acta Sci Agron* 2018;**40**, e35355.
61. Carneiro GDOP, Bontempo AF, Guimãraes FAR, et al. Carryover of tembotrione and atrazine in sugar beet. *Cienc Investig Agrar* 2019;**46**(3):319–24.
62. Magné C, Saladin G, Clément C. Transient effect of the herbicide flazasulfuron on carbohydrate physiology in *Vitis vinifera* L. *Chemosphere* 2006;**62**(4):650–7. https://doi.org/10.1016/j.chemosphere.2005.04.119.
63. Zaller JG, Cantelmo C, Dos Santos G, et al. Herbicides in vineyards reduce grapevine root mycorrhization and alter soil microorganisms and the nutrient composition in grapevine roots, leaves, xylem sap and grape juice. *Environ Sci Pollut Res* 2018; **25**(23):23215–26. https://doi.org/10.1007/s11356-018-2422-3.
64. Fuhlendorf SD, Engle DM, Arnold DC, Bidwell TG. Influence of herbicide application on forb and arthropod communities of North America tallgrass prairies. *Agric Ecosyst Environ* 2001;**92**:251–9.
65. Tunnel SJ, Stubbendieck J, Palazzolo S, Masters RA. Forb response to herbicides in a degraded tallgrass prairie. *Nat Areas J* 2006;**26**:72–7.
66. Browne M, Pagad S, Poorter MD. The crucial role of information exchange and research for effective responses to biological invasions. *Weed Res* 2009;**49**:6–18.
67. Rook EJ, Fischer DG, Seyferth RD, Kirsch JL, LeRoy CJ. Responses of prairie vegetation to fire, herbicide, and invasive species legacy. *Northwest Sci* 2011;**85**(2):288–302.
68. Stanley AG, Dunwiddie PW, Kaye TN. Restoring invaded Pacific Northwest prairies: management recommendations from a region-wide experiment. *Northwest Sci* 2011;**85**:233–46.
69. Weidlich EWA, Flórido FG, Sorrini TB, Brancalion PHS. Controlling invasive plant species in ecological restoration: a global review. *J Appl Ecol* 2020. https://doi.org/10.1111/1365-2664.13656.
70. OECD. *OECD guideline for the testing of chemicals. No. 217 soil microorganisms: carbon transformation test*; 2000 *https://wwwoecdorg/chemicalsafety/risk-assessment/1948325pdf.* 10 pp.
71. Bünemann EK, Schwenke GD, Van Zwieten L. Impact of agricultural inputs on soil organismsa review. *Aust J Soil Res* 2006;**44**(4):379–406. https://doi.org/10.1071/SR05125.
72. Imfeld G, Vuilleumier S. Measuring the effects of pesticides on bacterial communities in soil: a critical review. *Eur J Soil Biol* 2012;**49**:22–30. https://doi.org/10.1016/j.ejsobi.2011.11.010.
73. Van Bruggen AHC, He MM, Shin K, et al. Environmental and health effects of the herbicide glyphosate. *Sci Total Environ* 2018;**616-617**:255–68. https://doi.org/10.1016/j.scitotenv.2017.10.309.

74. Wurst S, Sonnemann I, Zaller JG. Soil macro-invertebrates- their impact on plants and associated aboveground communities in temperate regions. In: Ohgushi T, Wurst S, Johnson SN, editors. *Aboveground-belowground community ecology.* Springer; 2018. p. 175–200. [chapter 8]. *Ecological studies.*
75. van den Hoogen J, Geisen S, Routh D, et al. Soil nematode abundance and functional group composition at a global scale. *Nature* 2019;**572**(7768):194–8. https://doi.org/10.1038/s41586-019-1418-6.
76. Brady NC, Weil RR. *Elements of the nature and properties of soils.* 4th ed. Pearson; 2019.
77. Liphadzi KB, Al-Khatib K, Bensch CN, et al. Soil microbial and nematode communities as affected by glyphosate and tillage practices in a glyphosate-resistant cropping system. *Weed Sci* 2005;**53**(4):536–45.
78. Sánchez-Moreno S, Ferris H. Suppressive service of the soil food web: effects of environmental management. *Agric Ecosyst Environ* 2007;**119**(1):75–87. https://doi.org/10.1016/j.agee.2006.06.012.
79. Ishibashi N, Kondo E, Ito S. Effects of application of certain herbicides on soil nematodes and aquatic invertebrates in rice paddy fields in Japan. *Crop Prot* 1983;**2**(3):289–304. https://doi.org/10.1016/0261-2194(83)90003-0.
80. Zhao J, Neher D, Fu S, Li Z, Wang K-L. Non-target effects of herbicides on soil nematode assemblages. *Pest Manag Sci* 2013;**69**. https://doi.org/10.1002/ps.3505.
81. Cheng Z, Grewal PS, Stinner BR, Hurto KA, Hamza HB. Effects of long-term turfgrass management practices on soil nematode community and nutrient pools. *Appl Soil Ecol* 2008;**38**(2):174–84. https://doi.org/10.1016/j.apsoil.2007.10.007.
82. Hopkin SP. *Biology of the springtails (Insecta: Collembola).* Oxford University Press; 1997.
83. Lins VS, Santos HR, Gonçalves MC. The effect of glyphosate, 2,4-D, atrazine e nicosulfuron herbicides upon the Edaphic collembola (Arthropoda: Ellipura) in a no tillage system. *Neotrop Entomol* 2007;**36**(2):261–7.
84. Griffiths BS, Caul S, Thompson J, et al. Soil microbial and faunal responses to herbicide tolerant maize and herbicide in two soils. *Plant Soil* 2008;**308**(1):93–103. https://doi.org/10.1007/s11104-008-9609-1.
85. Bitzer RJ, Buckelew LD, Pedigo LP. Effects of transgenic herbicide-resistant soybean varieties and systems on surface-active springtails (Entognatha: Collembola). *Environ Entomol* 2002;**31**(3):449–61. https://doi.org/10.1603/0046-225x-31.3.449.
86. Fiera C, Ulrich W, Popescu D, et al. Tillage intensity and herbicide application influence surface-active springtail (Collembola) communities in Romanian vineyards. *Agric Ecosyst Environ* 2020;**107006**:300. https://doi.org/10.1016/j.agee.2020.107006.
87. Römbke J, Schmelz RM, Pélosi C. Effects of organic pesticides on enchytraeids (Oligochaeta) in agroecosystems: laboratory and higher-tier tests. *Front Environ Sci* 2017;**5**:20. https://doi.org/10.3389/fenvs.2017.00020.
88. Römbke J. *Enchytraeus albidus* (Oligochaeta, Annelida) as a test organism in terrestrial laboratory systems. In: Chambers PL, Chambers CM, Greim H, editors. *Biological monitoring of exposure and the response at the subcellular level to toxic substances. Archives of toxicology*, Springer; 1989. p. 402–5.
89. Novais SC, Amorim MJB. Changes in cellular energy allocation in *Enchytraeus albidus* when exposed to dimethoate, atrazine, and carbendazim. *Environ Toxicol Chem* 2013;**32**(12):2800–7. https://doi.org/10.1002/etc.2368.
90. Yang DL, Zhu J, Shen GX, et al. The acute toxicity of single and combined exposure of mercury and bromoxynil on *Fridericia bulbosa*. *Appl Mech Mater* 2011;**137**:280–5.

91. Scoriza RN, de Paula Silva A, MEF C, dos Santos Leles PS, de Resende AS. Efeito de herbicidas sobre a biota de invertebrados do solo em área de restauração florestal. *Rev Bras Ciênc Solo* 2015;**39**:1576–84.
92. Amorim MJB, Novais S, Römbke J, Soares AMVM. Enchytraeus albidus (Enchytraeidae): a test organism in a standardised avoidance test? Effects of different chemical substances. *Environ Int* 2008;**34**(3):363–71. https://doi.org/10.1016/j.envint.2007.08.010.
93. Salminen J, Eriksson I, Haimi J. Effects of terbuthylazine on soil fauna and decomposition processes. *Ecotoxicol Environ Saf* 1996;**34**(2):184–9. https://doi.org/10.1006/eesa.1996.0062.
94. Lavelle P. Earthworm activities and the soil system. *Biol Fertil Soils* 1988;**6**:237–51.
95. Edwards CA, Bohlen PJ. *Biology and ecology of earthworms*. 3rd ed. Chapman & Hall; 1996.
96. Jouquet P, Dauber J, Lagerlöf J, Lavelle P, Lepage M. Soil invertebrates as ecosystem engineers: intended and accidental effects on soil and feedback loops. *Appl Soil Ecol* 2006;**32**(2):153–64. https://doi.org/10.1016/j.apsoil.2005.07.004.
97. Cortet J, Gomot-De Vauflery A, Poinsot-Balaguer N, Gomot L, Texier C, Cluzeau D. The use of invertebrate soil fauna in monitoring pollutant effects. *Eur J Soil Biol* 1999;**35**(3):115–34.
98. Paoletti MG. The role of earthworms for assessment of sustainability and as bioindicators. *Agric Ecosyst Environ* 1999;**74**:137–55. https://doi.org/10.1016/S0167-8809(99)00034-1.
99. Pelosi C, Barot S, Capowiez Y, Hedde M, Vandenbulcke F. Pesticides and earthworms. A review. *Agron Sustain Dev* 2014;**34**(1):199–228.
100. Lee KE. *Earthworms. Their ecology and relationships with soils and land use*. Academic Press; 1985.
101. OECD. *Test No. 207: earthworm, acute toxicity tests*. OECD Publishing; 1984.
102. EEC. *EEC directive 79/831. Annex V. Part C: methods for the determination of ecotoxicity. Level I. C (L1) 4: toxicity for earthworms*; 1985.
103. Greene JC, Bartels CL, Warren-Hicks WJ, et al. *Protocols for short-term toxicity screening of hazardous waste sites*. Vol. EPA 600/3-88/029. ERLC; 1989.
104. Van Gestel CAM, Dirven-Van Breemen EM, Baerselman R, et al. Comparison of sublethal and lethal criteria for nine different chemicals in standardized toxicity tests using the earthworm Eisenia andrei. *Ecotoxicol Environ Saf* 1992;**23**(2):206–20. https://doi.org/10.1016/0147-6513(92)90059-C.
105. Wang YH, Wu SG, Chen LP, et al. Toxicity assessment of 45 pesticides to the epigeic earthworm *Eisenia fetida*. *Chemosphere* 2012;**88**(4):484–91. https://doi.org/10.1016/j.chemosphere.2012.02.086.
106. Lydy MJ, Linck SL. Assessing the impact of triazine herbicides on organophosphate insecticide toxicity to the earthworm Eisenia fetida. *Arch Environ Contam Toxicol* 2003;**45**(3):343–9.
107. Mosleh YY, Ismail SMM, Ahmed MT, Ahmed YM. Comparative toxicity and biochemical responses of certain pesticides to the mature earthworm *Aporrectodea caliginosa* under laboratory conditions. *Environ Toxicol* 2003;**18**(5):338–46. https://doi.org/10.1002/tox.10134.
108. Xiao NW, Jing B, Ge F, Liu XH. The fate of herbicide acetochlor and its toxicity to *Eisenia fetida* under laboratory conditions. *Chemosphere* 2006;**62**(8):1366–73. https://doi.org/10.1016/j.chemosphere.2006.07.043.

109. Dalby PR, Baker GH, Smith SE. Glyphosate, 2,4-DB and dimethoate: effects on earthworm survival and growth. *Soil Biol Biochem* 1995;**27**:1661–2.
110. Yasmin S, D'Souza D. Effect of pesticides on the reproductive output of *Eisenia fetida. Bull Environ Contam Toxicol* 2007;**79**(5):529–32. https://doi.org/10.1007/s00128-007-9269-5.
111. Correia FV, Moreira JC. Effects of glyphosate and 2,4-D on earthworms (*Eisenia foetida*) in laboratory tests. *Bull Environ Contam Toxicol* 2010;**85**:264–8.
112. Casabé N, Piola L, Fuchs J, et al. Ecotoxicological assessment of the effects of glyphosate and chlorpyrifos in an Argentine soya field. *J Soils Sediments* 2007;**8**:1–8.
113. Verrell P, Van Buskirk E. As the worm turns: *Eisenia fetida* avoids soil contaminated by a glyphosate-based herbicide. *Bull Environ Contam Toxicol* 2004;**72**:219–24.
114. Zaller JG, Heigl F, Ruess L, Grabmaier A. Glyphosate herbicide affects belowground interactions between earthworms and symbiotic mycorrhizal fungi in a model ecosystem. *Sci Rep* 2014;**4**:5634. https://doi.org/10.1038/srep05634.
115. Kaneda S, Okano S, Urashima Y, Murakami T, Nakajima M. Effects of herbicides, glyphosate, on density and casting activity of earthworm, *Pheretima* (*Amynthas*) *carnosus*. *Japanese J Soil Sci Plant Nutr* 2009;**80**:469–76.
116. Stellin F, Gavinelli F, Stevanato P, Concheri G, Squartini A, Paoletti MG. Effects of different concentrations of glyphosate (Roundup 360®) on earthworms (*Octodrilus complanatus*, *Lumbricus terrestris* and *Aporrectodea caliginosa*) in vineyards in the North-East of Italy. *Appl Soil Ecol* 2017. https://doi.org/10.1016/j.apsoil.2017.07.028.
117. Pochron S, Choudhury M, Gomez R, et al. Temperature and body mass drive earthworm (*Eisenia fetida*) sensitivity to a popular glyphosate-based herbicide. *Appl Soil Ecol* 2019;**139**:32–9. https://doi.org/10.1016/j.apsoil.2019.03.015.
118. Pochron S, Simon L, Mirza A, Littleton A, Sahebzada F, Yudell M. Glyphosate but not Roundup® harms earthworms (*Eisenia fetida*). *Chemosphere* 2020;**241**. https://doi.org/10.1016/j.chemosphere.2019.125017, 125017.
119. Treder K, Jastrzębska M, Kostrzewska MK, Makowski P. Do long-term continuous cropping and pesticides affect earthworm communities? *Agronomy* 2020;**10**(4):586.
120. Chelinho S, Moreira-Santos M, Lima D, et al. Cleanup of atrazine-contaminated soils: ecotoxicological study on the efficacy of a bioremediation tool with *Pseudomonas* sp. ADP. *J Soils Sediments* 2010;**10**(3):568–78. https://doi.org/10.1007/s11368-009-0145-2.
121. Frampton GK, Jänsch S, Scott-Fordsmand JJ, Römbke J, Van den Brink PJ. Effects of pesticides on soil invertebrates in laboratory studies: a review and analysis using species sensitivity distributions. *Environ Toxicol Chem* 2006;**25**(9):2480–9. https://doi.org/10.1897/05-438r.1.
122. Mosleh YY, Paris-Palacios S, Couderchet M, Vernet G. Effects of the herbicide isoproturon on survival, growth rate, and protein content of mature earthworms (*Lumbricus terrestris* L.) and its fate in the soil. *Appl Soil Ecol* 2003;**23**(1):69–77. https://doi.org/10.1016/S0929-1393(02)00161-0.
123. Marques C, Pereira R, Goncalves F. Using earthworm avoidance behaviour to assess the toxicity of formulated herbicides and their active ingredients on natural soils. *J Soils Sediments* 2009;**9**(2):137–47. https://doi.org/10.1007/s11368-009-0058-0.
124. Pelosi C, Toutous L, Chiron F, et al. Reduction of pesticide use can increase earthworm populations in wheat crops in a European temperate region. *Agric Ecosyst Environ* 2013;**181**:223–30. https://doi.org/10.1016/j.agee.2013.10.003.
125. Zhou CF, Wang YJ, Yu YC, et al. Does glyphosate impact on Cu uptake by, and toxicity to, the earthworm *Eisenia fetida*? *Ecotoxicology* 2012;**21**(8):2297–305. https://doi.org/10.1007/s10646-012-0986-0.

126. FAO, ITPS. *Global assessment of the impact of plant protection products on soil functions and soil ecosystems*; 2017. 40 pp.
127. Sannino F, Gianfreda L. Pesticide influence on soil enzymatic activities. *Chemosphere* 2001;**45**(4):417–25. https://doi.org/10.1016/S0045-6535(01)00045-5.
128. Hernández M, Jia Z, Conrad R, Seeger M. Simazine application inhibits nitrification and changes the ammonia-oxidizing bacterial communities in a fertilized agricultural soil. *FEMS Microbiol Ecol* 2011;**78**(3):511–9. https://doi.org/10.1111/j.1574-6941.2011.01180.x.
129. Wang J, Zhang H, Zhang X, Qin S, Tan H, Li X. Effects of long-term chlorimuron-ethyl application on the diversity and antifungal activity of soil *Pseudomonas* spp. in a soybean field in Northeast China. *Ann Microbiol* 2013;**63**(1):335–41. https://doi.org/10.1007/s13213-012-0479-7.
130. Wu M, Zhang X, Zhang H, et al. Soil pseudomonas community structure and its antagonism towards *Rhizoctonia solani* under the stress of acetochlor. *Bull Environ Contam Toxicol* 2009;**83**(3):313–7. https://doi.org/10.1007/s00128-009-9731-7.
131. Panettieri M, Lazaro L, López-Garrido R, Murillo JM, Madejón E. Glyphosate effect on soil biochemical properties under conservation tillage. *Soil Tillage Res* 2013;**133**:16–24. https://doi.org/10.1016/j.still.2013.05.007.
132. Mandl K, Cantelmo C, Gruber E, Faber F, Friedrich B, Zaller JG. Effects of glyphosate-, glufosinate- and flazasulfuron-based herbicides on soil microorganisms in a vineyard. *Bull Environ Contam Toxicol* 2018;**101**(5):562–9. https://doi.org/10.1007/s00128-018-2438-x.
133. Motta EVS, Raymann K, Moran NA. Glyphosate perturbs the gut microbiota of honey bees. *Proc Natl Acad Sci* 2018;**115**(41):10305–10. https://doi.org/10.1073/pnas.1803880115.
134. Sessitsch A, Mitter B. 21st century agriculture: integration of plant microbiomes for improved crop production and food security. *Microb Biotechnol* 2015;**8**(1):32–3. https://doi.org/10.1111/1751-7915.12180.
135. Zhan J, Liang Y, Liu D, et al. Antibiotics may increase triazine herbicide exposure risk via disturbing gut microbiota. *Microbiome* 2018;**6**(1):224. https://doi.org/10.1186/s40168-018-0602-5.
136. Martinez B, Tomkins J, Wackett LP, Wing R, Sadowsky MJ. Complete nucleotide sequence and organization of the atrazine catabolic plasmid pADP-1 from *Pseudomonas* sp. strain ADP. *J Bacteriol* 2001;**183**(19):5684–97. https://doi.org/10.1128/JB.183.19.5684-5697.2001.
137. Gomez E, Ferreras L, Lovotti L, Fernandez E. Impact of glyphosate application on microbial biomass and metabolic activity in a Vertic Argiudoll from Argentina. *Eur J Soil Biol* 2009;**45**(2):163–7. https://doi.org/10.1016/j.ejsobi.2008.10.001.
138. Nguyen DB, Rose MT, Rose TJ, van Zwieten L. Effect of glyphosate and a commercial formulation on soil functionality assessed by substrate induced respiration and enzyme activity. *Eur J Soil Biol* 2018;**85**:64–72. https://doi.org/10.1016/j.ejsobi.2018.01.004.
139. Arango L, Buddrus-Schiemann K, Opelt K, et al. Effects of glyphosate on the bacterial community associated with roots of transgenic Roundup Ready® soybean. *Eur J Soil Biol* 2014;**63**:41–8. https://doi.org/10.1016/j.ejsobi.2014.05.005.
140. Banks ML, Kennedy AC, Kremer RJ, Eivazi F. Soil microbial community response to surfactants and herbicides in two soils. *Appl Soil Ecol* 2014;**74**:12–20. https://doi.org/10.1016/j.apsoil.2013.08.018.
141. Zabaloy MC, Carné I, Viassolo R, Gómez MA, Gomez E. Soil ecotoxicity assessment of glyphosate use under field conditions: microbial activity and community structure of

Eubacteria and ammonia-oxidising bacteria. *Pest Manag Sci* 2016;**72**(4):684–91. https://doi.org/10.1002/ps.4037.

142. Nakatani AS, Fernandes MF, de Souza RA, et al. Effects of the glyphosate-resistance gene and of herbicides applied to the soybean crop on soil microbial biomass and enzymes. *Field Crop Res* 2014;**162**:20–9. https://doi.org/10.1016/j.fcr.2014.03.010.
143. Al-Ani MAM, Hmoshi RM, Kanaan IA, Thanoon AA. Effect of pesticides on soil microorganisms. *J Phys Conf Ser* 2019;**1294**. https://doi.org/10.1088/1742-6596/1294/7/072007, 072007.
144. Means NE, Kremer RJ, Ramsier C. Effects of glyphosate and foliar amendments on activity of microorganisms in the soybean rhizosphere. *J Environ Sci Health B* 2007;**42**(2):125–32. https://doi.org/10.1080/03601230601123227.
145. Araujo ASF, Monteiro RTR, Abarkeli RB. Effect of glyphosate on the microbial activity of two Brazilian soils. *Chemosphere* 2003;**52**(5):799–804. https://doi.org/10.1016/s0045-6535(03)00266-2.
146. Bruckner A, Schmerbauch A, Ruess L, Heigl F, Zaller J. Foliar Roundup application has minor effects on the compositional and functional diversity of soil microorganisms in a short-term greenhouse experiment. *Ecotoxicol Environ Saf* 2019;**174**:506–13. https://doi.org/10.1016/j.ecoenv.2019.02.073.
147. Aristilde L, Reed ML, Wilkes RA, et al. Glyphosate-induced specific and widespread perturbations in the metabolome of soil *Pseudomonas* species. *Front Environ Sci* 2017;**5**(34). https://doi.org/10.3389/fenvs.2017.00034.
148. Cycoń M, Piotrowska-Seget Z. Changes in bacterial diversity and community structure following pesticides addition to soil estimated by cultivation technique. *Ecotoxicology* 2009;**18**(5):632–42. https://doi.org/10.1007/s10646-009-0321-6.
149. Kumar A, Nayak AK, Shukla AK, et al. Microbial biomass and carbon mineralization in agricultural soils as affected by pesticide addition. *Bull Environ Contam Toxicol* 2012;**88**(4):538–42. https://doi.org/10.1007/s00128-012-0538-6.
150. Pampulha ME, Ferreira MA, Oliveira A. Effects of a phosphinothricin based herbicide on selected groups of soil microorganisms. *J Basic Microbiol* 2007;**47**(4):325–31. https://doi.org/10.1002/jobm.200610274.
151. Gyamfi S, Pfeifer U, Stierschneider M, Sessitsch A. Effects of transgenic glufosinate-tolerant oilseed rape (*Brassica napus*) and the associated herbicide application on eubacterial and *Pseudomonas* communities in the rhizosphere. *FEMS Microbiol Ecol* 2002;**41**(3):181–90. https://doi.org/10.1111/j.1574-6941.2002.tb00979.x.
152. Schmalenberger A, Tebbe CC. Bacterial diversity in maize rhizospheres: conclusions on the use of genetic profiles based on PCR-amplified partial small subunit rRNA genes in ecological studies. *Mol Ecol* 2003;**12**(1):251–62. https://doi.org/10.1046/j.1365-294X.2003.01716.x.
153. Ernst D, Rosenbrock-Krestel H, Kirchhof G, et al. Molecular investigations of the soil, rhizosphere and transgenic glufosinate-resistant rape and maize plants in combination with herbicide (Basta) application under field conditions. *Z Naturforsch C J Biosci* 2008;**63**(11-12):864–72. https://doi.org/10.1515/znc-2008-11-1214.
154. Zhang X, Li X, Zhang C, Li X, Zhang H. Ecological risk of long-term chlorimuron-ethyl application to soil microbial community: an in situ investigation in a continuously cropped soybean field in Northeast China. *Environ Sci Pollut Res* 2011;**18**(3):407–15. https://doi.org/10.1007/s11356-010-0381-4.
155. Widenfalk A, Bertilsson S, Sundh I, Goedkoop W. Effects of pesticides on community composition and activity of sediment microbes—responses at various levels of microbial community organization. *Environ Pollut* 2008;**152**(3):576–84. https://doi.org/10.1016/j.envpol.2007.07.003.

156. Crouzet O, Wiszniowski J, Donnadieu F, Bonnemoy F, Bohatier J, Mallet C. Dose-dependent effects of the herbicide mesotrione on soil cyanobacterial communities. *Arch Environ Contam Toxicol* 2012;**64**:23–31.
157. van Hoesel W, Tiefenbacher A, König N, et al. Single and combined effects of pesticide seed dressings and herbicides on earthworms, soil microorganisms, and litter decomposition. *Front Plant Sci* 2017;**8**:215. https://doi.org/10.3389/fpls.2017.00215.
158. Köberl M, Wagner P, Müller H, et al. Unraveling the complexity of soil microbiomes in a large-scale study subjected to different agricultural management in styria. *Front Microbiol* 2020;**11**(1052). https://doi.org/10.3389/fmicb.2020.01052.
159. Thiour-Mauprivez C, Martin-Laurent F, Calvayrac C, Barthelmebs L. Effects of herbicide on non-target microorganisms: towards a new class of biomarkers? *Sci Total Environ* 2019;**684**:314–25. https://doi.org/10.1016/j.scitotenv.2019.05.230.
160. Nosanchuk JD, Ovalle R, Casadevall A. Glyphosate inhibits melanization of cryptococcus neoformans and prolongs survival of mice after systemic infection. *J Infect Dis* 2001;**183** (7):1093–9. https://doi.org/10.1086/319272.
161. Roberts C, Roberts F, Lyons R, et al. The shikimate pathway and its branches in apicomplexan parasites. *J Infect Dis* 2002;**185**(Suppl. 1):S25–36. https://doi.org/10.1086/338004.
162. Hage-Ahmed K, Rosner K, Steinkellner S. Arbuscular mycorrhizal fungi and their response to pesticides. *Pest Manag Sci* 2019;**75**(3):583–90. https://doi.org/10.1002/ps.5220.
163. Druille M, Cabello MN, Omacini M, Golluscio RA. Glyphosate reduces spore viability and root colonization of arbuscular mycorrhizal fungi. *Appl Soil Ecol* 2013;**64**:99–103. https://doi.org/10.1016/j.apsoil.2012.10.007.
164. Druille M, Omacini M, Golluscio RA, Cabello MN. Arbuscular mycorrhizal fungi are directly and indirectly affected by glyphosate application. *Appl Soil Ecol* 2013;**72**:143–9. https://doi.org/10.1016/j.apsoil.2013.06.011.
165. Helander M, Saloniemi I, Omacini M, Druille M, Salminen JP, Saikkonen K. Glyphosate decreases mycorrhizal colonization and affects plant-soil feedback. *Sci Total Environ* 2018;**642**:285–91. https://doi.org/10.1016/j.scitotenv.2018.05.377.
166. Sánchez-Bayo F, Wyckhuys KAG. Worldwide decline of the entomofauna: a review of its drivers. *Biol Conserv* 2019;**232**:8–27. https://doi.org/10.1016/j.biocon.2019.01.020.
167. Cardoso P, Barton PS, Birkhofer K, et al. Scientists' warning to humanity on insect extinctions. *Biol Conserv* 2020;**242**. https://doi.org/10.1016/j.biocon.2020.108426, 108426.
168. Kraus EC, Direct SMJ. Indirect effects of herbicides on insect herbivores in rice, *Oryza sativa*. *Sci Rep* 2019;**9**(1):6998. https://doi.org/10.1038/s41598-019-43361-w.
169. Evans SC, Shaw EM, Rypstra AL. Exposure to a glyphosate-based herbicide affects agrobiont predatory arthropod behaviour and long-term survival. *Ecotoxicology* 2010;**19**(7):1249–57. https://doi.org/10.1007/s10646-010-0509-9.
170. Ahn YJ, Kim YJ, Yoo JK. Toxicity of the herbicide glufosinate-ammonium to predatory insects and mites of *Tetranychus urticae* (Acari: Tetranychidae) under laboratory conditions. *J Econ Entomol* 2001;**94**(1):157–61.
171. Saska P, Skuhrovec J, Lukáš J, Chi H, Tuan S-J, Honěk A. Treatment by glyphosate-based herbicide alters life history parameters of the rose-grain aphid *Metopolophium dirhodum*. *Sci Rep* 2016;**6**(1). https://doi.org/10.1038/srep27801, 27801.
172. Marcus SR, Fiumera AC. Atrazine exposure affects longevity, development time and body size in *Metopolophium dirhodum*. *J Insect Physiol* 2016;**91-92**:18–25. https://doi.org/10.1016/j.jinsphys.2016.06.006.

173. Böhme F, Bischoff G, Zebitz CPW, Rosenkranz P, Wallner K. Pesticide residue survey of pollen loads collected by honeybees (*Apis mellifera*) in daily intervals at three agricultural sites in South Germany. *PLoS One* 2018;**13**(7). https://doi.org/10.1371/journal.pone.0199995, e0199995.
174. David A, Botías C, Abdul-Sada A, et al. Widespread contamination of wildflower and bee-collected pollen with complex mixtures of neonicotinoids and fungicides commonly applied to crops. *Environ Int* 2016;**88**:169–78. https://doi.org/10.1016/j.envint.2015.12.011.
175. Mullin CA, Frazier M, Frazier JL, et al. High Levels of Miticides and agrochemicals in North American apiaries: implications for honey bee health. *PLoS One* 2010;**5**(3). https://doi.org/10.1371/journal.pone.0009754, e9754.
176. Moffett JO, Morton HL, MacDonald RH. Toxicity of some herbicidal sprays to honey bees. *J Econ Entomol* 1972;**65**:32–6.
177. Vázquez DE, Balbuena MS, Chaves F, Gora J, Menzel R, Farina WM. Sleep in honey bees is affected by the herbicide glyphosate. *Sci Rep* 2020;**10**(1):10516. https://doi.org/10.1038/s41598-020-67477-6.
178. Farina WM, Balbuena MS, Herbert LT, Mengoni Goñalons C, Vázquez DE. Effects of the herbicide glyphosate on honey bee sensory and cognitive abilities: individual impairments with implications for the hive. *Insects* 2019;**10**(10):354.
179. Faita MR, Oliveira EM, Alves VV, Orth AI, Nodari RO. Changes in hypopharyngeal glands of nurse bees (*Apis mellifera*) induced by pollen-containing sublethal doses of the herbicide Roundup®. *Chemosphere* 2018;**211**:566–72. https://doi.org/10.1016/j.chemosphere.2018.07.189.
180. Cullen MG, Thompson LJ, Carolan JC, Stout JC, Stanley DA. Fungicides, herbicides and bees: a systematic review of existing research and methods. *PLoS One* 2019;**14**(12). https://doi.org/10.1371/journal.pone.0225743, e0225743.
181. Herbert LT, Vázquez DE, Arenas A, Farina WM. Effects of field-realistic doses of glyphosate on honeybee appetitive behaviour. *J Exp Biol* 2014;**217**(19):3457–64. https://doi.org/10.1242/jeb.109520.
182. Hladik ML, Vandever M, Smalling KL. Exposure of native bees foraging in an agricultural landscape to current-use pesticides. *Sci Total Environ* 2016;**542**:469–77. https://doi.org/10.1016/j.scitotenv.2015.10.077.
183. Belsky J, Joshi NK. Effects of fungicide and herbicide chemical exposure on apis and non-apis bees in agricultural landscape. *Front Environ Sci* 2020;**8**(81). https://doi.org/10.3389/fenvs.2020.00081.
184. Abraham J, Benhotons GS, Krampah I, Tagba J, Amissah C, Abraham JD. Commercially formulated glyphosate can kill non-target pollinator bees under laboratory conditions. *Entomol Exp Appl* 2018;**166**(8):695–702. https://doi.org/10.1111/eea.12694.
185. Drapela T, Frank T, Heer X, Moser D, Zaller JG. Landscape structure affects activity density, body size and fecundity of *Pardosa* wolf spiders (Araneae, Lycosidae) in winter oilseed rape. *Eur J Entomol* 2011;**108**:609–14.
186. Niedobová J, Skalský M, Ouředníčková J, Michalko R, Bartošková A. Synergistic effects of glyphosate formulation herbicide and tank-mixing adjuvants on *Pardosa* spiders. *Environ Pollut* 2019;**249**:338–44. https://doi.org/10.1016/j.envpol.2019.03.031.
187. Godfrey JA, Rypstra AL. Impact of an atrazine-based herbicide on an agrobiont wolf spider. *Chemosphere* 2018;**201**:459–65. https://doi.org/10.1016/j.chemosphere.2018.03.023.

188. Schmidt-Jeffris RA, Cutulle MA. Non-target effects of herbicides on *Tetranychus urticae* and its predator, *Phytoseiulus persimilis*: implications for biological control. *Pest Manag Sci* 2019;**75**(12):3226–34. https://doi.org/10.1002/ps.5443.
189. Druart C, Millet M, Scheifler R, Delhomme O, de Vaufleury A. Glyphosate and glufosinate-based herbicides: fate in soil, transfer to, and effects on land snails. *J Soils Sediments* 2011;**11**:1373–84.
190. EFSA, Ockleford C, Adriaanse P, et al. Scientific Opinion on the state of the science on pesticide risk assessment for amphibians and reptiles. *EFSA J* 2018;**16**(2). https://doi.org/10.2903/j.efsa.2018.5125, e05125.
191. Jahn T, Hötker H, Oppermann R, Bleil R, Vele L. *Protection of biodiversity of free living birds and mammals in respect of the effects of pesticides. Vol. 30/2014*; 2014.
192. Todd BD, Bergeron CM, Hepner MJ, Hopkins WA. Aquatic and terrestrial stressors in amphibians: a test of the double jeopardy hypothesis based on maternally and trophically derived contaminants. *Environ Toxicol Chem* 2011;**30**(10):2277–84. https://doi.org/10.1002/etc.617.
193. Fryday S, Thompson H. *Toxicity of pesticides to aquatic and terrestrial life stages of amphibians and occurrence, habitat use and exposure of amphibian species in agricultural environments. EFSA Supporting Publications*; 2012. EN-343:348 pp.
194. Quaranta A, Bellantuono V, Cassano G, Lippe C. Why amphibians are more sensitive than mammals to xenobiotics. *PLoS One* 2009;**4**(11). https://doi.org/10.1371/journal.pone.0007699, e7699.
195. Brühl CA, Pieper S, Weber B. Amphibians at risk? Susceptibility of terrestrial amphibian life stages to pesticides. *Environ Toxicol Chem* 2011;**30**(11):2465–72. https://doi.org/10.1002/etc.650.
196. Storrs Méndez SI, Tillitt DE, Rittenhouse TAG, Semlitsch RD. Behavioral response and kinetics of terrestrial atrazine exposure in American toads (*Bufo americanus*). *Arch Environ Contam Toxicol* 2009;**57**(3):590–7. https://doi.org/10.1007/s00244-009-9292-0.
197. Dinehart SK, Smith LM, McMurry ST, Anderson TA, Smith PN, Haukos DA. Toxicity of a glufosinate- and several glyphosate-based herbicides to juvenile amphibians from the Southern High Plains, USA. *Sci Total Environ* 2009;**407**(3):1065–71. https://doi.org/10.1016/j.scitotenv.2008.10.010.
198. Takahashi M. Oviposition site selection: pesticide avoidance by gray treefrogs. *Environ Toxicol Chem* 2007;**26**(7):1476–80. https://doi.org/10.1897/06-511r.1.
199. Hatch AC, Belden LK, Scheessele E, Blaustein AR. Juvenile amphibians do not avoid potentially lethal levels of urea on soil substrate. *Environ Toxicol Chem* 2001;**20**(10):2328–35. https://doi.org/10.1002/etc.5620201027.
200. Smith PN, Cobb GP, Godard-Codding C, et al. Contaminant exposure in terrestrial vertebrates. *Environ Pollut* 2007;**150**(1):41–64. https://doi.org/10.1016/j.envpol.2007.06.009.
201. Brühl CA, Schmidt T, Pieper S, Alscher A. Terrestrial pesticide exposure of amphibians: an underestimated cause of global decline? *Sci Rep* 2013;**3**:1135. https://doi.org/10.1038/srep01135.
202. Lenhardt PP, Brühl CA, Leeb C, Theissinger K. Amphibian population genetics in agricultural landscapes: does viniculture drive the population structuring of the European common frog (*Rana temporaria*)? *PeerJ* 2017;**5**. https://doi.org/10.7717/peerj.3520, e3520.
203. Leeb C, Brühl C, Theissinger K. Potential pesticide exposure during the post-breeding migration of the common toad (*Bufo bufo*) in a vineyard dominated landscape. *Sci Total Environ* 2020;**706**:134430. https://doi.org/10.1016/j.scitotenv.2019.134430.

204. Lenhardt PP, Bruehl CA, Berger G. Temporal coincidence of amphibian migration and pesticide applications on arable fields in spring. *Basic Appl Ecol* 2015;**16**(1):54–63. https://doi.org/10.1016/j.baae.2014.10.005.
205. Van Meter RJ, Glinski DA, Henderson WM, Garrison AW, Cyterski M, Purucker ST. Pesticide uptake across the amphibian dermis through soil and overspray exposures. *Arch Environ Contam Toxicol* 2015;**69**(4):545–56. https://doi.org/10.1007/s00244-015-0183-2.
206. Berger G, Graef F, Pfeffer H. Glyphosate applications on arable fields considerably coincide with migrating amphibians. *Sci Rep (Nature)* 2013;**3**:2622. https://doi.org/10.1038/srep02622.
207. Carpenter JK, Monks JM, Nelson N. The effect of two glyphosate formulations on a small, diurnal lizard (*Oligosoma polychroma*). *Ecotoxicology* 2016;**25**(3):548–54. https://doi.org/10.1007/s10646-016-1613-2.
208. Weir S, Yu S, Knox A, Talent LG, Monks J, Salice C. Acute toxicity and risk to lizards of rodenticides and herbicides commonly used in New Zealand. *N Z J Ecol* 2016;**40**:342–50. https://doi.org/10.20417/nzjecol.40.43.
209. Hoffman DJ, Spann JW, LeCaptain LJ, Bunck CM, Rattner BA. Developmental toxicity of diphenyl ether herbicides in nestling American kestrels. *J Toxicol Environ Health* 1991;**34**(3):323–36. https://doi.org/10.1080/15287399109531571.
210. Marshall EJP, Brown V, Boatman N, Lutman P, Squire G. *The impact of herbicides on weed abundance and biodiversity. PN0940. A report for the UK Pesticides Safety Directorate*; 2001.
211. Bautista SL. *A summary of acute risk of four common herbicides to birds and mammals. United States Department of Agriculture Forest Service General Technical Report*; PNW 694:77 pp; 2007.
212. WHO. *Glyphosate environmental health criteria No. 159*; 1994. 177 pp http://www.inchem.org/documents/ehc/ehc/ehc159.htm.
213. Mineau P. A comprehensive re-analysis of pesticide dermal toxicity data in birds and comparison with the rat. *Environ Toxicol Pharmacol* 2012;**34**(2):416–27. https://doi.org/10.1016/j.etap.2012.05.010.
214. Tassin de Montaigu C, Goulson D. Identifying agricultural pesticides that may pose a risk for birds. *PeerJ* 2020;**8**. https://doi.org/10.7717/peerj.9526, e9526.
215. Jarrell ZR, Ahammad MU, Benson AP. Glyphosate-based herbicide formulations and reproductive toxicity in animals. *Vet Anim Sci* 2020;**10**. https://doi.org/10.1016/j.vas.2020.100126, 100126.
216. Shaner DL. Herbicide safety relative to common targets in plants and mammals. *Pest Manag Sci* 2004;**60**(1):17–24. https://doi.org/10.1002/ps.782.
217. Ross WE, Block ER, Chang R-Y. Paraquat-induced DNA damage in mammalian cells. *Biochem Biophys Res Commun* 1979;**91**(4):1302–8. https://doi.org/10.1016/0006-291X(79)91208-7.
218. Yadawa AK, Richa R, Chaturvedi CM. Herbicide paraquat provokes the stress responses of HPA axis of laboratory mouse, Mus musculus. *Pestic Biochem Physiol* 2019;**153**:106–15. https://doi.org/10.1016/j.pestbp.2018.11.008.
219. Želježić D, Žunec S, Bjeliš M, et al. Effects of the chloro-s-triazine herbicide terbuthylazine on DNA integrity in human and mouse cells. *Environ Sci Pollut Res* 2018;**25**(19):19065–81. https://doi.org/10.1007/s11356-018-2046-7.
220. Ren R, Sun D-J, Yan H, Wu Y-P, Zhang Y. Oral Exposure to the herbicide simazine induces mouse spleen immunotoxicity and immune cell apoptosis. *Toxicol Pathol* 2013;**41**(1):63–72. https://doi.org/10.1177/0192623312452488.

221. Mayer M, Duan X, Sunde P, Topping CJ. European hares do not avoid newly pesticide-sprayed fields: overspray as unnoticed pathway of pesticide exposure. *Sci Total Environ* 2020;**715**:136977. https://doi.org/10.1016/j.scitotenv.2020.136977.
222. de Snoo GR, Scheidegger NM, de Jong FMW. Vertebrate wildlife incidents with pesticides: a European survey. *Pestic Sci* 1999;**55**:47–54.
223. Edwards PJ, Fletcher MR, Berny P. Review of the factors affecting the decline of the European brown hare, *Lepus europaeus* (Pallas, 1778) and the use of wildlife incident data to evaluate the significance of paraquat. *Agric Ecosyst Environ* 2000;**79**:95–103.
224. EFSA, Hernández-Jerez A, Adriaanse P, et al. Scientific statement on the coverage of bats by the current pesticide risk assessment for birds and mammals. *EFSA J* 2019;**17**(7). https://doi.org/10.2903/j.efsa.2019.5758, e05758.
225. Stahlschmidt P, Brühl CA. Bats at risk? Bat activity and insecticide residue analysis of food items in an apple orchard. *Environ Toxicol Chem* 2012;**31**(7):1556–63. https://doi.org/10.1002/etc.1834.
226. Vandenberg LN, Colborn T, Hayes TB, Heindel JJ, Jacobs DR, Lee DH. Hormones and endocrine-disrupting chemicals: low-dose effects and nonmonotonic dose responses. *Endocr Rev* 2012;**33**:378–455. https://doi.org/10.1210/er.2011-1050.
227. Mnif W, Hassine AIH, Bouaziz A, Bartegi A, Thomas O, Roig B. Effect of endocrine disruptor pesticides: a review. *Int J Environ Res Public Health* 2011;**8**(6):2265–303. https://doi.org/10.3390/ijerph8062265.
228. Guillette LJ. Endocrine disrupting contaminants—beyond the dogma. *Environ Health Perspect* 2006;**114**(Suppl. 1):9–12. https://doi.org/10.1289/ehp.8045.
229. Hayes T, Haston K, Tsui M, Hoang A, Haeffele C, Vonk A. Feminization of male frogs in the wild. Water-borne herbicide threatens amphibian populations in parts of the United States. *Nature* 2002;**419**:895–6.
230. Hayes TB, Khoury V, Narayan A, et al. Atrazine induces complete feminization and chemical castration in male African clawed frogs (*Xenopus laevis*). *Proc Natl Acad Sci* 2010;**107**(10):4612–7. https://doi.org/10.1073/pnas.0909519107.
231. Moore A, Waring CP. Mechanistic effects of a triazine pesticide on reproductive endocrine function in mature male Atlantic salmon (*Salmo salar* L.) parr. *Pestic Biochem Physiol* 1998;**62**(1):41–50. https://doi.org/10.1006/pest.1998.2366.
232. Rey F, González M, Zayas MA, et al. Prenatal exposure to pesticides disrupts testicular histoarchitecture and alters testosterone levels in male *Caiman latirostris*. *Gen Comp Endocrinol* 2009;**162**(3):286–92. https://doi.org/10.1016/j.ygcen.2009.03.032.
233. Trentacoste SV, Friedmann AS, Youker RT, Breckenridge CB, Zirkin BR. Atrazine effects on testosterone levels and androgen-dependent reproductive organs in peripubertal male rats. *J Androl* 2001;**22**(1):142–8. https://doi.org/10.1002/j.1939-4640.2001.tb02164.x.
234. Friedmann AS. Atrazine inhibition of testosterone production in rat males following peripubertal exposure. *Reprod Toxicol* 2002;**16**(3):275–9. https://doi.org/10.1016/S0890-6238(02)00019-9.
235. Kniewald J, Jakominić M, Tomljenović A, et al. Disorders of male rat reproductive tract under the influence of atrazine. *J Appl Toxicol* 2000;**20**(1):61–8. https://doi.org/10.1002/(sici)1099-1263(200001/02)20:1<61::Aid-jat628>3.0.Co;2-3.
236. Cook LE, Chen Y, Renfree MB, Pask AJ. Long-term maternal exposure to atrazine in the drinking water reduces penis length in the tammar wallaby *Macropus eugenii*. *Reprod Fertil Dev* 2020;**32**(13):1099–107. https://doi.org/10.1071/RD20158.
237. Ruuskanen S, Rainio MJ, Gómez-Gallego C, et al. Glyphosate-based herbicides influence antioxidants, reproductive hormones and gut microbiome but not reproduction: a long-term experiment in an avian model. *Environ Pollut* 2020;**266**. https://doi.org/10.1016/j.envpol.2020.115108, 115108.

238. Ruuskanen S, Rainio MJ, Uusitalo M, Saikkonen K, Helander M. Effects of parental exposure to glyphosate-based herbicides on embryonic development and oxidative status: a long-term experiment in a bird model. *Sci Rep* 2020;**10**(1):6349. https://doi.org/10.1038/s41598-020-63365-1.
239. Alarcón R, Rivera OE, Ingaramo PI, et al. Neonatal exposure to a glyphosate-based herbicide alters the uterine differentiation of prepubertal ewe lambs. *Environ Pollut* 2020;**265**. https://doi.org/10.1016/j.envpol.2020.114874, 114874.
240. Zanardi MV, Schimpf MG, Gastiazoro MP, et al. Glyphosate-based herbicide induces hyperplastic ducts in the mammary gland of aging Wistar rats. *Mol Cell Endocrinol* 2020. https://doi.org/10.1016/j.mce.2019.110658, 501110658.
241. Ingaramo P, Alarcón R, Muñoz-de-Toro M, Luque EH. Are glyphosate and glyphosate-based herbicides endocrine disruptors that alter female fertility? *Mol Cell Endocrinol* 2020. https://doi.org/10.1016/j.mce.2020.110934, 110934.
242. Rossetti MF, Stoker C, Ramos JG. Agrochemicals and neurogenesis. *Mol Cell Endocrinol* 2020. https://doi.org/10.1016/j.mce.2020.110820, 510110820.
243. Zaller JG. *Daily Poison. Pesticides—an underestimated danger.* Springer International; 2020.
244. Amano T, Kusumoto Y, Okamura H, et al. A macro-scale perspective on within-farm management: how climate and topography alter the effect of farming practices. *Ecol Lett* 2011;**14**(12):1263–72. https://doi.org/10.1111/j.1461-0248.2011.01699.x.
245. Köhler HR, Triebskorn R. Wildlife ecotoxicology of pesticides: can we track effects to the population level and beyond? *Science* 2013;**341**(6147):759–65. https://doi.org/10.1126/science.1237591.
246. Topping CJ, Elmeros M. Modeling exposure of mammalian predators to anticoagulant rodenticides. *Front Environ Sci* 2016;**4**:80. https://doi.org/10.3389/fenvs.2016.00080.
247. McKinlay R, Plant JA, Bell JNB, Voulvoulis N. Endocrine disrupting pesticides: implications for risk assessment. *Environ Int* 2008;**34**(2):168–83. https://doi.org/10.1016/j.envint.2007.07.013.

Indirect herbicide effects on biodiversity, ecosystem functions, and interactions with global changes

Carsten A. Brühl[a] *and Johann G. Zaller*[b]

[a]iES Landau, Institute for Environmental Sciences, University of Koblenz-Landau, Landau, Germany, [b]Institute of Zoology, University of Natural Resources and Life Sciences (BOKU), Vienna, Austria

Chapter outline

Introduction

When herbicides affect nontarget organisms, they inevitably also influence the functions of these organisms for ecosystem processes and the services they provide for humans. However, these aspects are less often addressed, partly because they are not mandatory within official environmental risk assessments (ERAs) with endpoints mainly at the organism level, and partly because of the inherent complexity of disentangling direct lethal or sublethal effects from indirect herbicide effects that act through disruptions of the food web.

Herbicides. https://doi.org/10.1016/[illegible]

Although we have an increasing understanding of a herbicide's mode of action on nontarget species, we still largely do not understand the knock-on consequences for ecosystem processes this might trigger, especially when effects are sublethal, multigenerational, or subtle. Here we try to provide an integrated view of the existing knowledge. It is obvious that herbicides at least reduce plant biodiversity in fields, in comparison to less effective mechanical or other nonchemical weed control measures. Consequences on food webs and associated biodiversity within the crop field or agricultural landscape are not well studied and are influenced by numerous stress factors (Fig. 8.1).

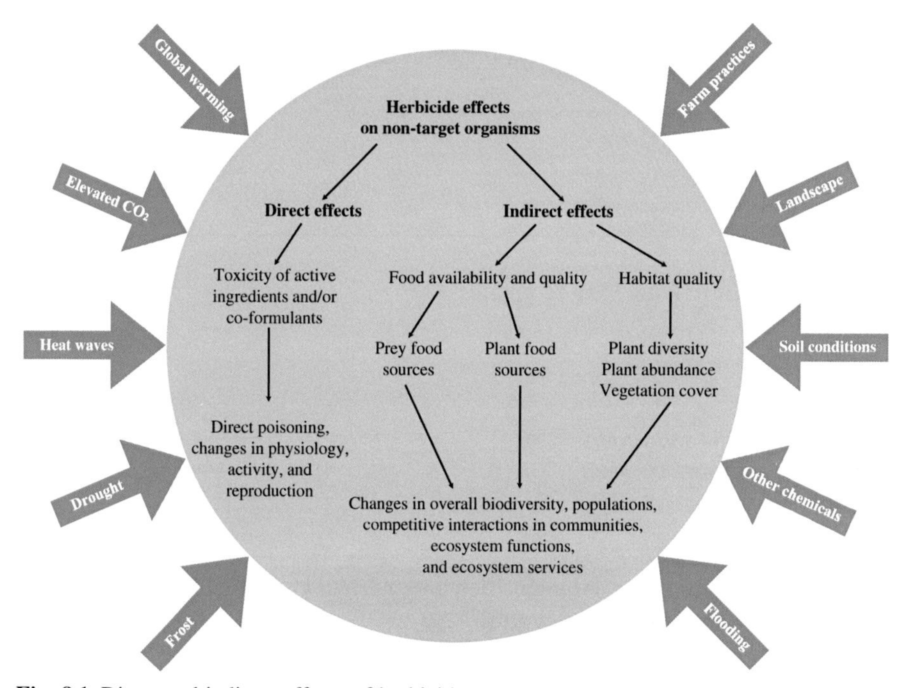

Fig. 8.1 Direct and indirect effects of herbicides on nontarget organisms, and stress factors that can influence them.

Herbicide use in industrialized agriculture with large-scale monocultures in many countries results in an overall reduction in landscape structure and heterogeneity, decreased crop diversity, decreased in-field plant diversity, and decreased landscape biodiversity because of degraded suitability for wildlife, including pollinators and pest antagonists that provide ecosystem services.

Herbicide transport: From local to global

Herbicides, like all pesticides, do not stay on the site where they are applied–the crop fields. The closest areas, such as field margins, receive overspray. Spray drift into neighboring noncropping areas exposes plants and animals there. Rain runoff exposes adjacent terrestrial habitats and streams. Herbicides can be transported over longer distances when volatilized or bound to blown soil particles. Field edges, old fields, and other seminatural habitats in agricultural landscapes support diverse plant communities that help sustain pollinators, predators, and other beneficial arthropods as well as many vertebrate species. It is likely that biodiversity in seminatural habitats is consistently exposed to low doses of various herbicides in combination with other pesticides.

Overspray

Overspray occurs because spray nozzles are mounted on the spray arm in such a way that the spray cones of two nozzles overlap, which is necessary to assure a 100% application rate (Fig. 8.2). The last nozzle of the spray arm is placed above the field edge, and as a result, parts of the adjacent field margin (depending on the field cultivation and the corresponding height of the spray arm) are oversprayed.[1] This can affect entire margins, most of which in Germany are between 1 and 3 m wide for 85% of their

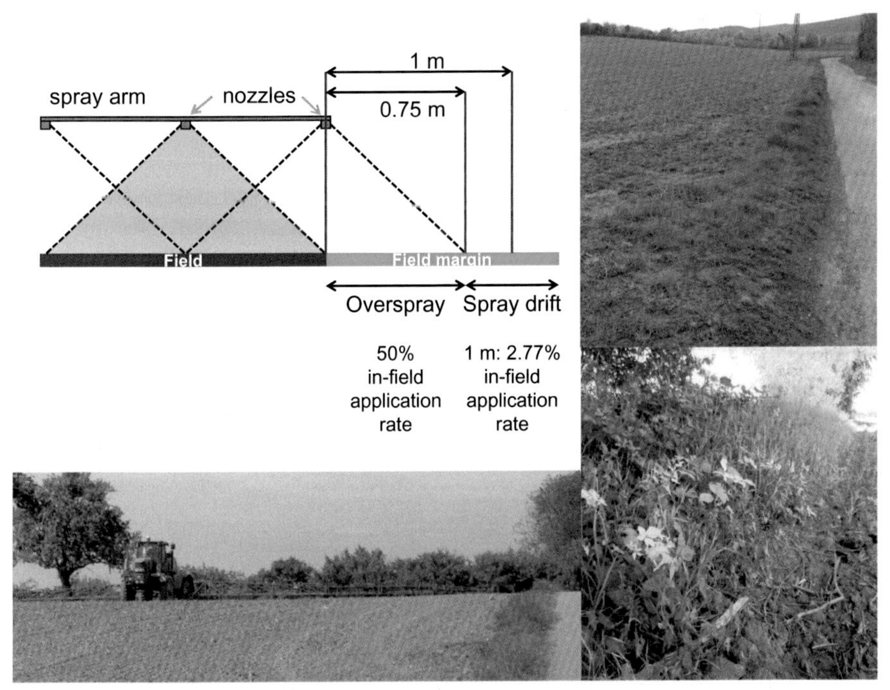

Fig. 8.2 Pesticide input via overspray and spray drift in cereal field margins. The grey area in the upper left panel illustrates the spray cone of one nozzle.[1] A herbicide application with the spray boom extending toward the field edge is in the lower left. The effects of herbicide application with yellow or white leaves of wild plants can be observed in the field margins (right).
(Images: C. Bruhl.)

length.[2] The precise impact of overspraying and the following drift in the first meter of field margin is highly relevant; it can affect plant composition in the narrow field margins of Europe. The overspray can reach 50% of the recommended application rate at the field edge. A mitigation option was recently suggested in Germany by the Federal Ministry of Food and Agriculture: special edge nozzles.[3] Whether this measure will be accepted and used successfully is not yet known.

Spray drift

Herbicides can cause serious injury to neighboring crops and wild plants when applied improperly. Spray drift occurs when small droplets emitted from a ground or aerial sprayer are transported by the wind.[4, 5] Spray drift values as 95th percentiles of collected data are used for environmental risk assessment.[6] According to this measure, 2.77% of the field application rate is present at 1 m for an arable crop but at 20 m for an early fruit orchard application without leaves. Differences between two- and three-dimensional sprayers are evident, and spray drift is therefore generally higher in permanent crops such as vineyards and orchards. With aerial application, similar levels of drift are reached at greater distances downwind. Averaged over a wide range of conditions, approximately 2% of the field-applied rate with oil-based applications and 1% with water-based applications were deposited 500 m downwind of a cotton field boundary.[7]

Spray drift depends on droplet formation and weather conditions. The highest potential is when the weather is hot, dry, and windy. Wind speed and direction during and after application determine the drift pattern.[8] Generally, herbicides and other pesticides should not be sprayed when the wind speed is more than 5 m/s (18 km or 11 miles/h), according to the Good Agricultural Practices recommendation in Germany.[3] If wind speed or direction changes during application, the buffer size (distance to the adjacent field or noncrop area) or location should be adjusted or the application stopped. Drift is also influenced by temperature inversion, occurring when warm air rises and cool air settles near the ground. With warm above cool, there is no mixing, and droplets are not dispersed but stay in a concentrated mass, move with any subtle airflow, and may land off-target. Inversions typically start at dusk and break up with sunrise because of vertical mixing.

A practical measure to mitigate and reduce herbicide drift is the use of drift-reducing nozzles because the nozzle directly affects the size of the spray droplet. Often, herbicide applications require the use of specific nozzles that will produce a coarse- or medium-sized droplet. Coarse droplets resist drift. Farmers are commonly required to have their spray equipment regularly inspected for blockage or wear and calibrated. In addition, the boom height of the spray equipment is a factor, with a higher boom increasing drift.

Herbicide/pesticide applicators must be adjustable to particular circumstances because the user is legally responsible for spray drift in neighboring crops. Besides potential effects on nontarget plants and wildlife, of course, the health and safety of the user and people nearby must be considered.

Runoff

Herbicides can be transported off crop fields in surface or subsurface water flow. An experiment evaluated the offsite transport of glyphosate through retention, leaching, and runoff under rainfall simulation.[9] Eighty-eight percent of the applied glyphosate was retained in the upper 9 cm of the surface soil layer, and leaching was negligible compared to runoff (4%) and spray drift (7%). Water erosion could lead to a higher loss of glyphosate and AMPA, with an estimated maximum export of 9.8 kg/ha/year in soils with low to medium herbicide content and of 47.7 kg/ha/year in soils with higher content. Permanent crops and cereals present the highest exports through water erosion.[10] The solubility of the herbicide is most important to describe its potential for water transport. Compounds with high (pyrithiobac sodium) to moderate solubility (diuron and metolachlor) were studied in simulator plots.[11] Pyrithiobac sodium was transported >90% in the water phase in rainfall and irrigation runoff. In contrast, diuron was transported 55% in sediment at 2 days after treatment and about 85% in sediment at 34 days.

Long-range transport

Transport also occurs when products volatilize or evaporate. Certain compounds (e.g., clomazone, 2,4-D, and dicamba) move by vapor drift, and the volatility of some products increases at temperatures above 25°C. Some can move into the atmosphere attached to soil particles and be deposited in rainfall, reaching concentrations high enough to injure plants.[12, 13]

Glyphosate is supposed to degrade rapidly in the atmosphere by photochemical oxidation,[14] but results from air and rain analyses indicate that it and AMPA can persist and be washed out and redistributed by rain (wet deposition). Glyphosate residue levels across Europe were found in areas with low to medium glyphosate or AMPA content in soil (0.05–0.5 mg/kg). Estimated glyphosate and AMPA removal by wind erosion reaches 1.9 kg/ha/year, and in some areas exceeds 3.0 kg/ha/year.[10] Wind tunnel experiments show that the contents of glyphosate and AMPA were particularly high (>15 and > 0.6 μg/g) in the finest soil particle fractions (<10 μm), which can be inhaled by humans directly.[15] Both compounds were often (>50%) detected in air samples collected from agricultural areas in the United States.[16] AMPA, the degradation product of glyphosate, is formed in soil; therefore, particle-bound wind transport through erosion processes is the only source. In a comprehensive environmental survey conducted in the United States,[17] glyphosate and AMPA were present in more than 70% of the precipitation samples, at maximum concentrations of 2.5 and 0.5 μg/L. In Europe, lower frequencies of detection in rainwater were reported–glyphosate in 10% of samples, AMPA in 13%–but with higher maximum concentrations, 6.2 and 1.2 μg/L.[18]

In Germany, 116 sites across the country were recently monitored for 500 chemicals using passive samplers, filters from ventilation systems, bee bread (a mixture of pollen and honey) out of beehives, and samples of tree bark.[19] Within 1 year (2019), 138 active ingredients from agricultural pesticides were found, among them the herbicides glyphosate, metolachlor, pendimethalin, terbuthylazin, prothioconazoldesthio (metabolite of prothioconazols), dimethenamid, prosulfocarb, AMPA, flufenacet, aclonifen, chlorflurenol, and MCPA. Glyphosate was found in 100% of the samples. Thirty percent of the detected pesticides had no current market approval or had never been approved in Germany.

Long-range transport results in deposition and detection in remote areas far from any agricultural activity. Herbicides are, for example, transported from the agricultural areas in Northern Italy to alpine glaciers, where terbuthylazine was measured in the meltwater at 1.98 ng/L.[20] Simazine was most frequently detected with eight other pesticides inside Pacific chorus frogs (*Pseudacris regilla*) collected high in California's Sierra Nevada mountains.[21] These compounds originated in the Central Valley agricultural area, where use is high in the cultivation of vegetables for distribution throughout the country. The effects of this agrochemical contamination on the biological processes of *P. regilla* and other frog species are unknown. Herbicides were also detected in tropical mountains in Costa Rica. Specifically, elevated concentrations of dacthal on the volcanoes Barva and Poas in the central cordillera, downwind of the extensive banana plantations of the Caribbean lowland, indicate atmospheric transport and wet deposition at high altitudes.[22] In all these montane areas, deposition in rain is occurring along the slopes, and degradation is slowed by the colder temperatures at high altitudes, resulting in accumulation.

Long-range transport occurs on a global scale, from temperate cropping areas to the Arctic. Monitoring in air, water, sediment, and biota detected, among 10 current use pesticides, the herbicides dacthal and trifluralin.[23] The authors conclude that a more systematic approach is needed to assess whether other pesticides might be accumulating in the Arctic, and to assess whether that has biological significance or results in risks for human consumers. Analysis of a Svalbard ice core detected dacthal and trifluralin frequently and also recorded metribuzin.[24] Dacthal, metribuzin, pendimethalin, quizalofop ethyl, triallate, and trifluralin were the most frequently recorded in the last decade in the Arctic.[25] Although concentrations were low, only a few measurements in this ecosystem are available, and potential effects are poorly understood in this environment where organisms might be especially sensitive to additional stress.

Plant diversity and food webs

Between 1993 and 2019, about 100 publications addressed the association between herbicides and plant diversity, more than half of which considered weed communities. Herbicides and food webs were the focus of more than 100 publications between 1986 and 2019, though the majority studied aquatic environments. Very little is known about how herbicides affect the nutritional quality of crop plants and the competitive interactions between weed species. The effects of herbicides on plant competitive interactions, and the effect of intraspecific competition on plant responses, might be such as to merit inclusion in the ecological risk assessment of this class of compounds.

Diversity of target plants or "weeds"

Mechanization and intensification of agriculture, including the use of herbicides, have major roles in determining the composition, diversity, and abundance of weed flora. Applying herbicides is a common method to maximize crop yield by suppressing the growth of unwanted wild species competing for the same resources (space, light, nutrients). In conventional agriculture, herbicides are now used in many arable crops at least once per season in a pre- or postemergent application aimed at eradication (Fig. 8.3). There is evidence that weed flora in Europe have changed over the past century, with many species declining in abundance with a few increasing.[26] There is also strong evidence for a decline in the size of arable weed seedbanks. A 50-year monitoring in Danish arable fields from 1964 to 2014 showed that the number of seeds in the bank varied but the species number was reduced by 50%.[27] Relatively few species dominated and were able to adapt to modern agricultural practice.

In recent decades, a decline in the diversity and abundance of wild plants in agrosystems has been reported in Europe and North America.[28, 29] This is a worrying trend, as these systems harbor a significant proportion of overall plant biodiversity, especially in central Europe.[30] One cause is the increased use of herbicides.[28] In a pan-European study of nine areas ranging from Spain to Sweden, weed density was

Fig. 8.3 Eradication of "weeds" using herbicides in a cereal field (upper left and right) and a vineyard (lower left). No weeds are visible in a maize field in May (lower right). (Images. C. Bruhl.)

measured in 30 cereal fields. In low-intensity cultivation fields, more than 20 species were recorded; in high-intensity fields, the number fell below five, with some devoid of weeds. Of the 13 agricultural intensification variables, herbicide application frequency (but also field size, insecticide application frequency, and amounts of fungicides) was identified as a significant factor for the observed reduction in weed density.

Weeds grow on the cropped land that constitutes the majority of the land area in an agricultural landscape. They grow in agricultural fields. This represents 22% of the land area in Europe, reaching more than 30% in Germany and France.[31] Therefore, weed species are, just by area contribution, key components of the food web in agricultural landscapes, providing food to many organisms from insects (including honeybees and wild bees) to vertebrates.

Diversity of nontarget plants

Plants growing outside the crop field are called "nontarget" with respect to herbicide application. This distinction is not related to a species but to the place where it grows. The same species identified as a weed in the field is a nontarget or wild species elsewhere.

Agricultural landscapes are typically land use mosaics, with fragments of seminatural habitat including pastures, grasslands, and forest interspersed within a matrix of arable fields. In contrast to the heavily managed environment of crop fields, where growers actively minimize plant diversity to enhance production, habitat fragments are the reservoir of plant biodiversity in many agricultural landscapes.[32–35] The nontarget plants occurring at these sites need to be protected in view of their functions in ecosystems, as described in the plant protection goals published by the European Food Safety Authority.[36] Plant biodiversity supports several ecosystem services that are crucial for sustainable agriculture, including pollination and biological pest control, but this diversity is at risk from herbicides applied in nearby crop fields.[37] Overspray and spray drift are the principal sources of exposure of nontarget terrestrial plants adjacent to crop fields.

Risk assessment for nontarget plants is based on single species phytotoxicity tests measuring direct effects on seed germination and plant vegetative vigor.[38–40] This approach does not reflect relevant interactions between plant species in terrestrial ecosystems. The current risk assessment scheme is based on endpoints measured at the organismic level, and the assessment of ecological effects relies on extrapolation from one species to another or from a single species to a community. Most importantly, the required single-species tests do not allow addressing of interactions between species (e.g., intraspecific competition for light, water, or nutrients). A current review shows that potential higher-tier approaches for terrestrial nontarget plants are limited.[39] Sixteen studies were found that assessed the effects of herbicides on nontarget plant communities by performing microcosm, mesocosm, or field analysis. These showed that microcosms might provide useful data and help to reduce uncertainties associated with single-species tests, but development and standardization are needed. More field experimentation was suggested to establish the required baseline knowledge on the effects of drift onto natural plant communities and to compare data generated in tiered testing with information obtained from natural systems. This would allow the establishment of science-based uncertainty factors to be used in the environmental risk assessment of herbicides.

Sublethal doses of glyphosate and metsulfuron methyl can also alter the competition between nontarget species (*Centaurea cyanus*, *Silene noctiflora*) because they respond differently to herbicides.[41] Direct effects can also lead to reduced seed germination (see Chapter 7), causing a shift in species composition and succession of the vegetation. In the long term, susceptible species disappear and tolerant species gain abundancy.

Sublethal and indirect effects that change interactions between plant species take a longer time to become obvious and measurable. Study durations restricted to one growing season will in most cases not detect them. Data based on multiple years are needed and provide a better estimate of potential effects. In a field study addressing realistic herbicide inputs to a surrogate field margin, the effects of a herbicide on plant community composition became stronger over time and were only changing the communities significantly after 3 years.[42] In addition to effects of direct toxicity on single species, indirect effects such as resultant changes in shading come into play. The elimination of keystone species in grassland communities, due to higher sensitivity, can drive these shifts. In the mentioned field study, the common rattle (*Rhinanthus alectorolophus*) was specifically sensitive to the sulfonyl-urea herbicide product Atlantis.[1] The hemiparasitic plant decreases the growth of its grass hosts. Therefore, the reduction of the common rattle leads to higher and denser grasses and a reduction of light availability for herbal species, resulting in community composition shifts. Consequently, distinct communities could be identified after 3 years, revealing the separate effects of fertilizer and herbicide and their combination. The loss of species and effect on frequencies caused a significantly lower species diversity in these treatments than in the control plots. Herbicide treatment also significantly reduced the biomass of plants, by more than 50% in some years. It is important to note that the field experiment used a realistic herbicide input of 30% of the recommended field rate relevant for narrow field margins (Fig. 8.4).

In an 11-year experiment on seminatural plant communities, the glyphosate formulation Roundup Bio and fertilizer were applied at different rates and combinations.[43] Glyphosate led to a decrease in the average specific leaf area and canopy height at the community level and in species richness. However, the phylogenetic diversity of plant communities increased when herbicide and fertilizer were applied together, likely because functional traits facilitating plant success in those conditions were not phylogenetically conserved. The authors conclude that predicting the cumulative effects of agrochemicals is more complex than anticipated due to their distinct selection of traits.

Pollinators

More than three-quarters of the leading types of global food crops rely to some extent on animal pollination for yield and/or quality. Pollinator-dependent crops contribute to 35% of global crop production volume.[44] Indeed, pollination provides

Fig. 8.4 Effects of realistic off-field herbicide inputs on a plant community. A randomized design was used to evaluate herbicide, insecticide, and fertilizer effects (upper left). The flower density of the common buttercup was reduced in herbicide plots (upper right). The plant community composition was shifted toward lower species numbers and reduced flowers (control plot in front and herbicide plot in back after 3 years, lower left). The flowers of the common buttercup were affected by the herbicide and were smaller and less attractive to insects (left compared to control flowers, lower right).
(Images: C. Brühl.)

several ecosystem services such as enabling crop and honey production and regulating weeds. In agricultural landscapes, flowers are provided by mass-flowering crops such as oilseed (rape, sunflower) and fruit orchards, by weeds in annual crops, and by plants in hedgerows and field margins. Wild plants in seminatural habitats are important resources for flower-visiting insects, which provide a natural pollination service for crops.[45] A reduction or change in the availability of floral resources due to herbicide exposure can have a negative impact on flower-visiting insect populations (Fig. 8.5). In Northern Europe, the decline of wildflowers and the consequent shortage for flower-visiting insects is a major concern and one of the drivers of historical pollinator decline.[46, 47] Hence, a change in flowering phenologies and the floral availability of wild plants secondarily affect pollinator populations and, ultimately, crop pollination.

Records of bees, hoverflies, and plants were analyzed over multiple decades in Britain and the Netherlands.[46] The study revealed declines (post- versus pre-1980) in local bee diversity in both countries but divergent trends in hoverflies. In both countries, about 30% fewer species account for half the post-1980 records.

Fig. 8.5 Insect plant interactions. If plants are killed by herbicides, the nutritional basis for the associated insects is removed and alterations in their populations are inevitable. Pollinators of the common buttercup are in the upper row and other plants in the lower row. Herbivorous insects of different life stages are in the center row.
(Images: C. Brühl.)

Pollinator declines were most frequent in habitat and flower specialists, in univoltine (one generation per year) species, and in nonmigrants. Flowering plants declined relative to other plant species in parallel. Although the authors do not mention herbicides as a causal factor for the reduction of flowering plant diversity, they probably were.

Sublethal doses of dicamba approximating particle drift events can delay, reduce, or prevent the flowering of plant species found in agricultural landscapes and lead to reduced visitation by pollinators.[48] To understand the influence of a herbicide on pollinators, several sublethal, drift-level rates ($\approx 1\%$ of the field application rate) were applied, and plant flowering and floral visitation on alfalfa (*Medicago sativa*) and the common boneset (*Eupatorium perfoliatum*) by pollinators was evaluated. Dicamba drift delayed the onset of flowering and reduced the number of flowers of each plant species but did not affect pollen quality (protein concentrations). However, plants affected by drift were visited less often by pollinators. Because plants exposed to sublethal levels of herbicides may produce fewer floral resources and be less frequently visited, drift can lead to disturbances of plant and beneficial insect communities in agricultural landscapes.

The application of a sulfonylurea product (Atlantis) at a realistic overspray rate of 30% for a narrow field margin suppressed flower intensity and reduced the density of the common buttercup (*Ranunculus acris*) by 85% (Fig. 8.4). Many flower-visiting insects may be affected by a reduced density of pollen plants. This food source decrease could be especially severe for specialist pollinators such as the large scissor-bee (*Chelostoma florisomne*), which depends entirely on *Ranunculus* pollen (Fig. 8.5). However, the pollen of *R. acris* is consumed by many insects, and a total of 117 flower-visiting insects on this plant species alone was recorded.[1, 49]

A very prominent pollinating insect species in North America is the monarch butterfly (*Danaus plexippus*), whose populations are in steady decline. Although there are likely multiple contributing factors, such as climate and resource-related effects on breeding, migrating, and overwintering populations, the key landscape-level change appears to be associated with the widespread use of genetically modified herbicide-resistant crops that dominate the summer breeding range. A review showed that glyphosate-tolerant soybean and maize have enabled the extensive use of this herbicide, generating widespread loss of milkweed (*Asclepias* spp.), the only host plants for monarch caterpillars.[50]

Not only mass-flowering crops such as oilseed rape or sunflowers but also flowering weeds are attractive to visiting insects. Most weed species are associated with generalist insect pollinators.[47] Yet, due to herbicide use in arable fields, many insect-pollinated weed species are rare or declining. In Europe, this is also because they depend on flower visitors for the seed set.[46, 51, 52] Especially rare species have difficulty attracting insects, accelerating their decline,[53] and it is suggested that the decline in pollinators is highly linked to weed decline.[54] The link seems logical, as the area for weeds, the agricultural crops, represents a major part of the European land cover and has been targeted for decades by herbicides. Moreover, the herbicide resistance of some weed species leads to increasing herbicide use on crops,[55–57] resulting in a further decrease of general weed density. Once flowering weed density is below a threshold, pollination is reduced because of the reduced attraction of insects, leading to a reduction in the seed set and therefore further decline. The decline of the food source also affects the flower-visiting insects; therefore, parallel declines in both groups are to be expected and are observed.[46]

Agricultural practice, especially the use of herbicides, can have detrimental effects on weed diversity and bees. In a study in Germany in three regions with seven pairs of conventionally and organically cultivated wheat fields along a gradient from heterogeneous to homogeneous landscapes, bee and plant diversity was recorded.[58] Higher bee diversity, flower cover, and diversity of flowering plants were recorded in organic fields where herbicide use is not allowed, compared with conventional fields. Bee diversity was related to both flower cover and the diversity of flowering plants.

Food web interactions and biodiversity

Plants form the energetic basis of terrestrial ecosystems, hence herbicide effects on the diversity, abundance, and fitness of plants inevitably affect all organisms depending on these plants. Arable weed species and wild plants in the seminatural habitats of agricultural landscapes support a high diversity of insect species; therefore, reductions in the abundance of their host plants may affect the associated insects. Some insect groups and farmland birds have shown marked population declines in recent decades, and these declines are associated with changes in agricultural practices, including the use of herbicides.[28] Certainly, reductions in food availability in winter and for nestling birds in spring are implicated, notably in the grey partridge (*Perdix perdix*). Thus, wild plants as weeds or nontarget plants are supporting biodiversity in agroecosystems (Fig. 8.6).

The reduction of plant density and diversity related to herbicide application or drift was documented in the field for weed communities and in neighboring seminatural habitats such as field margins and hedges. Herbivorous insects occur in many groups, such as caterpillars (of butterflies and moths), crickets, grasshoppers, and bugs. In addition to being food for them, these plants provide cover and reproduction sites. An analysis of the importance of European weeds as food for phytophagous insects showed that the number of insects recorded varies markedly among weed species. Some weeds had very few records; for example, just one insect species was recorded for the birdeye speedwell (*Veronica persica*) with more than 70 for the common chickweed (*Stellaria media*).[26] The authors noted that several insect species are dependent on particular weeds to complete their life cycle, so these weeds are of specific importance.

Autumn herbicide application on flora and invertebrate fauna was investigated in winter wheat (*Triticum aestivum*) crops over 3 years. This showed that untreated plots had greater floral cover and weed diversity and significantly higher numbers of many invertebrate taxa, notably those that are important in the diets of farmland birds.[59]

In a 7-year field experiment, soil-associated arthropods were positively correlated with weed biomass and negatively with crop plant biomass; also, community structures of arthropods and plants were correlated.[60] Herbicide treatments facilitated high numbers of some taxa in maize when this coincided with plot invasion by herbicide-tolerant weeds. A recent paper synthesized data from seven UK studies over a period of 18 years, providing information on arable weeds and invertebrates in winter wheat.

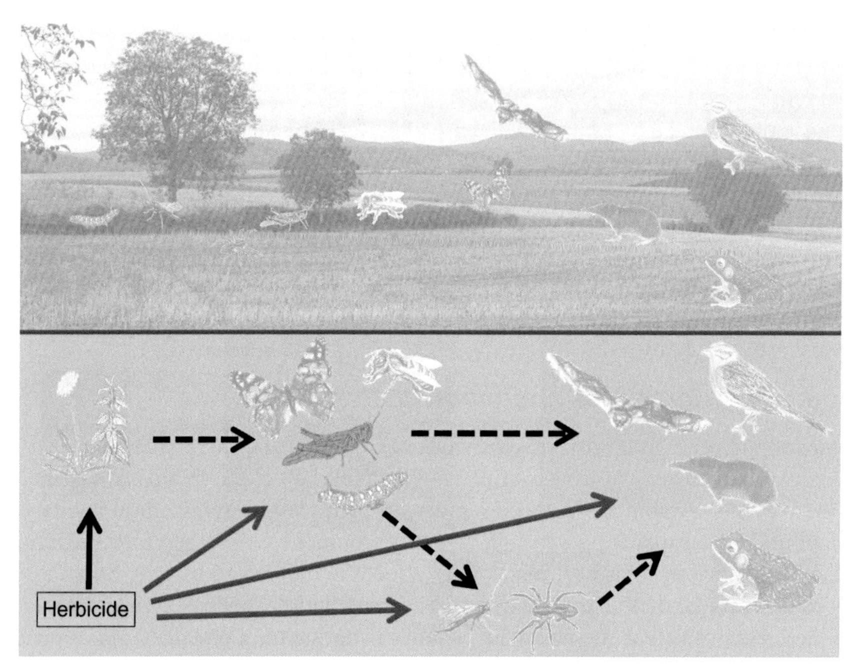

Fig. 8.6 Direct (*solid arrows*) and indirect (*dotted arrows*) effects of herbicides on food web interactions in an agroecosystem. Black indicates a strong effect, grey a weaker effect.

It focused on the role of arable weeds in supporting invertebrate populations and the selected ecosystem services they deliver.[61] Both phytophagous and predatory invertebrates responded to weed cover, but to different degrees. Phytophages showed a stronger positive relationship with weed cover than the predators because they rely on the resources provided by the weeds, whereas the predatory species response is likely to be mediated by its prey.

Herbicides affect not only plant density and diversity but also plant nutrient levels and hormone pathways used in defense, both of which may influence susceptibility to herbivorous insects.[48] For instance, aphids perform better on herbicide-stressed plants. The interaction effects of herbicides on plants and on related insect performance have been poorly explored. The application of a combination of 2,4-DB and MCPA (Clovermax) negatively affected herbivory in polyphagous grasshoppers (*Pseudochorthippus parallelus*) while a higher grass diversity generally increased herbivory.[62]

These indirect effects are the consequence of herbicides modifying basic biochemical processes in plants, such as electron transport and amino acid synthesis. For example, glyphosate inhibits an enzyme of the shikimate pathway, and sulfonylurea herbicides inhibit acetolactate synthase. Both enzymes are necessary for the synthesis of specific amino acids in plants. Amino acids and proteins are crucial to the development of herbivorous insects (and other organisms), and therefore herbicide-treated plants may not meet their dietary requirements. In addition, herbicides cause stress in plants and can induce genes involved in plant defense, triggering the release of an entire battery of toxic compounds.[63] Many plants can synthesize secondary metabolites as a stress reaction, such as glucosinolates, a defense mechanism that may deter or be toxic to herbivores.[64, 65] Some of these plant secondary metabolites, such as pyrethrum, are also used as insecticides.[66] Therefore, even if a plant is not directly killed by a herbicide, the herbicide might trigger its defense mechanisms and the plant might become unsuitable or less nutritious to herbivores.

The few studies that focus on the effects of herbicides on host plant quality have yielded mixed results. The observed responses include no effects,[65] increased numbers of aphids (*Rhopalosiphum maidis*), and heavier corn borer pupae (*Ostrinia nubilalis*) on herbicide-treated maize plants[67]; reduced longevity of caterpillars of the soybean looper (*Pseudoplusia includens*) on soybean plants[68]; and extended development times and reduced female egg numbers of the green dock beetle (*Gastrophysa viridula*) on the broad-leaved dock *Rumex obtusifolius*.[69] In another study, black bindweed (*Polygonum convolvulus*) plants were treated with sublethal rates of chlorsulfuron following the introduction of the larvae of the leaf-eating beetle *G. polygoni* (Coleoptera).[70] The larvae showed up to 80% mortality on those plants receiving the highest dosage. Although chlorsulfuron was not directly toxic to the larvae, the authors argue that the pesticide might enhance a herbivore-induced plant response. Aphids and other species living in meristematic tissues appear to respond positively to herbicide-treated plants, whereas foliar feeders (and other feeding guilds) tend to be negatively affected.

To study the effects of herbicides on the host plant quality of a moth, three plant species, English plantain (*Plantago lanceolata*), greater plantain (*P. major*), and common buttercup (*R. acris*), were treated with sublethal rates of either a sulfonylurea (Atlantis WG, Bayer CropScience) or a glyphosate (Roundup LB Plus, Monsanto) herbicide, and the development of caterpillars of the cabbage moth (*Mamestra brassicae*) feeding on these plants was observed.[71] Of the six tested plant-herbicide combinations, one (*R. acris* and the sulfonylurea) resulted in considerable effects. Caterpillars had a significantly lower weight after 3 weeks, increased time to pupation (approximately one-third longer), and increased overall development time (6 days longer) compared with larvae that were fed unsprayed plants (Fig. 8.7).

Fig. 8.7 Caterpillars of the cabbage moth (*M. brassicae*) 3 weeks after feeding on the common buttercup (*R. acris*) untreated control or treated with 10% of the field rate of a sulfonylurea-based herbicide (Atlantis WG).[71]

These observed effects may have been caused by a lower nutritional value of the host plants or increased concentrations of secondary metabolites that are involved in plant defense. The results suggest potential risks to herbivores feeding on host plants treated with sublethal rates of herbicides. It is difficult to evaluate the impact of this indirect effect in nature because the effects of herbicides on host plant quality appear to be species-specific. For example, in Germany, there are approximately 3500 Lepidoptera species, 4200 flowering plant and fern species, and more than 580 registered herbicide products. This situation results in a vast number of possible combinations, making it difficult to estimate the overall risks of one or more herbicides to herbivorous insects. Relationships between herbicides, plants, and insects were mostly studied in-field, where herbicides reduce weed cover significantly and insects are affected due to a loss of food plants. The subtler changes in off-crop, nontarget plant communities as highlighted above are rarely addressed in a food web approach.

The indirect effects of herbicides by eliminating or decreasing plant food or its quality lead to reductions in insect abundance and biomass. Insects, in turn, are substantial food sources for species at higher trophic levels such as birds, bats, and other mammals.[72, 73] The decline of farmland birds was noticed for decades and is now recorded thoroughly. The European Union farmland bird index, including 39 selected common species, declined by 17% from 2000 to 2018, whereas an increase of more than 7% occurred for forest birds.[74] The indirect effects of herbicides in industrialized agriculture represent one of the causal factors. Several reviews and comprehensive studies on indirect herbicide effects with continuing impacts on populations of farmland bird species are available.[75]

The food web effects of herbicides were already studied almost 50 years ago for their consequences on the grey partridge (*P. perdix*), which declined markedly in the United Kingdom and all around Europe. The British population had been dropping since 1945; the key factor was the mortality of chicks. The mortality of partridge chicks was then suspected of being related to insect availability because the chicks of this seed-eating bird require insects as an essential protein source for the first weeks of life. The hypothesis was tested in a remarkable field experiment in the early 1980s in Hampshire.[76] A 6 m strip at the edge of cereal fields was either sprayed with herbicide and in some instances also fungicide or left unsprayed as conservation headland. The insect abundance and breeding success of partridges were recorded. Unsprayed areas showed significantly higher abundances of true bugs (Heteroptera), leaf beetles (Chrysomelidae), and weevils (Curculionidae) than sprayed areas. Sawfly larvae (Hymenoptera) and butterfly/moth (Lepidoptera) caterpillar abundance did not vary significantly, though the method of recording them quantitatively (sweep netting) was unsuitable. The grey partridge brood size (number of chicks) was reduced by 80% in sprayed compared with unsprayed headlands. The density of male grey partridges was negatively related to the number of herbicide applications per field and positively related to the mean number of dicotyledonous weed species.[77]

Another study focused on insects eaten by ring-necked pheasant and grey partridge chicks and on beneficial arthropods that prey on insect pests and weed seeds.[78] Beneficial arthropods and vegetation-dwelling chick-food insects were more common in weedy fields than in the more monocultural fields. The study demonstrates that herbicides do affect arthropods that serve as avian food resources as well as beneficial predators in agroecosystems, and that these effects are most likely mediated by changes in the plant community.

Herbicides not only affect birds by reducing weed and related herbivorous insect densities, but they also affect weed seed production, potentially reducing the availability of food for adult granivorous birds.[73] Relationships between bird feeding densities and seed densities have been demonstrated for several species. Yellowhammers (*Emberiza citrinella*) responded to the provision of supplementary seeds in a field experiment, and greater settling densities were noted the following spring in areas where seed was supplied. The linnet (*Carduelis cannnabina*) population decline coincided with reductions in the abundance of key weed species sensitive to herbicides that provided the bulk of the seeds for chicks. Turtle doves (*Streptopelia turtur*) also feed and may rely extensively on weed seed during the breeding season. Several of the important weed species in the turtle dove diet (e.g., chickweed, *S. media*) have undergone widespread declines on farmland in Britain and are susceptible to the frequency of herbicide use. The densities of singing corn buntings (*Emberiza calandra*) were also significantly lower in fields with high numbers of herbicide and fungicide applications.[77]

Environmental risk assessment (ERA) focusses on environmental effects that can occur in seminatural structures outside agricultural fields. Currently, no ERA for in-field risk is mandated, but the scientific opinion for nontarget arthropods mentions that biodiversity "has to be supported to a certain degree in the in-field areas in order to provide important ecosystem services."[79] The current guideline does not address in-field biodiversity effects at all. Negative effects on biodiversity are therefore deliberately accepted in the cropping area where pesticides are directly applied at biologically effective rates.[80] One has to consider that in an ideal conventional, industrialized cropping system, a farmer aims to eliminate all weeds to guarantee complete resource (nutrients, water, and light) availability to the crop plant. Biodiversity in these systems is therefore ideally reduced to exactly one species, the crop. Additionally, all insects that would potentially feed on this crop are regarded as pests and therefore targeted with insecticides. Because no herbivorous insect species are tolerated (at least in theory), higher trophic levels are also reduced and total biodiversity declines. The ERA required for pesticide regulation in most cases does not address the impact of pesticide use in agricultural fields and does not include food web-related ecosystem effects. This fundamental misconception leads to an ERA scheme and a resulting pesticide regulation that is not protective for biodiversity.[80, 81] If we continue working with the ERA scheme in place, further declines of many groups of organisms such as farmland birds and insects in the agricultural landscape will be inevitable. Understanding the relationships between herbicide use for weed management, biodiversity, ecosystem functions, and agricultural practice is required for developing more sustainable agricultural systems.[82]

For mammals, long-term studies on herbicide effects in species on the population level are lacking. Individuals are negatively affected directly by poisoning, but also indirectly by effects on habitat quality and food availability. Due to the general lack of knowledge about population size and trends, but also about ecological features such as habitat occurrence and diet choice of small mammal species, the actual impact is difficult to estimate. Feeding on wild plant parts and simultaneously needing a high proportion of ground cover to avoid predation makes small mammals vulnerable to indirect herbicide effects. Although few studies addressed this issue, several authors mention the possibility–for example, for the brown hare (*Lepus europaeus*), the common shrew (*Sorex araneus*), the wood mouse (*Apodemus sylvaticus*), and the European badger (*Meles meles*)–by removing plant food resources and changing the microclimate.[75] In a comprehensive review of the impacts on small mammal populations in North American grasslands, including shifts in diet with subsequent decreased survival, lower reproductive success and increasing foraging dispersal have been described.[37]

The population abundance of small mammal assemblages living along a gradient of agricultural land-use intensification has been studied in northeast Italy.[83] Population abundance, species presence, and species diversity were affected by agricultural intensification, herbicide use, and landscape naturalness, but no specific herbicide effect was seen. A study on the habitat selection of small mammals in relation to herbicide application investigated winter wheat fields under different treatments in the United Kingdom.[84] The reduced application of herbicides in experimental plots led to increased floral and invertebrate abundance, and wood mice actively sought those plots in preference to normally sprayed plots. Unsprayed field edges have been found to have positive effects on the presence and diversity of plant species and therewith on small mammal abundance.[85] Brown hares inhabiting agricultural landscapes need species-rich field margins with herbal undergrowth because the diet offered by large crops is too one-sided due to the rapid decline of herbs on arable land, which is derived from the application of herbicides (and fertilizer). The application of 2,4-D reduced foliar cover by forbs and shrubs and thus the diet of prairie dogs (*Cynomys ludovicianus*).[86] However, despite a drastic change in diet composition, there was little evidence for negative effects on squirrels, which remained in good condition and showed no significant difference in activity from those in an untreated area. Decreased abundance and a change in the sex ratio of meadow voles (*Microtus pennsylvanicus*) were found in an experimental study investigating the response of small mammals to vegetation changes after the application of 2,4-D.[87] The herbicide significantly altered the plant community structure and reduced plant species diversity, changing the area into a "monoculture-type" giant foxtail (*Setaria faberii*) habitat. Voles were exposed to food quality differences, which reduced the trapped population by half.

Changes in rodent populations after the application of 2,4-D by reduced coverage of forbs and sages were reported in the late 1960s in North American rangeland.[88] Densities of pocket gophers (*Thomomys talpoides*) and least chipmunks (*Eutamias minimus*) were decreased by altered food availability. In contrast, the widely distributed and polyphagous deer mouse (*Peromyscus maniculatus*) was not significantly affected in density or litter size. The numbers of montane voles (*Microtus montanus*) even increased, benefiting from the increased grass cover that usually followed the herbicide treatment of perennial forb and shrub ranges. Populations of pocket gophers and voles were re-established after a recovery of the forbs.

For both farmland bird and small mammal species, changes in food availability seem to be the most important negative consequence of herbicide application.

Ecosystem functions: Litter decomposition and nutrient cycling

The decomposition of organic matter is an important ecosystem function provided by various soil biota important for nutrient cycling. Herbicides and litter decomposition or nutrient cycling were documented in more than 25 and 100 publications, between 1979 and 2019. Soil organic matter also strongly influences soil water-holding capacity and nutrient retention, and affects herbicide fate and mobility. Organic matter turnover also liberates nutrients for crop growth and plays a key role in greenhouse gas efflux and carbon sequestration.

The effects of herbicides in litter decomposition have been examined in several recent studies. Glyphosate applied at 1.1 kg/ha to vineyard soil stimulated litter decomposition measured by bait-lamina feeding activity.[89] A comparison of feeding activity after treatment with paraquat, simazine, glyphosate, and a glyphosate-terbutylazine mixture on different soil plots without cover crops was also undertaken. It showed that, with the exception of simazine, the herbicides increased feeding activity.

From the review papers available, the emerging picture is one of compound-specific effects on particular soil functions. With regard to herbicides, adverse effects on phosphatase activity by glyphosate,[90] inhibition of nitrification by simazine,[91] and adverse effects on pathogen-antagonistic *Pseudomonas* bacteria by acetochlor and chlorimuron-ethyl[92, 93] are just some of the examples of the potential effects of herbicides on soil health as related to plant nutrition and disease.

The litter decomposition rate in soil was found to be insensitive to glyphosate.[94] However, the stabilization factor of litter in soil tended to increase, suggesting conversion from labile into more recalcitrant compounds.[95] Decomposition was unaffected by glyphosate, glufosinate, or flazasulfuron under field conditions.[96] Three glyphosate-based herbicides (Roundup LB Plus, Touchdown Quattro, Roundup PowerFlex) or their corresponding active ingredients (isopropylammonium, diammonium, or potassium salts) at recommended rates in a greenhouse study showed no effect on litter decomposition or different soil organic matter levels.[97]

In contrast to the previously mentioned studies, recent research found that glyphosate application to a black oat cover crop slowed the breakdown of plant residues.[98] The authors hypothesized that this resulted from a change in the C:N content of the residues during the glyphosate-induced plant senescence, rather than the herbicide-inhibiting decomposer organisms. Glyphosate inhibited the breakdown of broad bean (*Vicia faba*) residues, but accelerated the decomposition of wheat biomass.[99] This further emphasizes the complexity of interactions, and that the impact of glyphosate on plant-residue breakdown may be regulated by litter quality.

Alachlor and metolachlor applied at recommended and high rates delayed the decomposition process, which recovered when there was no residual activity in the soil.[100] The authors stress the importance of compound- and site-specific interactions, which make generalization difficult even within a chemical class.

Glyphosate application led to increased soil concentration of nitrate by 1592% and phosphate by 127% in comparison to no treatment. The authors explain this by reduced plant uptake after herbicide application, and point to potential risks for nutrient leaching into streams, lakes, and groundwater.[95] Glyphosate also adds phosphorus to agricultural landscapes, influencing the accumulation and cycling of phosphorus in soil and nearby surface waters. Yet pesticides have been largely ignored when monitoring anthropogenic sources of phosphorus in agricultural watersheds. Across the United States, the mean inputs of glyphosate-derived phosphorus increased from 160 kg/ha in 1993 to 940 kg/ha in 2014, with values frequently exceeding 2000 kg/ha in areas planted with glyphosate-resistant crops.[101] Compared to fertilizers, this is still a minor source. However, phosphorus inputs from glyphosate use have now reached levels comparable to those from sources for which phosphorus regulations were initiated in the past and require greater recognition in watershed research and management. The soil biology also influences the availability of phosphorus and other elements to crops. Of special importance is the symbiotic association of plants with mycorrhizal fungi. These associations mutually benefit both partners through a flow of reduced carbon substrates from the plant to the fungi in return for other nutrients, especially phosphorus and microelements.

Ecosystem service: Crop disease control

The interaction of herbicides, pathogens, and crop plants has the potential to either increase or decrease the incidence of disease and subsequent yields through a number of mechanisms. Herbicides can affect organisms that cause direct crop damage, including insects, nematodes, fungi, and bacteria (Chapter 7). Less well known are the mechanisms by which herbicides interact with soil processes that suppress the occurrence of disease, despite the presence of disease-causing organisms.[102] Herbicides and biocontrol were addressed in more than 400 publications between 1985 and 2019; however, only 22 address herbicides and crop diseases since 1995. Kortekamp recently reviewed these interactions, and the reader is directed to this review for a more in-depth discussion, particularly with respect to glyphosate and glufosinate.[103] The following is a brief summary of the pertinent findings, and an update.

As with other nontarget organisms, it is difficult to distinguish between direct and indirect herbicide effects on crop diseases. For instance, the impact on plants can also indirectly affect soil microbes by reduced carbon input into the rhizosphere. When plants die following herbicide application, the remaining plant debris provides a resource to support microbial growth and activity. Further, a reduction in plant cover may result in an increase in soil temperature and a decrease in water content, both of which affect rates of microbial activity. Moreover, herbicide impacts on plant community composition may also affect soil biota via the plant-mediated selection of distinct microbial communities.

The effects of glyphosate on crop diseases are most often studied. A review article by Van Bruggen and colleagues refers to, among other things, the promotion of plant pathogenic soil-borne fungi by glyphosate.[104] This is documented, for instance, in an increase in *Pythium* infections in sunflowers,[105] and in a peanut (*Arachis hypogaea*)-corn (*Zea mays*)-soybean (*Glycine* max) rotation system.[106] Generally, *Pythium* fungi are a very diverse group that can (but may not always) cause root rot in several crops.

Fusarium fungal pathogens cause important diseases, such as root or crown rot and head blight, in cereal crops. There are no commercially available cereal cultivars with good resistance, so agronomic practices that reduce these pathogens need to be identified. Several studies report a promotion of *Fusarium* occurrence in soils after glyphosate treatment,[107] especially after application in glyphosate-resistant soybean and maize.[108] Plant pathogens are promoted by the treatment-related reduction of secondary plant constituents (the shikimate metabolism provides important precursors in the synthesis pathway of these constituents) and the resulting increased susceptibility of the crop plants, or by a limited intake of micronutrients.[109] An increased incidence of *Fusarium* sp. deafness was observed in wheat when minimal tillage was applied to glyphosate-treated areas compared to nonglyphosate-treated areas.[110] However, this effect was not observed in conventional tillage and no-tillage.

In greenhouse studies using glyphosate-resistant sugar beets, increased disease severity was observed following glyphosate application and inoculation with *Rhizoctonia solani* and *Fusarium oxysporum*.[111] Thus, the herbicide seemed to reduce the plant's ability to protect itself against pathogens. Even sublethal doses of glyphosate inhibited the expression of resistance in soybeans to *Phytophthora megasperma* f.sp. glycinea,[112] in beans to *Colletotrichum lindemuthianum*,[113] and in tomatoes to *Fusarium* spp.[114] Furthermore, glyphosate applied to the soil increased the disease symptoms caused by *Cylindrocarpon* sp. in grapevines.[115]

It is suggested that glyphosate can upset the balance of the soil microbial community and reduce the innate suppressiveness of the soil to pathogen dominance.[103] For example, glyphosate at recommended field rates inhibited the growth of *Pseudomonads* and indole-acetic acid-producing microorganisms, concomitant with increasing *Fusarium* infection in soybean roots in a dose-dependent manner.[107]

Also important are interactions between soil biota and soil-borne plant diseases such as *Fusarium*[116] and *Sclerotinia*.[117] Earthworms have been shown to affect or control these soil-borne pathogens. If herbicides affect the activity of earthworms as documented (see Chapter 7), this important biocontrol activity of earthworms is compromised.

Besides reports indicating that herbicides inhibit fungal species involved in soil-borne diseases,[103] others found no negative effect of glyphosate on the vegetative growth of several *R. solani* isolates and anastomosis groups.[118] However, the herbicide influenced the production of fruiting bodies of this pathogen.

Herbicides have been shown to affect not only pathogenic microorganisms but also symbiotic fungi such as *Rhizobia* (bacteria inside root nodules of legumes that fix nitrogen). Most data are available for glyphosate. Most soybean (*G. max*) cultivars grown in the United States are Roundup Ready, genetically modified for resistance to glyphosate. In greenhouse experiments, glyphosate-treated glyphosate-resistant soybeans had lower *Rhizobia* root nodule mass, root mass, chlorophyll content, total plant N, and nitrogenase activity than the untreated non-GM cultivar.[119] So, it appears that glyphosate exerts stress to the glyphosate-resistant soybean. However, further research is needed to verify the greenhouse experiment findings and possible yield effects under field conditions.

In a large-scale study including 116 composite samples from vineyards, orchards, and other crops from Austria, agricultural management as well as distinct soil parameters were identified as drivers of the indigenous microbial communities in soils.[120] While the microbiome of vineyard soils maintained a higher stability when herbicides were applied, orchard soils exhibited drastic shifts within community composition. Herbicides decreased the abundance of the bacterial order Chthoniobacterales and antagonists of phytopathogens (*Flavobacterium*, *Monographella*). Moreover, the soils of herbicide-treated orchards revealed a significantly higher presence of potential apple pathogenic fungi (*Nectria*, *Thelonectria*).

Several other herbicides have been assessed. Acetochlor at rates of 50–250 mg/kg altered the structure of soil fungal communities, with a temporary increase in pathogens and a reduction in common nonpathogens.[121] Experiments with the plant pathogenic fungi *R. solani* show that chlorsulfuron at the equivalent of 2.5 g/ha significantly increased root disease caused by *R. solani* in wheat and barley, but did not increase the incidence of the disease take-all, caused by *Gaeumannomyces graminis* var. *tritici*, in wheat.[122] Sublethal doses ($<20\%$ of recommended rate) of imazamox and propoxycarbazone-Na reduced barley growth and increased *R. solani* disease symptoms.[123] These results suggest that farmers using sulfonylurea or other acetolactate synthase-inhibiting herbicides should monitor previously treated crops for disease symptoms.

The following major herbicide effects may lead to increased disease in crop plants.[103, 124, 125]:

- Reduction in the biochemical defenses of the host against the pathogen.
- Reduction of the structural defenses of the host.
- Stimulation of increased exudation from host plants.
- Stimulation of pathogen growth and/or production of chemicals that damage the plant.
- Inhibition of microflora competing with potential pathogens.
- Interference with rhizosphere microbial ecology by enhancing the population and/or virulence of some phytopathogenic microbial species in the crop rhizosphere.
- The as yet incompletely elucidated reduction in the uptake and utilization of nutrient metals by crops.

Many of the reported effects of herbicides on plant diseases were detected in field studies that do not allow the determination of whether the observed effect is due to a direct herbicide-pathogen interaction or to an indirect effect of making the plant less or more resistant to the pathogen.[109] The particular effect is a function of many factors, including the herbicide class and its formulation, the disease species, the plant species, the timing of herbicide application and infection, and environmental factors. These secondary effects of herbicides need to be addressed to fully understand their environmental toxicology implications and for the success of integrated pest management.

Global change and other environmental stressors

Herbicide application stresses ecological systems by reducing food plants and plant cover and interacting with other stress factors. Therefore, some ecotoxicologists state that ecotoxicology should not be seen as a subdiscipline of toxicology but rather as a case of stress ecology.[126] Nevertheless, interactions between effects of herbicides and environmental factors are rarely investigated. The results of herbicide effects in conjunction with global change were published in fewer than 40 studies in the last 10 years. Nine studies have addressed the interaction of herbicides and microplastics since 2015.

The ERAs of pesticides are based predominantly on the results of laboratory studies where test organisms are exposed to a range of concentrations of single compounds. This approach is useful for the generation of dose-response relationships and the derivation of toxicity data such as the concentration causing 50% impairment of a life history trait, in most cases mortality. In such laboratory experiments, the test organisms are kept under ideal conditions (temperature, moisture, food, etc.) to optimize performance in the control condition and isolate the effects of the chemical in question. However, in their natural settings, organisms rarely experience optimal conditions; for most of their lifespan, they are forced to cope with suboptimal conditions and frequent exposure to severe environmental stress. These added environmental stressors may or may not alter the effects of herbicides.

Among the most often investigated aspects of global change is whether a climate-change-induced temperature increase will alter herbicide effects on nontarget organisms. It has been suggested that climate change itself does not strongly affect amphibians, but rather acts in combination with biotic and abiotic factors to increase their effects.[127] However, a phenological shift of some amphibian species toward earlier reproduction due to climate change might expose them less to GBHs than to those that do not show a shift in reproduction time, although glyphosate applications changed over time and might also change in the future.[128, 129]

Pesticide toxicity to various organisms at above-optimum temperatures has been investigated in several studies.[130] This is important because the seasonal pattern of pesticide application during the summer months makes exposure under heat stress a realistic scenario. However, most studies were conducted with insecticides, and none mentioned interaction with herbicides. No study with herbicides is available for freezing temperatures, and none have investigated interactions between herbicides and drought.[130] It could be expected that organisms that had been subjected to previous drought would be much more sensitive to herbicides than control organisms, as was shown for the insecticide lindane on springtails (*Onychiurus quadriocellatus*).[131]

Studies on interactions between climatic parameters and the toxicity of herbicides have been conducted mainly in aquatic environments. For example, temperature modulated atrazine toxicity in microalgae.[132] Interactions between GBH concentrations (Roundup LB Plus at 0.0, 0.5, 1.0, 1.5, or 2.5 mg AI/L) and temperature (15 °C versus 20 °C) altered egg development in European toads (*Bufo bufo*), with more pronounced effects at the lower temperature.[133] In another instance,

Roundup PowerFlex interacted with temperature to produce tail deformation and increased mortality in the tadpoles of common toads.[134] It seems likely that temperature is also affecting herbicide toxicity in terrestrial organisms.

Climate change will certainly interact with the spatial distribution and effects of herbicides in nature. Models show that climate change leads to increased pesticide and herbicide use in Europe.[135] Elevated water temperature may change the metabolite pattern of herbicides via alterations in biotransformation processes, and changes in precipitation may result in changes in volatilization and deposition. Global warming is expected to affect the toxicological potency of herbicides because most studies on the combined effects of elevated temperature and pesticide exposure have revealed the synergistic action of both factors.[130] Herbicide interactions with global warming will probably influence the direction in which selection acts upon biota, a factor that will be particularly problematic for populations or species living at the edge of their physiological tolerance. Further problems in a warming world may result from temperature interactions with the metabolic rates of heterothermic organisms and, for endocrine-disruptive compounds, with physiological processes involved in temperature-dependent sex determination, as is known for reptile species.[136] In addition, changes in the geographic range and incidence of many infectious diseases that may be fostered by pesticide-exerted immunotoxicity have been predicted.[137] Higher-level herbicide effects, such as changes in plant communities, will probably interfere with the effects of global change on biodiversity and thus affect ecosystem function.

Moreover, erosion processes can strongly influence the dissipation of herbicides such as glyphosate and AMPA applied with a GBH in agricultural soils. The structural soil state shortly before erosive rainfall can be a key parameter for the distribution of glyphosate and its metabolites. Field rain simulation experiments showed that severe erosion processes immediately after application of Roundup Max can lead to serious unexpected glyphosate loss, even in soils with a high presumed adsorption, if their structure is unfavorable.[138] In one of the no-tillage plots where the herbicide formulation was used, up to 47% of the glyphosate dissipated with surface runoff.

In comparison to other natural stressors, the effects of nutritional conditions on the toxicity of herbicides are poorly studied.[130] Interactions between herbicides and starvation may be complex. On the one hand, there is evidence that exposed organisms may be less resistant to starvation than unexposed while on the other hand, organisms exposed to contaminated food frequently decrease their consumption rate. The latter mechanism allows organisms to avoid excessive consumption of herbicides if there is no alternative food source, and starvation may result. Consequently, increased mortality in animals exposed to contaminated food may occur due to not only intoxication but also starvation.

Another aspect of the effect of food limitation on toxicity is that nutritional conditions during an organism's development may have a profound influence on overall life history characteristics.[139] Even if an organism recovers from transient food deprivation in the early life stages, it may still have a permanent effect on the adult individual, and even on offspring, raising the possibility that epigenetic mechanisms may play a role in determining effects. Despite the limited literature on interactions between the nutritional status of organisms and the toxicity of chemicals, in most cases significant interactions were found. However, the effect of starvation is hard to predict, as it can provoke both an increase and a decrease in sensitivity to pollutants. This, together with the scarcity of hard data, calls for further studies on this issue, which represents a highly relevant ecological scenario.

Very little is known about herbicide effects on greenhouse gas emissions. Butachlor (5–100 mg/kg) dose-dependently reduced CH_4 production in alluvial rice soil, even at the lowest concentration.[140] This is interesting, as metolachlor at 5.7 kg/ha had no significant effect on CH_4 emissions in aerobic shortgrass steppe soil.[141] The effect that herbicides and pesticides have on trace gas production and consumption in agricultural soils is often overlooked. Field and laboratory experiments were used to measure the effects that the commonly used herbicides prosulfuron (0.46 L/ha) and metolachlor (5.7 L/ha) have on trace gas fluxes (CO_2, N_2O, and CH_4) from shortgrass steppe soils.[141] During an initial 1-year study, prosulfuron stimulated N_2O emissions and CH_4 consumption by as much as 1600% and 1300%, respectively. After 2 years, prosulfuron application led to an average of ~50% reduction in the global warming potential from N_2O and CH_4 fluxes while metolachlor application did not significantly affect the trace gas fluxes measured.

By changing the soil biota and functions, herbicides may affect the production of greenhouse gases. A single dose of bensulfuron-methyl or pretilachlor at recommended rates has been shown to reduce N_2O and CH_4 emissions; however, when they were applied together, this effect was absent or reversed.[142]

The use of agrochemicals such as mineral fertilizers and herbicides in agricultural systems may affect the potential of soil to act as a sink for methane. Typically, the effect of each agrochemical on soil methane oxidation is investigated separately, whereas in the field these agrochemicals are used together to form one comprehensive land management system. In a field experiment, the combined effects of multiple fertilizer and herbicide (nicosulfuron, dimethenamide, atrazine) applications on the soil methanotrophic community were examined.[143] While organic fertilizer increased methane oxidation rates, herbicides did not alter it. The methanotrophic community structure was more affected by soil type than by herbicides. In another study, the effects of quizalofop-p-ethyl and bentazone on methane (CH_4), carbon dioxide (CO_2), and nitrous oxide (N_2O) emissions from soil planted with a forage crop (alfalfa, *M. sativa*) were measured over 2 years.[144] Soil CO_2 emissions and the soil uptake of CH_4 increased in both years following herbicide treatment, although CO_2 emissions differed between years. N_2O emission decreased relative to control and showed no significant difference between the years. Overall, bentazone and moreso quizalofop-p-ethyl increased CO_2 emissions, which contributed to a significant increase in greenhouse gas after herbicide application. Because there are hundreds of herbicide products and combinations and a multitude of soil parameters and climatic conditions, the effects of herbicides on greenhouse gases need to be investigated in more detail and addressed in nationwide plans to reduce climate change effects.

The increased use of plastic films and herbicides on agricultural soil leads to the accumulation of plastic debris and herbicide residue in soil. This accumulation has become a serious environmental issue. In a soil incubation experiment, three glyphosate levels (0, 3.6 kg/ha, 7.2 kg/ha) were applied in different settings with microplastics (homopolymer polypropylene powder) addition.[145] An interaction between glyphosate and low microplastic content became evident, leading to the loss of bioavailable carbon and phosphorus. When glyphosate was applied to soils with a high microplastic content, dissolved organic nitrogen was reduced compared to treatments with glyphosate alone. In other experiments it was shown that microplastics can be carriers of the antibiotics amoxicillin and phenol but not for the herbicides atrazine or diuron.[146] Further studies are necessary to disentangle the effects of plastic debris and herbicides.

Adaptations, multiple applications, and interactions with other pesticides

Although herbicides show mainly moderate to low half-lives (see Chapter 7, Table 7.2), residues can have long-term effects on nontarget organisms and their contribution to ecosystem functioning.

An aspect that is rarely studied is the adaptation of terrestrial organisms to herbicides. Such adaptations have been shown, for instance, in earthworm populations (*Aporrectodea caliginosa*) with different chemical exposure histories.[147] Worms originating from 20 years of conventional farming using fungicides (epoxiconazole) showed acclimation to fungicides in their burrowing behavior. Another study suggests that this physiological adaptation is a general response pattern. The exposure history of endogeic earthworms (*A. caliginosa*, *Allolobophora chlorotica*) modified their responses to herbicides.[148] Acute exposure of preadapted and naive *A. caliginosa* to the GBH Roundup Flash (2.5 μg AI/g dry soil) and the fungicide epoxiconazole and their mixture revealed that environmental pre-exposure accelerated the activation of detoxification enzymes (sGST) toward epoxiconazole.

Indications of adaptation have also been documented for cellulolytic microbes. The application of bensulfuron-methyl to virgin soil inhibited cellulolytic microbes, but did not do so in soil that had received historical applications.[149] In this study, only the higher dose of bensulfuron-methyl in the second soil (which had previously received bensulfuron treatment) substantially inhibited nitrification, possibly because of the greater persistence of the herbicide in the soil.

The effects of simulated multiple applications were studied by applying four different herbicides (bentazone, isoproturon, fluchloralin, 2,4-D) on chickpeas (*Cicer arietinum*) in a greenhouse experiment.[150] Each rate of the four herbicides except the normal dose of bentazone reduced rhizobial populations within single nodules of the chickpea plant. The authors caution that growers should avoid accumulation of these herbicides in soils to prevent yield loss.

The effects of continued herbicide use were especially studied for glyphosate, used in Roundup Ready crops year by year. A study tested the effect of single versus multiple applications on the microbial and potassium properties of soils.[151] Multiple applications may create a selection pressure in soil microbial communities that could affect nutrient dynamics. Roundup PowerMAX significantly increased microbial respiration, especially in soils with a history of receiving glyphosate. But there was no significant effect of repeated glyphosate application on soil microbial structure or microbial biomass potassium content. The authors conclude that glyphosate stimulates microbial respiration, particularly on soils with a history of glyphosate application, but with no effect on functional diversity or microbial biomass nutrient content. In another study, in contrast, glyphosate (95% technical grade, no information on salt form) increased microbial respiration in a grassland soil without a glyphosate history but had no effect on soils with years of glyphosate applications.[152] This was explained as a stress response of glyphosate-sensitive microorganisms in newly treated soil, and an adapted and depressed respiration in the chronically exposed soil. These contrasting results demand further studies investigating the responses of sites with different herbicide application histories.

Little studied is the translocation of a herbicide from plants to soil. The translocation of GBH Roundup Bio (applied: 360 g/L of AI; 720 g/ha) from quinoa (*Chenopodium quinoa*) to sandy loam soil was studied in a greenhouse pot experiment.[153] Eight days after application, 4% of the glyphosate was detected in soil and about 12% in roots. About 1.5 months later, 12% was found in soil and 8% in roots. The authors conclude that in field studies, glyphosate residues found in soil must originate from plant roots, and that such translocation processes should be included in both leaching assessments and pesticide fate models.

Mixtures of different herbicides are often used to target multiple weed species with a single application or to respond to weed resistance against certain active ingredients. The sheer number of possible combinations of different herbicides makes an assessment of the toxicity of specific mixtures extremely difficult. However, more than 400 publications addressed the topic between 1962 and 2019. Several herbicide combinations (including 2,4-D, glyphosate, ethalfluralin, sethoxydim, clopyralid, and ethametsulfuron) in canola cropping systems showed a number of significant differences in the functional diversity of soil bacteria.[154] However, no general conclusions about why particular herbicide combinations caused greater effects on soil bacteria could be made.

Another complex aspect is the sequence of different pesticide classes, for example between seed dressing (the treatment of crop seeds with insecticides and/or fungicides) and herbicide application. While pesticides in seed dressings aim to protect seeds from pests and diseases, those crop fields in most cases also receive herbicide applications. In a greenhouse pot experiment, the single and interactive effects of the seed dressing of winter wheat (*T. aestivum* L. var. Capo) with neonicotinoid insecticides and/or strobilurin and triazolinthione fungicides and an additional one-time application of a GBH (Roundup LB Plus with about 1.5 times the recommended dosage) on soil biota and processes were examined.[155] Seed dressings significantly reduced the surface activity of earthworms (no difference between insecticides and fungicides), and subsequent herbicide application intensified this effect. Neither seed dressings nor herbicide affected litter decomposition, soil basal respiration, microbial biomass, or specific respiration. This study points to the need to investigate interactive effects of different pesticide classes that are commonly applied agriculturally.

There are few reports of the additive or synergistic effects of the combined application of herbicides with fungicides,[156, 157] even though effects can be expected.

The simultaneous use of these agents to control diseases and weeds could lead to antagonistic interactions[156, 158] that reduce the efficacy of both. It would be useful to determine potential herbicide-fungicide interactions in distinct plant-pathogen combinations, and to use herbicides that interact synergistically with fungicides; thus, they could be used to lower the amount of fungicide necessary to prevent disease.[103] Unfortunately, not much data on the environmental effects of pesticide combinations exist, even though a considerable amount has been published on the effects of individual agrochemicals toward nontarget organisms and ecological processes.

Winter honeybees, which have a longer lifespan than summer bees, were exposed through food to the insecticide imidacloprid, the fungicide difenoconazole, and the herbicide glyphosate, alone or in mixtures at environmental concentrations (0.1, 1, and 10 μg/L).[159] Survival was significantly reduced after exposure to these three pesticides individually and in combination, relative to untreated controls. Overall, the combinations had a higher impact than the agents alone, with a maximum mortality of 53% after 20 days of exposure to the insecticide-fungicide binary mixture at 1 μg/L. In surviving bees, the pesticide combinations had a systemic impact on the physiological state. These results demonstrate the importance of studying the effects of chemical cocktails based on realistic exposure levels, and of developing long-term tests to reveal possible lethal and adverse sublethal interactions in honeybees and other insect pollinators.

Some evidence suggests that the long-term application of atrazine can induce significant changes in the soil microbial population. Shifts in the bacterial community structure in soil under a maize monoculture after 18 years of annual use of atrazine (0.75 kg/ha) and metolachlor (2 kg/ha) have been found.[160] Twenty-year-long applications of atrazine and metalochlor with different fertilizer types on the endophytic community of maize plants grown in different field experiments were assessed. Atrazine (0.75 kg/ha) and metalochlor (2 kg/ha) had no effect on the root endophytic community structure.[161] Taken together, it is difficult to make general conclusions about the long-term effect of repeated herbicide applications because too few studies addressed this issue.[162]

When reporting the nontarget effects of pesticide formulations, the side effects of numerous nondeclared surfactants also need to be considered, as they might be more toxic than the active ingredient.[163–166] See also Chapter 4.

Conclusions

Maintaining healthy agroecosystems with closed nutrient cycles and intact natural biocontrol at high biodiversity is essential for sustainable food production. The current scheme for ERAs of herbicides is inadequate to halt the decline in biodiversity. The main points that are often raised are the inclusion of new test or surrogate species, the extension of studies to more realistic scenarios, the validity of selected uncertainty (assessment) factors, the lack of including sublethal endpoints in risk assessments, and the need to address ignored groups of organisms.[80] The consideration of interactions of herbicide effects with stressors such as nutrient shortage and climate change is also fundamental.

Herbicides applied with the ultimate goal of maximizing productivity and economic returns are not only active on the crop fields but also on neighboring seminatural structures. Transport to remote ecosystems such as montane and arctic regions is documented, but the resulting effects on the biota of this long-term chronic exposure are currently unknown. Herbicides change plant community composition, resulting in a homogenization and reduction in biomass and biodiversity. This effect is especially noted in the cropping area. Because agricultural cropping areas cover a considerable proportion of the terrestrial land area (up to 30% in the EU), the re-establishment of a diverse weed community seems to be central to restore biodiversity in agricultural landscapes. Herbicides also act at the expense of ecosystem functions. Although not immediately obvious, ecosystem services also contribute to crop health by promoting stubble turnover, pathogen suppression, nutrient cycling, and maintenance of soil structure.

A reduction or change in the availability of floral resources due to herbicide exposure can have negative impacts on flower-visiting insect populations. The decline of wildflowers and the consequent shortage for flower-visiting insects is a major concern and one of the potential drivers of historical and worldwide decline in pollinators. Hence, a change in flowering phenologies and the floral availability of nontarget plants may secondarily affect pollinator populations and, ultimately, crop pollination. Plant biomass as food for herbivorous insects or birds and mammals is also reduced by herbicides, leading to declines at higher trophic levels. Additionally, changes in the food quality of host plants affect insect development, and shifts in plant diet affect the population persistence of herbivorous mammals.

Conclusions and recommendations for specific protection goals have been outlined by European authorities.[79] If the provision of a certain level of ecosystem function (e.g., food web support, pollination, pest control) is to be maintained, ERAs should account for multiple stresses caused by normal agricultural practices (e.g., sequential use of different pesticides) and the impacts of climate change. Especially, the indirect effects of in-field herbicide applications need to be addressed.

To better understand the full range of indirect effects of herbicides on biodiversity and ecological interaction in agricultural landscapes, there is a strong need for carefully planned and executed field experiments that also include aquatic ecosystems within the landscape (see the following close-up).

Close-up: Indirect herbicide effects in aquatic food webs: Looking left and right is just as important as up and down

Pollutants transported from terrestrial to aquatic ecosystems through wastewater or runoff have been at the center of ecotoxicology for many years. Those pollutants have been assessed for their effects predominantly by using single-species short-term toxicity tests that focus on the direct effects on nontarget organisms such as aquatic primary producers, invertebrates, or fish.

However, aquatic ecosystems harbor complex food webs in which various species at different trophic levels interact. Consequently, the effects of pollutants on one species may have a cascading impact on others. The impacts may be bottom-up or top-down (vertical) or may affect the interactions among organisms at the same trophic level (horizontal). As species differ in their sensitivity to a certain pollutant, cascading impacts may modify the structure of the food web, although this pollutant may not have caused meaningful reactions in a species when exposed directly. In other words, if environmental risk assessments of pollutants are based only on direct effects, potential side effects (= indirect effects) remain unnoticed.[167]

Indirect effects have attracted increasing attention over the past few decades but are still underrepresented. Fig. 8.8 illustrates potential vertical and horizontal interactions among groups of organisms belonging to different or the same trophic levels within a hypothetical autotrophic food web. In this schematic web, a pollutant such as a herbicide likely affects primary producers, which have receptors targeted by herbicides. These effects may translate to changes at higher trophic levels in a bottom-up manner. For example, algal biofilms growing on macrophytes can have their biomass reduced by a herbicide. Consequently, the macrophytes, which may be less sensitive than algae, harvest more energy from sunlight and ultimately show a higher accrual of biomass. In another example, the biofilm composition and biomass are negatively affected, leading to lower fitness in organisms grazing on this biofilm. In this way, herbicides can theoretically affect organisms that would show little response when exposed directly.

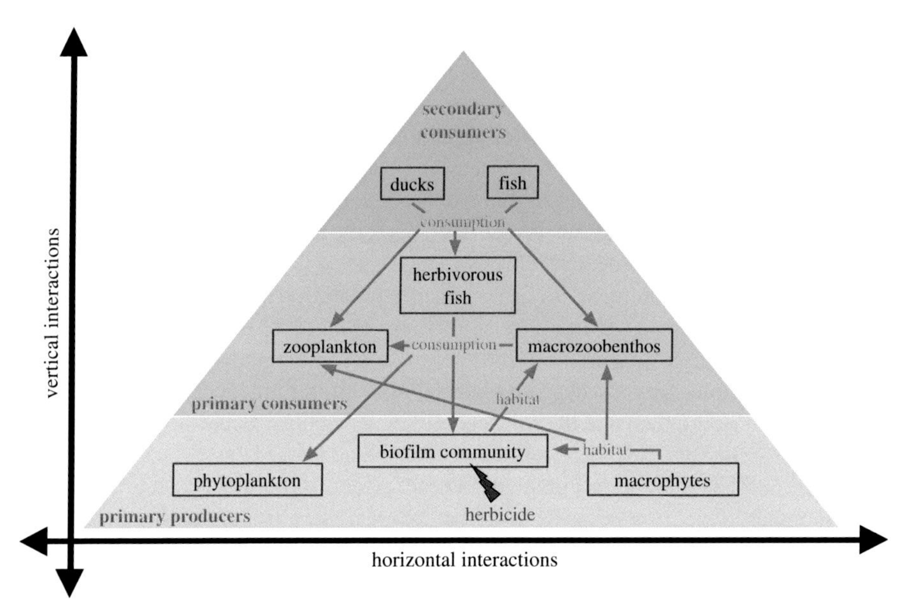

Fig. 8.8 Simplified overview of horizontal and vertical interactions among aquatic organisms within a hypothetical autotrophic food web affected by a herbicide.

In lotic and lentic ecosystems, where primary production is realized by autotrophic biofilms, macrophytes, and phytoplankton,[168] herbicides might, as theoretically explained above, disturb the energy transfer in those food webs. Releya, for instance, examined the impact of two globally used herbicides, glyphosate and 2,4-D, on the biodiversity and productivity of model ecosystems.[167] Within 2 weeks, he found unexpected indirect negative effects: a complete elimination of tadpoles caused by glyphosate in concert with a decrease in predator biomass. This observation suggests that glyphosate induced bottom-up effects initiated by changes in primary producers, leading to impacts in primary and ultimately secondary consumers.

In a recent article, Zhao et al. studied the influence of changing the horizontal (the number of species within trophic levels) and vertical (the number of trophic levels) compositions on the effects of the herbicide linuron on primary producers.[169] Specifically, next to the randomly selected primary producer *Raphidocelis subcapitata*, the authors added up to four competing species of primary producers to increase diversity, and up to four herbivores in the presence and absence of one predator species. Linuron decreased the biomass of primary producers (measured by a cell counter) less when surrounded by a higher diversity. This finding was explained by the compensation effect, balancing for the loss of sensitive producers. By adding primary consumers, linuron further decreased the biomass of primary producers with increasing consumer diversity. The indirect negative effects of linuron on the herbivore biomass were also recorded. These effects were increased in magnitude by adding predators, namely secondary consumers. It can be concluded that horizontal and vertical interactions within food webs can indeed mitigate or potentiate pollution-induced effects.

In a similar study, the direct and indirect effects of the herbicides diuron and hexazinone were assessed within model aquatic communities, including 95 phytoplankton, 18 zooplankton, and 14 macroinvertebrate species in outdoor mesocosms.[170] Biodiversity and biomass were the main points of interest. Herbicides led to decreases in the total abundance of all species, with indirect effects on higher trophic organisms at concentrations inducing limited effects under laboratory conditions. Consequently, the indirect effects induced by herbicides may be relevant under field conditions, as this group of pesticides is among the most frequently detected in pesticide-monitoring programs of surface waters.[171]

To our knowledge, studies on the indirect effects of herbicides in aquatic food webs have not considered the changes of food quality induced at lower trophic levels (here primary producers), except for Rybicki and colleagues. These authors examined the effects of the herbicide terbutryn on biofilm communities and extended this experiment by addressing the potential direct and indirect effects on a primary consumer, a grazing mayfly.[172] This mayfly (*Rhithrogena semicolorata*) showed reduced energy reserves (triglycerides), which was interpreted as an indication of moderate starvation. A decrease in energy reserves might ultimately affect grazer populations (and higher trophic levels) in the long term.

Understanding the impact of herbicides on primary producers' diversity, the consequences for the food quality of the biofilms for higher trophic levels, and further bottom-up or top-down indirect effects are among the future research challenges. These insights will not only support our basic understanding on how ecosystems and food webs function, but also whether the assessment of risks performed prior to the authorization of such pollutants to the market is protective.

Sophie Oster and Mirco Bundschuh

iES Landau, Institute for Environmental Sciences, University of Koblenz-Landau, Landau, Germany

References

1. Schmitz J, Schäfer K, Brühl CA. Agrochemicals in field margins—assessing the impacts of herbicides, insecticides, and fertilizer on the common buttercup (*Ranunculus acris*). *Environ Toxicol Chem* 2013;**32**(5):1124–31. https://doi.org/10.1002/etc.2138.
2. Hahn M, Lenhardt PP, Brühl CA. Characterization of field margins in intensified agro-ecosystems—why narrow margins should matter in terrestrial pesticide risk assessment and management. *Integr Environ Assess Manag* 2014;**10**:456–62.

3. BMEL. *Anwendung von Pflanzenschutzmitteln. Verhalten in unmittelbarer Nähe zu Wohnbebauungen*. Gärten oder Personen; 2018.*https://wwwbmelde/SharedDocs/Downloads/DE/Broschueren/Flyer-Poster/Flyer-Pflanzenschutzanwendungpdf?__blob=publicationFile&v=3.*
4. de Jong FMW, de Snoo GR, van de Zande JC. Estimated nationwide effects of pesticide spray drift on terrestrial habitats in the Netherlands. *J Environ Manag* 2008;**86**(4):721–30. https://doi.org/10.1016/j.jenvman.2006.12.031.
5. Wang M, Rautmann D. A simple probabilistic estimation of spray drift—factors determining spray drift and development of a model. *Environ Toxicol Chem / SETAC* 2008;**27**:2617–26. https://doi.org/10.1897/08-109.1.
6. Rautmann D, Streloke M, Winkler R. New basic drift values in the authorization procedure for plant protection products, In: *Workshop on risk management and risk mitigation measures in the context of authorization of plant protection products*; 1999. p. 133–41.
7. Woods N, Craig IP, Dorr G, Young B. Spray drift of pesticides arising from aerial application in cotton. *J Environ Qual* 2001;**30**(3):697–701. https://doi.org/10.2134/jeq2001.303697x.
8. Linhart C, Niedrist GH, Nagler M, et al. Pesticide contamination and associated risk factors at public playgrounds near intensively managed apple and wine orchards. *Environ Sci Eur* 2019;**31**(1):28. https://doi.org/10.1186/s12302-019-0206-0.
9. Lupi L, Bedmar F, Puricelli M, et al. Glyphosate runoff and its occurrence in rainwater and subsurface soil in the nearby area of agricultural fields in Argentina. *Chemosphere* 2019;**225**:906–14. https://doi.org/10.1016/j.chemosphere.2019.03.090.
10. Silva V, Montanarella L, Jones A, et al. Distribution of glyphosate and aminomethylphosphonic acid (AMPA) in agricultural topsoils of the European Union. *Sci Total Environ* 2018;**621**:1352–9. https://doi.org/10.1016/j.scitotenv.2017.10.093.
11. Silburn DM, Foley JL, de Voil RC. Managing runoff of herbicides under rainfall and furrow irrigation with wheel traffic and banded spraying. *Agric Ecosyst Environ* 2011;https://doi.org/10.1016/j.agee.2011.08.018.
12. Hill BD, Harker KN, Hasselback P, Moyer JR, Inaba DI, Byers SD. Phenoxy herbicides in Alberta rainfall: potential effects on sensitive crops. *Can J Plant Sci* 2002;**82**:481–4.
13. Tuduri L, Harner T, Blanchard P, et al. A review of currently used pesticides (CUPs) in Canadian air and precipitation. Part 2: regional information and perspectives. *Atmos Environ* 2006;**40**(9):1579–89. https://doi.org/10.1016/j.atmosenv.2005.11.020.
14. EFSA. *Glyphosate renewal assessment report of 18 December 2013. Rapporteur member state (RMS): Germany*. Co-RMS: Slovakia available on request athttp://darefsaeuropaeu/dar-web/provision; 2013.
15. Bento CPM, Yang X, Gort G, et al. Persistence of glyphosate and aminomethylphosphonic acid in loess soil under different combinations of temperature, soil moisture and light/darkness. *Sci Total Environ* 2016;**572**:301–11. https://doi.org/10.1016/j.scitotenv.2016.07.215.
16. Chang F-C, Simcik MF, Capel PD. Occurrence and fate of the herbicide glyphosate and its degradate aminomethylphosphonic acid in the atmosphere. *Environ Toxicol Chem* 2011;**30**(3):548–55. https://doi.org/10.1002/etc.431.
17. Battaglin WA, Meyer MT, Kuivila KM, Dietze JE. Glyphosate and its degradation product AMPA occur frequently and widely in U.S. soils, surface water, groundwater, and precipitation. *J Am Water Resour Assoc* 2014;**50**(2):275–90. https://doi.org/10.1111/jawr.12159.
18. Quaghebeur D, Smet BD, Wulf ED, Steurbaut W. Pesticides in rainwater in Flanders, Belgium: results from the monitoring program 1997–2001. *J Environ Monit* 2004;**6**(3):182–90. https://doi.org/10.1039/B312558K.

19. BEL. *Pestizid-Belastung der Luft. Eine deutschlandweite Studie zur Ermittlung der Belastung der Luft mit Hilfe von technischen Sammlern, Bienenbrot, Filtern aus Be- und Entlüftungsanlagen und Luftgüte-Rindenmonitoring hinsichtlich des Vorkommens von Pestizid-Wirkstoffen, insbesondere Glyphosat.* 140pp.
20. Ferrario C, Finizio A, Villa S. Legacy and emerging contaminants in meltwater of three Alpine glaciers. *Sci Total Environ* 2017;**574**:350–7. https://doi.org/10.1016/j.scitotenv.2016.09.067.
21. Smalling KL, Fellers GM, Kleeman PM, Kuivila KM. Accumulation of pesticides in Pacific chorus frogs (*Pseudacris regilla*) from California's Sierra Nevada mountains, USA. *Environ Toxicol Chem* 2013;**32**:2026–34. https://doi.org/10.1002/etc.2308.
22. Daly GL, Lei YD, Teixeira C, DCG M, Castillo LE, Wania F. Accumulation of current-use pesticides in neotropical montane forests. *Environ Sci Technol* 2007;**41**(4):1118–23. https://doi.org/10.1021/es0622709.
23. Hoferkamp L, Hermanson MH, Muir DCG. Current use pesticides in Arctic media; 2000–2007. *Sci Total Environ* 2010;**408**(15):2985–94. https://doi.org/10.1016/j.scitotenv.2009.11.038.
24. Ruggirello RM, Hermanson MH, Isaksson E, et al. Current use and legacy pesticide deposition to ice caps on Svalbard, Norway. *J Geophys Res-Atmos* 2010;**115**(D18)https://doi.org/10.1029/2010jd014005.
25. Balmer JE, Morris AD, Hung H, et al. Levels and trends of current-use pesticides (CUPs) in the arctic: an updated review, 2010–2018. *Emerg Contam* 2019;**5**:70–88. https://doi.org/10.1016/j.emcon.2019.02.002.
26. Marshall EJP, Brown VK, Boatman ND, Lutman PJW, Squire GR, Ward LK. The role of weeds in supporting biological diversity within crop fields. *Weed Res* 2003;**43**(2):77–89. https://doi.org/10.1046/j.1365-3180.2003.00326.x.
27. Andreasen C, Jensen HA, Jensen SM. Decreasing diversity in the soil seed bank after 50 years in Danish arable fields. *Agric Ecosyst Environ* 2018;**259**:61–71. https://doi.org/10.1016/j.agee.2018.02.034.
28. Geiger F, Bengtsson J, Berendse F, et al. Persistent negative effects of pesticides on biodiversity and biological control potential on European farmland. *Basic Appl Ecol* 2010;**11**:97–105.
29. Boutin C, Strandberg B, Carpenter D, Mathiassen SK, Thomas PJ. Herbicide impact on non-target plant reproduction: what are the toxicological and ecological implications? *Environ Pollut* 2014;**185**:295–306. https://doi.org/10.1016/j.envpol.2013.10.009.
30. Robinson RA, Sutherland WJ. Post-war changes in arable farming and biodiversity in Great Britain. *J Appl Ecol* 2002;**39**(1):157–76. https://doi.org/10.1046/j.1365-2664.2002.00695.x.
31. Eurostat. *EU land cover statistics.* https://eceuropaeu/eurostat/statistics-explained/indexphp/Land_cover_statistics; 2019. Accessed 27 March 2019.
32. Boutin C, Jobin B. Intensity of agricultural practices and effects on adjacent habitats. *Ecol Appl* 1998;**8**(2):544–57.
33. Egan JF, Mortensen DA. Quantifying vapor drift of dicamba herbicides applied to soybean. *Environ Toxicol Chem* 2012;**31**(5):1023–31. https://doi.org/10.1002/etc.1778.
34. Liira J, Schmidt T, Aavik T, et al. Plant functional group composition and large-scale species richness in European agricultural landscapes. *J Veg Sci* 2008;**19**(1):3–14. https://doi.org/10.3170/2007-8-18308.
35. Phalan B, Onial M, Balmford A, Green RE. Reconciling food production and biodiversity conservation: land sharing and land sparing compared. *Science* 2011;**333**(6047):1289–91. https://doi.org/10.1126/science.1208742.

36. EFSA PPR Panel. Guidance on tiered risk assessment for plant protection products for aquatic organisms in edge-of-field surface waters. *EFSA J* 2013;**11**(7):3290. https://doi.org/10.2903/j.efsa.2013.3290.
37. Freemark K, Boutin C. Impacts of agricultural herbicide use on terrestrial wildlife in temperate landscapes—a review with special reference to North America. *Agric Ecosyst Environ* 1995;**52**(2–3):67–91. https://doi.org/10.1016/0167-8809(94)00534-l.
38. Boutin C, White AL, Carpenter D. Measuring variability in phytotoxicity testing using crop and wild plant species. *Environ Toxicol Chem* 2010;**29**(2):327–37. https://doi.org/10.1002/etc.30.
39. Schmitz J, Stahlschmidt P, Brühl CA. Assessing the risk of herbicides to terrestrial non-target plants using higher-tier studies. *Hum Ecol Risk Assess* 2015;**21**(8):2137–54.
40. Schmitz J, Stahlschmidt P, Brühl CA. *Protection of terrestrial non-target plant species in the regulation of environmental risks of pesticides*. Federal Environment Agency (Germany), TEXTE; 2015. 148pp*http://www.umweltbundesamt.de/publikationen/protection-of-terrestrial-non-target-plant-species*.
41. Boutin C, Montroy K, Mathiassen SK, Carpenter DJ, Strandberg B, Damgaard C. Effects of sublethal doses of herbicides on the competitive interactions between 2 nontarget plants, *Centaurea cyanus* L. and *Silene noctiflora* L. *Environ Toxicol Chem* 2019;**38**(9):2053–64. https://doi.org/10.1002/etc.4506.
42. Schmitz J, Hahn M, Brühl CA. Agrochemicals in field margins—an experimental field study to assess the impacts of pesticides and fertilizers on a natural plant community. *Agric Ecosyst Environ* 2014;**193**:60–9. https://doi.org/10.1016/j.agee.2014.04.025.
43. Pellissier L, Wisz MS, Strandberg B, Damgaard C. Herbicide and fertilizers promote analogous phylogenetic responses but opposite functional responses in plant communities. *Environ Res Lett* 2014;**9**(2):024016. https://doi.org/10.1088/1748-9326/9/2/024016.
44. IPBES. *The assessment report of the Intergovernmental Science-Policy Platform on Biodiversity and Ecosystem Services on pollinators, pollination and food production*. Secretariat of the Intergovernmental Science-Policy Platform on Biodiversity and Ecosystem Services; 2017 552 pp.
45. Carvalheiro LG, Seymour CL, Veldtman R, Nicolson SW. Pollination services decline with distance from natural habitat even in biodiversity-rich areas. *J Appl Ecol* 2010;**47**(4):810–20. https://doi.org/10.1111/j.1365-2664.2010.01829.x.
46. Biesmeijer JC, Roberts SPM, Reemer M, et al. Parallel declines in pollinators and insect-pollinated plants in Britain and the Netherlands. *Science* 2006;**313**(5785):351–4. https://doi.org/10.1126/science.1127863.
47. Rollin O, Benelli G, Benvenuti S, et al. Weed-insect pollinator networks as bio-indicators of ecological sustainability in agriculture. A review. *Agron Sustain Dev* 2016;**36**(1):8. https://doi.org/10.1007/s13593-015-0342-x.
48. Bohnenblust EW, Vaudo AD, Egan JF, Mortensen DA, Tooker JF. Effects of the herbicide dicamba on nontarget plants and pollinator visitation. *Environ Toxicol Chem* 2016;**35**(1):144–51. https://doi.org/10.1002/etc.3169.
49. Weiner CN, Werner M, Linsenmair KE, Blüthgen N. Land use intensity in grasslands: changes in biodiversity, species composition and specialisation in flower visitor networks. *Basic Appl Ecol* 2011;**12**(4):292–9. https://doi.org/10.1016/j.baae.2010.08.006.
50. Stenoien C, Nail KR, Zalucki JM, Parry H, Oberhauser KS, Zalucki MP. Monarchs in decline: a collateral landscape-level effect of modern agriculture. Review. *Insect Sci* 2018;**25**(4):528–41. https://doi.org/10.1111/1744-7917.12404.
51. Sutcliffe OL, Kay QON. Changes in the arable flora of central southern England since the 1960s. *Biol Conserv* 2000;**93**(1):1–8. https://doi.org/10.1016/S0006-3207(99)00119-6.

52. Motten AF. Pollination ecology of the spring wildflower community of a temperate deciduous forest. *Ecol Monogr* 1986;**56**(1):21–42. https://doi.org/10.2307/2937269.
53. Pontin DR, Wade MR, Kehrli P, Wratten SD. Attractiveness of single and multiple species flower patches to beneficial insects in agroecosystems. *Ann Appl Biol* 2006;**148**(1):39–47. https://doi.org/10.1111/j.1744-7348.2005.00037.x.
54. Kluser S, Peduzzi P. *Global pollinator decline: a literature review*. https://unepgrid.ch/storage/app/media/legacy/37/Global_pollinator_decline_literature_review_2007.pdf; 2007.
55. Andreasen C, Streibig JC. Evaluation of changes in weed flora in arable fields of Nordic countries—based on Danish long-term surveys. *Weed Res* 2011;**51**(3):214–26. https://doi.org/10.1111/j.1365-3180.2010.00836.x.
56. Heap I. Herbicide resistant weeds. In: Pimentel D, Peshin R, editors. *Integrated pest management*. Springer; 2014. p. 281–301.
57. Fernandez P, Gauvrit C, Barro F, Menendez J, De Prado R. First case of glyphosate resistance in France. *Agron Sustain Dev* 2015;**35**(4):1469–76. https://doi.org/10.1007/s13593-015-0322-1.
58. Holzschuh A, Steffan-Dewenter I, Tscharntke T. Agricultural landscapes with organic crops support higher pollinator diversity. *Oikos* 2008;**117**(3):354–61. https://doi.org/10.1111/j.2007.0030-1299.16303.x.
59. Moreby SJ, Southway SE. Influence of autumn applied herbicides on summer and autumn food available to birds in winter wheat fields in southern England. *Agric Ecosyst Environ* 1999;**72**(3):285–97. https://doi.org/10.1016/s0167-8809(99)00007-9.
60. Wardle DA, Nicholson KS, Bonner KI, Yeates GW. Effects of agricultural intensification on soil-associated arthropod population dynamics, community structure, diversity and temporal variability over a seven-year period. *Soil Biol Biochem* 1999;**31**(12):1691–706. https://doi.org/10.1016/S0038-0717(99)00089-9.
61. Smith BM, Aebischer NJ, Ewald J, Moreby S, Potter C, Holland JM. The potential of arable weeds to reverse invertebrate declines and associated ecosystem services in cereal crops. *Front Sustain Food Syst* 2020;**3**(118)https://doi.org/10. 3389/fsufs.2019.00118.
62. Gutiérrez Y, Ott D, Scherber C. Direct and indirect effects of plant diversity and phenoxy herbicide application on the development and reproduction of a polyphagous herbivore. Article. *Sci Rep* 2020;**10**(1):7300. https://doi.org/10.1038/s41598-020-64252-5.
63. Pasquer F, Ochsner U, Zarn J, Keller B. Common and distinct gene expression patterns induced by the herbicides 2,4-dichlorophenoxyacetic acid, cinidon-ethyl and tribenuron-methyl in wheat. *Pest Manag Sci* 2006;**62**(12):1155–67. https://doi.org/10.1002/ps.1291.
64. Ahuja I, Rohloff J, Bones AM. Defence mechanisms of Brassicaceae: implications for plant-insect interactions and potential for integrated pest management. A review. *Agron Sustain Dev* 2010;**30**(2):311–48. https://doi.org/10.1051/agro/2009025.
65. Kjaer C, Heimbach U. Relationships between sulfonylurea herbicide treatment of host plants and the performance of herbivorous insects. *Pest Manag Sci* 2001;**57**(12):1161–6.
66. Glynne-Jones A. Pyrethrum. *Pestic Outlook* 2001;**12**(5):195–8. https://doi.org/10.1039/B108601B.
67. Oka IN, Pimentel D. Herbicide (2,4-D) increases insect and pathogen pests on corn. *Science* 1976;**193**(4249):239–40. https://doi.org/10.1126/science.193.4249.239.
68. Agnello AM, van Duyn JW, Bradley Jr. JR. Influence of postemergence herbicides on populations of bean leaf beetle, *Cerotoma trifurcata* (Coleoptera: Chrysomelidae), and corn earworm, *Heliothis zea* (Lepidoptera: Noctuidae), in soybeans. *J Econ Entomol* 1986;**79**(1):261–5. https://doi.org/10.1093/jee/79.1.261.

69. Speight RI, Whittaker JB. Interactions between the chrysomelid beetle *Gastrophysa viridula*, the weed *Rumex obtusifolius* and the herbicide Asulam. *J Appl Ecol* 1987;**24**:119–29.
70. Kjær C, Elmegaard N. Effect of herbicide treatment on host plant quality for a leaf-eating beetle. *Pestic Sci* 1996;**47**(4):319–25. https://doi.org/10.1002/(sici)1096-9063(199608)47:4<319::Aid-ps421>3.0.Co;2-z.
71. Hahn M, Geisthardt M, Bruehl CA. Effects of herbicide-treatmed host plants on the development of *Mamestra brassicae* L. caterpillars. *Environ Toxicol Chem* 2014;**33**(11):2633–8. https://doi.org/10.1002/etc.2726.
72. Wilson JD, Morris AJ, Arroyo BE, Clark SC, Bradbury RB. A review of the abundance and diversity of invertebrate and plant foods of granivorous birds in northern Europe in relation to agricultural change. *Agric Ecosyst Environ* 1999;**75**(1–2):13–30.
73. Boatman ND, Brickle NW, Hart JD, et al. Evidence for the indirect effects of pesticides on farmland birds. *Ibis* 2004;**146**(s2):131–43. https://doi.org/10.1111/j.1474-919X.2004.00347.x.
74. Eurostat. *Bird populations on the decline*. https://eceuropaeu/eurostat/web/products-eurostat-news/product/–/asset_publisher/VWJkHuaYvLIN/content/EDN-20200605-1/pop_up?_101_INSTANCE_VWJkHuaYvLIN_viewMode=print&_101_INSTANCE_VWJkHuaYvLIN_languageId=en_GB.
75. Jahn T, Hötker H, Oppermann R, Bleil R, Vele L. *Protection of biodiversity of free living birds and mammals in respect of the effects of pesticides*. Vol. 30. .
76. Rands MRW. Pesticide use on cereals and the survival of Grey partridge chicks: a field experiment. *J Appl Ecol* 1985;**22**(1):49–54. https://doi.org/10.2307/2403325.
77. Ewald JA, Aebischer NJ. *Pesticide use, avian food resources and bird densities in Sussex*. Joint Nature Conservation Committee Report No 296; 1999 103pp.
78. Taylor RL, Maxwell BD, Boik RJ. Indirect effects of herbicides on bird food resources and beneficial arthropods. *Agric Ecosyst Environ* 2006;**116**(3/4):157–64. https://doi.org/10.1016/j.agee.2006.01.012.
79. EFSA. Scientific Opinion addressing the state of the science on risk assessment of plant protection products for non-target arthropods. *EFSA J* 2015;**13**(2):3996. https://doi.org/10.2903/j.efsa.2015.3996.
80. Brühl CA, Zaller JG. Biodiversity decline as a consequence of an inappropriate environmental risk assessment of pesticides. Opinion. *Front Environ Sci* 2019;**7**(177)https://doi.org/10.3389/fenvs.2019.00177.
81. Topping CJ, Aldrich A, Berny P. Overhaul environmental risk assessment for pesticides. *Science* 2020;**367**(6476):360–3.
82. Gaba S, Alignier A, Aviron S, et al. Ecology for sustainable and multifunctional agriculture. In: Gaba S, Smith B, Lichtfouse E, editors. *Sustainable agriculture reviews 28: Ecology for agriculture*. Springer International Publishing; 2018. p. 1–46.
83. Gentili S, Sigura M, Bonesi L. Decreased small mammals species diversity and increased population abundance along a gradient of agricultural intensification. Article. *Hystrix* 2014;**25**(1)https://doi.org/10.4404/hystrix-25.1-9246.
84. Tew TE, Macdonald DW, Rands MRW. Herbicide application affects microhabitat use by arable wood mice (*Apodemus sylvaticus*). *J Appl Ecol* 1992;**29**:532–9.
85. de Snoo GR. Unsprayed field margins: effects on environment, biodiversity and agricultural practice. *Landsc Urban Plan* 1999;**46**(1):151–60. https://doi.org/10.1016/S0169-2046(99)00039-0.
86. Fagerstone KA, Tietjen HP, Lavoie GK. Effects of range treatment with 2,4-D on prairie dog diet. *J Range Manag* 1977;**30**:57–60.

87. Spencer SR, Barrett GW. Meadow vole population response to vegetational changes resulting from 2,4-D application. *Am Midl Nat* 1980;**103**:32–46.
88. Johnson DR, Hansen RM. Effects of range treatment with 2,4-D on rodent populations. *J Wildl Manag* 1969;**33**:125–32.
89. Reinecke AJ, Helling B, Louw K, Fourie J, Reinecke SA. The impact of different herbicides and cover crops on soil biological activity in vineyards in the Western Cape, South Africa. *Pedobiology* 2002;**46**:475–84.
90. Sannino F, Gianfreda L. Pesticide influence on soil enzymatic activities. *Chemosphere* 2001;**45**(4):417–25. https://doi.org/10.1016/S0045-6535(01)00045-5.
91. Hernández M, Jia Z, Conrad R, Seeger M. Simazine application inhibits nitrification and changes the ammonia-oxidizing bacterial communities in a fertilized agricultural soil. *FEMS Microbiol Ecol* 2011;**78**(3):511–9. https://doi.org/10.1111/j.1574-6941.2011.01180.x.
92. Wang J, Zhang H, Zhang X, Qin S, Tan H, Li X. Effects of long-term chlorimuron-ethyl application on the diversity and antifungal activity of soil *Pseudomonas* spp. in a soybean field in Northeast China. *Ann Microbiol* 2013;**63**(1):335–41. https://doi.org/10.1007/s13213-012-0479-7.
93. Wu M, Zhang X, Zhang H, et al. Soil Pseudomonas community structure and its antagonism towards *Rhizoctonia solani* under the stress of acetochlor. *Bull Environ Contam Toxicol* 2009;**83**(3):313–7. https://doi.org/10.1007/s00128-009-9731-7.
94. Casabé N, Piola L, Fuchs J, et al. Ecotoxicological assessment of the effects of glyphosate and chlorpyrifos in an Argentine soya field. *J Soils Sediments* 2007;**8**:1–8.
95. Gaupp-Berghausen M, Hofer M, Rewald B, Zaller JG. Glyphosate-based herbicides reduce the activity and reproduction of earthworms and lead to increased soil nutrient concentrations. *Sci Rep* 2015;**5**:12886. https://doi.org/10.1038/srep12886.
96. Zaller JG, Cantelmo C, Dos Santos G, et al. Herbicides in vineyards reduce grapevine root mycorrhization and alter soil microorganisms and the nutrient composition in grapevine roots, leaves, xylem sap and grape juice. *Environ Sci Pollut Res* 2018;**25**(23):23215–26. https://doi.org/10.1007/s11356-018-2422-3.
97. Maderthaner M, Weber M, Takács E, et al. Commercial glyphosate-based herbicides effects on springtails (Collembola) differ from those of their respective active ingredients and vary with soil organic matter content. *Environ Sci Pollut Res* 2020;**27**(14):17280–9. https://doi.org/10.1007/s11356-020-08213-5.
98. Damin V, Trivelin PCO, de Godoy BT, de Carvalho SJP, Moraes MF. Mineralization and cor recovery of 15Nitrogen from black oats residues treated with herbicides. *J Plant Nutr* 2012;**35**(12):1830–42. https://doi.org/10.1080/01904167.2012.706679.
99. Abdel-Mallek AY, Abdel-Kader MIA, Shonkeir AMA. Effect of glyphosate on fungal population, respiration and the decay of some organic matters in Egyptian soil. *Microbiol Res* 1994;**149**(1):69–73. https://doi.org/10.1016/S0944-5013(11)80139-4.
100. Sahid IB, Yap MY. Effects of two acetanilide herbicides on microbial populations and their cellulolytic activities. *Bull Environ Contam Toxicol* 1994;**52**(1):61–8. https://doi.org/10.1007/BF00197358.
101. Hébert M-P, Fugère V, Gonzalez A. The overlooked impact of rising glyphosate use on phosphorus loading in agricultural watersheds. *Front Ecol Environ* 2019;**17**(1):48–56. https://doi.org/10.1002/fee.1985.
102. Mazzola M. Mechanisms of natural soil suppressiveness to soilborne diseases. *Antonie Van Leeuwenhoek* 2002;**81**(1):557–64. https://doi.org/10.1023/A:1020557523557.
103. Kortekamp A. Unexpected side effects of herbicides: modulation of plant-pathogen interactions. In: Kortekamp A, editor. *Herbicides and environment*. InTech; 2011. p. 85–104.

104. Van Bruggen AHC, He MM, Shin K, et al. Environmental and health effects of the herbicide glyphosate. *Sci Total Environ* 2018;**616–617**:255–68. https://doi.org/10.1016/j.scitotenv.2017.10.309.
105. Descalzo RC, Punja ZK, Lévesque CA, Rahe JE. Glyphosate treatment of bean seedlings causes short-term increases in *Pythium* populations and damping off potential in soils. *Appl Soil Ecol* 1998;**8**(1):25–33. https://doi.org/10.1016/S0929-1393(97)00069-3.
106. Meriles JM, Gil SV, Haro RJ, March GJ, Guzmán CA. Selected soil-borne fungi under glyphosate application and crop residues from a long-term field experiment. *Biol Agric Hortic* 2008;**26**(2):193–205. https://doi.org/10.1080/01448765.2008.9755080.
107. Zobiole LHS, Kremer RJ, Oliveira Jr. RS, Constantin J. Glyphosate affects microorganisms in rhizospheres of glyphosate-resistant soybeans. *J Appl Microbiol* 2011;**110**(1):118–27. https://doi.org/10.1111/j.1365-2672.2010.04864.x.
108. Kremer RJ, Means NE. Glyphosate and glyphosate-resistant crop interactions with rhizosphere microorganisms. *Eur J Agron* 2009;**31**(3):153–61. https://doi.org/10.1016/j.eja.2009.06.004.
109. Duke SO, Wedge DE, Cerdeira AL, Matallo MB. Herbicide effects on plant diseases. *Outlook Pest Manag* 2007;36–40.
110. Fernandez MR, Zentner RP, Basnyat P, Gehl D, Selles F, Huber DM. Glyphosate associations with cereal diseases caused by *Fusarium* spp. in the Canadian Prairies. *Eur J Agron* 2009;**31**:133–43. https://doi.org/10.1016/j.eja.2009.07.003.
111. Larson R, Hill A, Kniss A, Hanson L, Miller S. Influence of glyphosate on *Rhizoctonia* and *Fusarium* root rot in sugar beet. *Pest Manag Sci* 2006;**62**:1182–92. https://doi.org/10.1002/ps.1297.
112. Keen NT, Holliday MJ, Yoshikawa M. Effects of glyphosate on glyceollin production and the expression of resistance to *Phytophthora megasperma* f. sp. *glycinea* in soybean. *Phytopathology* 1982;**72**:1468–70.
113. Johal GS, Rahe JE. Role of phytoalexins in the suppression of resistance of *Phaseolus vulgaris* to *Colletotrichum lindemuthianum* by glyphosate. *Can J Plant Pathol* 1990;**12**:225–35.
114. Brammal RA, Higgins VJ. The effects of glyphosate on resistance of tomato to *Fusarium* crown rot and root diseases and on the formation of host structural defence barriers. *Can J Bot* 1988;**66**:1547–55.
115. Weckert MA. Interaction between *Cylindrocarpon* and glyphosate in young vine decline. *Phytopathol Mediterr* 2010;**49**:117–8.
116. Meyer-Wolfarth F, Schrader S, Oldenburg E, Weinert J, Brunotte J. Biocontrol of the toxigenic plant pathogen *Fusarium culmorum* by soil fauna in an agroecosystem. *Mycotox Res* 2017;**33**:237–44. https://doi.org/10.1007/s12550-017-0282-1.
117. Euteneuer P, Wagentristl H, Steinkellner S, Scheibreithner C, Zaller JG. Earthworms affect decomposition of soil-borne plant pathogen *Sclerotinia sclerotiorum* in a cover crop field experiment. *Appl Soil Ecol* 2019;**138**:88–93. https://doi.org/10.1016/j.apsoil.2019.02.020.
118. Harikrishnan R. Influence of herbicides on growth and sclerotia production in *Rhizoctonia solani*. *Weed Sci* 2009;**49**:241–7. https://doi.org/10.1614/0043-1745(2001)049[0241:IOHOGA]2.0.CO;2.
119. Fan L, Feng Y, Weaver DB, Delaney DP, Wehtje GR, Wang G. Glyphosate effects on symbiotic nitrogen fixation in glyphosate-resistant soybean. *Appl Soil Ecol* 2017;**121**:11–9. https://doi.org/10.1016/j.apsoil.2017.09.015.
120. Köberl M, Wagner P, Müller H, et al. Unraveling the complexity of soil microbiomes in a large-scale study subjected to different agricultural management in Styria. Original research. *Front Microbiol* 2020;**11**(1052)https://doi.org/10.3389/fmicb.2020.01052.

121. Xin-Yu L, Zhen-Cheng S, Xu L, Cheng-Gang Z, Hui-Wen Z. Assessing the effects of acetochlor on soil fungal communities by DGGE and clone library analysis. *Ecotoxicology (London, England)* 2010;**19**:1111–6. https://doi.org/10.1007/s10646-010-0493-0.
122. Rovira AD, McDonald HJ, Glen Osmond SA. Effects of the herbicide chlorsulfuron on *Rhizoctonia* bare patch and take-all of barley and wheat. *Plant Dis* 1986;**70**:879–82. https://doi.org/10.1094/PD-70-879.
123. Lee H, Ullrich SE, Burke IC, Yenish J, Paulitz TC. Interactions between the root pathogen *Rhizoctonia solani* AG-8 and acetolactate-synthase-inhibiting herbicides in barley. *Pest Manag Sci* 2012;**68**(6):845–52. https://doi.org/10.1002/ps.2336.
124. Duke SO, Wedge DE, Cerdeira AL, Matallo MB. Interactions of synthetic herbicides with plant disease and microbial herbicides. In: Vurro M, Gressel J, editors. *Novel biotechnologies for biocontrol agent enhancement and management. NATO Security through Science Series*Springer; 2007. p. 277–96.
125. Martinez DA, Loening UE, Graham MC. Impacts of glyphosate-based herbicides on disease resistance and health of crops: a review. *Environ Sci Eur* 2018;**30**(1):2. https://doi.org/10.1186/s12302-018-0131-7.
126. van Straalen NM. Peer reviewed: ecotoxicology becomes stress ecology. *Environ Sci Technol* 2003;**37**(17):324A–330A. https://doi.org/10.1021/es0325720.
127. López-Alcaide S, Macip-Ríos R. Effects of climate change in amphibians and reptiles. In: Grillo O, editor. *Biodiversity loss in a changing planet*. InTech; 2011. p. 163–84.
128. Lötters S, Filz KJ, Wagner N, Schmidt BR, Emmerling C, Veith M. Hypothesizing if responses to climate change affect herbicide exposure risk for amphibians. *Environ Sci Eur* 2014;**26**(1):31. https://doi.org/10.1186/s12302-014-0031-4.
129. Berger G, Graef F, Pallut B, Hoffmann J, Brühl CA, Wagner N. How does changing pesticide usage over time affect migrating amphibians: a case study on the use of glyphosate-based herbicides in german agriculture over 20 years. *Front Environ Sci* 2018;**6**:6. https://doi.org/10.3389/fenvs.2018.00006doi:10.3389/fenvs. 2018.00006.
130. Holmstrup M, Bindesbøl A-M, Oostingh GJ, et al. Interactions between effects of environmental chemicals and natural stressors: a review. *Sci Total Environ* 2010;**408**(18):3746–62. https://doi.org/10.1016/j.scitotenv.2009.10.067.
131. Demon A, Eijsackers H. The effects of lindane and azinphosmethyl on survival time of soil animals, under extreme or fluctuating temperature and moisture conditions. *Z Angew Entomol* 1985;**100**(1–5):504–10. https://doi.org/10.1111/j.1439-0418.1985.tb02812.x.
132. Chalifour A, Arts MT, Kainz MJ, Juneau P. Combined effect of temperature and bleaching herbicides on photosynthesis, pigment and fatty acid composition of *Chlamydomonas reinhardtii*. *Eur J Phycol* 2014;**49**(4):508–15. https://doi.org/10.1080/09670262.2014.977962.
133. Baier F, Jedinger M, Gruber E, Zaller JG. Temperature-dependence of glyphosate-based herbicide's effects on egg and tadpole growth of Common Toads. *Front Environ Sci* 2016;**4**:51. https://doi.org/10.3389/fenvs.2016.00051.
134. Baier F, Gruber E, Hein T, et al. Non-target effects of a glyphosate-based herbicide on common toad larvae (*Bufo bufo*, Amphibia) and associated algae are altered by temperature. *Peer J* 2016;**4**:e2641https://doi.org/10.7717/peerj.2641.
135. Kattwinkel M, Kühne J-V, Foit K, Liess M. Climate change, agricultural insecticide exposure, and risk for freshwater communities. *Ecol Appl* 2011;**21**(6):2068–81. https://doi.org/10.1890/10-1993.1.
136. Crain DA, Guillette LJ. Reptiles as models of contaminant-induced endocrine disruption. *Anim Reprod Sci* 1998;**53**(1):77–86. https://doi.org/10.1016/S0378-4320(98)00128-6.

137. Kiesecker JM. Global stressors and the global decline of amphibians: tipping the stress immunocompetency axis. *Ecol Res* 2011;**26**(5):897–908. https://doi.org/10.1007/s11284-010-0702-6.
138. Todorovic GR, Rampazzo N, Mentler A, Blum WEH, Eder A, Strauss P. Influence of soil tillage and erosion on the dispersion of glyphosate and aminomethylphosphonic acid in agricultural soils. *Int Agrophys* 2014;**28**:93–100.
139. Metcalfe NB, Monaghan P. Compensation for a bad start: grow now, pay later? *TREE* 2001;**16**(5):254–60. https://doi.org/10.1016/S0169-5347(01)02124-3.
140. Mohanty SR, Nayak DR, Babu YJ, Adhya TK. Butachlor inhibits production and oxidation of methane in tropical rice soils under flooded condition. *Microbiol Res* 2004;**159**(3):193–201. https://doi.org/10.1016/j.micres.2004.03.004.
141. Kinney C, Mosier A, Ferrer I, Furlong E, Mandernack K. Effects of the herbicides prosulfuron and metolachlor on fluxes of CO_2, N_2O, and NH_4 in a fertilized Colorado grassland soil. *J Geophys Res* 2004;**109**:https://doi.org/10.1029/2003JD003656.
142. Das S, Ghosh A, Adhya TK. Nitrous oxide and methane emission from a flooded rice field as influenced by separate and combined application of herbicides bensulfuron methyl and pretilachlor. *Chemosphere* 2011;**84**(1):54–62. https://doi.org/10.1016/j.chemosphere.2011.02.055.
143. Seghers D, Siciliano S, Top E, Verstraete W. Combined effect of fertilizer and herbicide applications on the abundance, community structure and performance of the soil methanotrophic community. *Soil Biol Biochem* 2005;**37**:187–93. https://doi.org/10.1016/j.soilbio.2004.05.025.
144. Shi L, Guo Y, Ning J, Lou S, Hou F. Herbicide applications increase greenhouse gas emissions of alfalfa pasture in the inland arid region of northwest China. *Peer J* 2020;**8**:e9231. https://doi.org/10.7717/peerj.9231.
145. Liu H, Yang X, Liang C, et al. Interactive effects of microplastics and glyphosate on the dynamics of soil dissolved organic matter in a Chinese loess soil. *Catena* 2019;**182104177**:https://doi.org/10.1016/j.catena.2019.104177.
146. Godoy V, Martín-Lara MA, Calero M, Blázquez G. The relevance of interaction of chemicals/pollutants and microplastic samples as route for transporting contaminants. *Process Saf Environ Prot* 2020;**138**:312–23. https://doi.org/10.1016/j.psep.2020. 03.033.
147. Givaudan N, Wiegand C, Le Bot B, et al. Acclimation of earthworms to chemicals in anthropogenic landscapes, physiological mechanisms and soil ecological implications. *Soil Biol Biochem* 2014;**73**:49–58. https://doi.org/10.1016/j.soilbio.2014.01.032.
148. Givaudan N, Binet F, Le Bot B, Wiegand C. Earthworm tolerance to residual agricultural pesticide contamination: field and experimental assessment of detoxification capabilities. *Environ Pollut* 2014;**192**:9–18. https://doi.org/10.1016/j.envpol.2014.05.001.
149. Gigliotti C, Allievi L, Salardi C, Ferrari F, Farini A. Microbial ecotoxicity and persistence in soil of the herbicide bensulfuron-methyl. *J Environ Sci Health B* 1998;**33**(4):381–98. https://doi.org/10.1080/03601239809373152.
150. Khan M, Zaidi A, Aamil M. Influence of herbicides on Chickpea-*Mesorhizobium* symbiosis. *Agronomie* 2004;**24**:https://doi.org/10.1051/agro:2004009.
151. Lane M, Lorenz N, Saxena J, Ramsier C, Dick RP. The effect of glyphosate on soil microbial activity, microbial community structure, and soil potassium. *Pedobiology* 2012;**55**(6):335–42. https://doi.org/10.1016/j.pedobi.2012.08.001.
152. Zabaloy MC, Gómez E, Garland JL, Gómez MA. Assessment of microbial community function and structure in soil microcosms exposed to glyphosate. *Appl Soil Ecol* 2012;**61**:333–9. https://doi.org/10.1016/j.apsoil.2011.12.004.

153. Laitinen P, Rämö S, Siimes K. Glyphosate translocation from plants to soil – does this constitute a significant proportion of residues in soil? *Plant Soil* 2007;**300**:51–60.
154. Lupwayi NZ, Harker KN, Clayton GW, O'Donovan JT, Blackshaw RE. Soil microbial response to herbicides applied to glyphosate-resistant canola. *Agric Ecosyst Environ* 2009;**129**(1):171–6. https://doi.org/10.1016/j.agee.2008.08.007.
155. van Hoesel W, Tiefenbacher A, König N, et al. Single and combined effects of pesticide seed dressings and herbicides on earthworms, soil microorganisms, and litter decomposition. *Front Plant Sci* 2017;**8**:215. https://doi.org/10.3389/fpls.2017.00215.
156. Hill TL, Stratton GW. Interactive effects of the fungicide chlorothalonil and the herbicide metribuzin towards the fungal pathogenAlternaria solani. *Bull Environ Contam Toxicol* 1991;**47**(1):97–103. https://doi.org/10.1007/BF01689459.
157. Schuster E, Schröder D. Side-effects of sequentially-applied pesticides on non-target soil microorganisms: field experiments. *Soil Biol Biochem* 1990;**22**(3):367–73. https://doi.org/10.1016/0038-0717(90)90115-G.
158. Heydari A, Misaghi IJ, Balestra G. Pre-emergence herbicides influence the efficacy of fungicides in controlling cotton seedling damping-off in the field. *Int J Agric Res* 2007;**2**:1049–53. https://doi.org/10.3923/ijar.2007.1049.1053.
159. Almasri H, Tavares DA, Pioz M, et al. Mixtures of an insecticide, a fungicide and a herbicide induce high toxicities and systemic physiological disturbances in winter *Apis mellifera* honey bees. *Ecotox Env Safety* 2020;**203111013**:https://doi.org/10.1016/j.ecoenv.2020.111013.
160. Seghers D, Verthé K, Reheul D, et al. Effect of long-term herbicide applications on the bacterial community structure and function in an agricultural soil. *FEMS Microbiol Ecol* 2003;**46**:139–46. https://doi.org/10.1016/S0168-6496(03)00205-8.
161. Seghers D, Wittebolle L, Top EM, Verstraete W, Siciliano SD. Impact of agricultural practices on the *Zea mays* L. endophytic community. *Appl Environ Microbiol* 2004;**70**(3):1475–82. https://doi.org/10.1128/aem.70.3.1475-1482.2004.
162. Rose MT, Cavagnaro TR, Scanlan CA, et al. Impact of herbicides on soil biology and function. In: Donald LS, editor. *Advances in agronomy*. Academic Press; 2016. p. 133–220.
163. Mullin CA, Chen J, Fine JD, Frazier MT, Frazier JL. The formulation makes the honey bee poison. *Pestic Biochem Physiol* 2015;**120**:27–35. https://doi.org/10.1016/j.pestbp.2014.12.026.
164. Mullin CA, Fine JD, Reynolds RD, Frazier MT. Toxicological risks of agrochemical spray adjuvants: organosilicone surfactants may not be safe. Mini review. *Front Public Health* 2016;**4**:92. https://doi.org/10.3389/fpubh.2016.00092.
165. Cuhra M, Bøhn T, Cuhra P. Glyphosate: too much of a good thing? Review. *Front Environ Sci* 2016;**4**:28. https://doi.org/10.3389/fenvs.2016.00028.
166. Moore LJ, Fuentes L, Rodgers JHJ, et al. Relative toxicity of the components of the original formulation of roundup to five north American anurans. *Ecotox Env Safety* 2012;**78**:128–33.
167. Relyea RA. The impact of insecticides and herbicides on the biodiversity and productivity of aquatic communities. *Ecol Appl* 2005;**15**(2):618–27.
168. Wu Y. *Periphyton: functions and application in environmental remediation*. 434pp.
169. Zhao Q, De Laender F, Van den Brink PJ. Community composition modifies direct and indirect effects of pesticides in freshwater food webs. *Sci Total Environ* 2020;**739**:139531. https://doi.org/10.1016/j.scitotenv.2020.139531.

170. Hasenbein S, Lawler SP, Connon RE. An assessment of direct and indirect effects of two herbicides on aquatic communities. *Environ Toxicol Chem* 2017;**36**(8):2234–44. https://doi.org/10.1002/etc.3740.
171. Schreiner VC, Szöcs E, Bhowmik AK, Vijver MG, Schäfer RB. Pesticide mixtures in streams of several European countries and the USA. *Sci Total Environ* 2016;**573**:680–9. https://doi.org/10.1016/j.scitotenv.2016.08.163.
172. Rybicki M, Jungmann D. Direct and indirect effects of pesticides on a benthic grazer during its life cycle. *Environ Sci Eur* 2018;**30**(1):35. https://doi.org/10.1186/s12302-018-0165-x.

Analytical strategies to measure toxicity and endocrine-disrupting effects of herbicides: Predictive toxicology

9

Robin Mesnage
King's College London, London, United Kingdom

Chapter outline

Introduction

Exposure to herbicides is frequently linked to adverse health outcomes in human populations. This includes acute intoxication but also the development of chronic conditions such as cancer, neurodegenerative diseases, and metabolic problems (see Chapter 6).

The potential human health effects of herbicides are not always correctly predicted from the battery of animal bioassays performed before the products are made available to consumers. There is a delay between the introduction of new synthetic chemicals and the detection of their health effects. Harmful chemicals can be regulated after decades of contradictory debates, as has been the case for some drugs (e.g., thalidomide, diethylstilboestrol), asbestos, plasticizers, and heavy metals, which were all considered to be safe while they were used on a large scale over the last century.[1] This is also the case now for pesticides. For instance, exposure to terbufos, deltamethrin, and glyphosate has been linked to an increased risk of developing non-Hodgkin

Herbicides. https://doi.org/10.1016/B978-0-12-823674-1.00011-0

lymphoid malignancies in cohorts of agricultural workers from France, Norway, and the United States.[2] Another example is the commonly used herbicide dicamba, which was also associated with an increased risk of liver and intrahepatic bile duct cancer.[3] Even though some animal studies suggested that dicamba could be a liver carcinogen,[4] it is not classified as such by regulatory authorities.

Debate on whether herbicides can be harmful to humans can be influenced by private, political, and economic interests.[5] Heated controversy arose after glyphosate applicators filed lawsuits claiming that they developed non-Hodgkin lymphoma.[6] Controversies about health effects can also originate in the inability of regulatory tests to identify risks. This chapter provides an overview of the strategies used to evaluate health effects, and cites case studies for common herbicides.

From food tasters to modern rodent bioassays

Food and chemical toxicity has long been a concern and was the subject of experiments before the implementation of regulations based on the results of rodent toxicity bioassays. Although animal testing probably dates to ancient times when hunter-gatherers tested the toxicity of plants on domesticated animals, humans seem to have been the early model of choice. Concerns about intoxication by chemicals can be traced to ancient Egypt, when slaves served as food tasters to prevent the assassination of royalty.[7] Assassination by poison—chemical terrorism—is still a major issue. For instance, the herbicide paraquat was used to lace a bottled beverage and kill 13 people in Japan in 1985.[8]

Rodents were the preferred species from the 16th century on. One of the first reports was by Joseph Priestley, who placed mice in jars of gases to study their toxicity in the 1770s.[9] The first known study of chemical carcinogenicity was John Hill's observation that individuals who used tobacco snuff had higher rates of nasal cancer.[10] Systematic animal studies started 150 years later, when Katsusaburo Yamagiwa experimentally reproduced occupational exposure to coal tar by applying it to the ears of rabbits.[11] Chemical toxicity was systematized in 1927 when J.W. Trevan proposed the 50% lethal dose (LD_{50}) as an index.[12]

Although toxicological studies were common at the beginning of the 20th century, the findings were not always translated into regulations. The first landmark chemical toxicity regulation was set when the Food, Drug, and Cosmetics Act of 1938 was signed in the United States to require premarket approval of drugs, after an untested formulation of sulfanilamide killed 73 people.[13] The first guidelines to standardize testing for acute, subacute, and chronic toxicity in animals appeared in the early 1950s,[14] when pesticides started to be used on a large scale. The 1947 Federal Insecticide, Fungicide, and Rodenticide Act (FIFRA) implemented a set of rules requiring the registration of pesticides before marketing.[15] Toxicity tests were incrementally improved over the next decades, and the modern version was set in 1981 when member countries of the Organization for Economic Co-operation and Development implemented the OECD Test Guidelines for Chemicals and OECD Principles of Good Laboratory Practice.[16]

OECD guidelines for the testing of chemicals

The guidelines and the good laboratory practice framework were introduced to ensure the reliability and reproducibility of safety tests, including toxicity tests, and the characterization of physicochemical properties. Implementation was prompted by a series of fraudulent practices in testing laboratories, the most famous being Industrial Bio-Test Laboratories (IBT), which was involved in the safety testing of many widely used household and industrial products, including 325 insecticides and herbicides.[17] IBT performed 35%–40% of all toxicology studies in the United States over its last decade of operation.[17]

Since their implementation in 1981, OECD guidelines have been continuously improved to ensure that they reflect state-of-the-art science and techniques. There are four sections to test physical chemical properties: effects on biotic systems, environmental fate, behavior, and health.

According to the FAO/WHO data requirements for the registration of pesticides,[18] the set of OECD test guideline (TG) bioassays used to evaluate health effects consists of the study of:

- **Absorption, distribution, metabolism, and excretion**. Measurement of the rate and extent of oral absorption, the distribution and potential for accumulation, the rate and extent of excretion, metabolism, and toxicologically relevant compounds (TG 417).
- **Acute toxicity**. LD_{50} for oral (TG 401), dermal (TG 402), and inhalation (TG 403) in rat; skin (TG 404) and eye irritation (TG 405) in rabbit; skin sensitization in guinea pig (TG 406).
- **Short-term toxicity**. Four-week study in rat (TG 407); 13-week feeding study in rodents (TG 408) or nonrodents (dog, TG 409).
- **Genotoxicity**. Bacterial reverse mutation assay (Ames test, TG 471); in vitro mammalian cell mutation (e.g., mouse lymphoma TK assay, TG 476), and chromosome aberration assay (TG 473 or 487); in vivo cytogenetics (micronucleus rodent bone marrow assay, TG 474 preferred).
- **Chronic toxicity and carcinogenicity**. Chronic toxicity (1 year) alone (TG 452); 2-year carcinogenicity (TG 451); combined chronic toxicity/carcinogenicity (TG 453) in two rodent species (rat and mouse preferred).
- **Reproductive toxicity**. Prenatal developmental toxicity (TG 414, rat and rabbit preferred); reproduction and fertility effects in rat (TG 421).
- **Neurotoxicity**. Acute delayed neurotoxicity in hen (TG 418); acute neurotoxicity in rat (TG 424); developmental neurotoxicity in rat (TG 426).

Although this battery is suitable to predict general toxic effects, it is not always sufficient to detect chronic metabolic or endocrine diseases, attention deficit disorder, autism, diabetes, obesity, and childhood cancers. These diseases can occur after in utero exposure to a combination of chemicals disrupting hormone signalling and appear only in adulthood. In the case of endocrine disruption, although OECD guidelines exist to detect alterations of endocrine function,[19] they are not always required by the European Union or US regulations.[20] In addition, 48 human nuclear receptors exist, but effects are mostly focused on estrogen, androgen, and thyroid receptors.[21]

An increasing number of researchers have called for the modification of testing guidelines to better reflect real-life exposure, specifically sequential and combined exposure to low levels of multiple chemicals.[22]

Designing animal bioassays to inform human health risk assessment

As stated above, chronic and repeated-dose exposure studies may not accurately predict effects in human populations because they do not reflect real-life exposure. This weakness can be addressed by changes in the following aspects of experimental design:

- **Dose range**. Current animal studies use high doses to predict low-dose effects by starting with a maximum tolerable level group and decreasing the level in sequential groups until a no-observable-adverse-effect level is found. This assumes that effects are linear. In some cases, effects are nonlinear, and metabolic perturbations can be detected at lower environmental levels that are rarely assessed in regulatory tests.[23] A classic example is the effect of the herbicide atrazine on amphibian development.[24]
- **Period of exposure**. Exposure to environmental chemicals during early life can have short- and long-term consequences on human health.[25] Animal studies starting chemical exposure in utero and continuing until the end of life have revealed the carcinogenic effects of chemicals that were not detectable with current OECD guidelines.[26]. A known example is the carcinogenicity testing strategy implemented by the Ramazzini Institute (Fig. 9.1) conducting lifetime cancer bioassays in rodents since the 1970s.[26] Effects could also be passed to descendants through epigenetic changes. For instance, exposure to the herbicide Agent Orange during the Vietnam War had effects on sperm, which could transmit adverse health to the next generation.[27] The fungicide vinclozolin can even have transgenerational effects through the male germ line to males up to the fourth generation.[28]
- **Exposure route**. Most short-term animal bioassays use intragastric gavage to mimic the effects of oral exposure. The tube bypasses surfaces in the oral cavity, which can be highly permeable. In some cases, this has been shown to bias the generalization of animal results to human exposure.[29] Absorption can also be influenced by the presence of lipophilic compounds, a mixture effect.[30]
- **Cocktail effects**. Humans are rarely exposed to single compounds. Predicting the effects of chemical mixtures is a challenge. Pesticides can interact, as in the potentiation of paraquat induction of Parkinson's disease by the fungicide maneb,[31] or with other compounds such as

Fig. 9.1 Experimental toxicology at the Cesare Maltoni Cancer Research Center, Ramazzini Institute, Bologna, Italy. The Ramazzini Institute has worked for years with governmental institutions to develop long-term carcinogenesis bioassays,[26] testing more than 200 compounds, including herbicides such as glyphosate.

the increased triazine herbicide exposure caused by antibiotic-mediated disturbances of the gut microbiome.[32] Combined effects can also be expected with bioaccumulated pollutants, which increase the susceptibility to other pollutants.[33] Solutions have been proposed,[34] such as the prioritization of the study of combinations of compounds sold together or occurring together in the environment.

- **Species model**. Some current animal models fail to translate accurately to humans. This happens for complex chronic diseases, notably behavioral[35] and endocrine-mediated disorders.[36] Disease modeling in human cells or organoids is increasingly reliable and more accurate than whole animal models for understanding human physiological responses.[37] Organ-on-a-chip technologies even reproduce functional units of human organs and their interactions, which can be used to model liver mechanistic toxicity,[38] kidney drug transport,[39] and intestinal function,[40] moving toward human body-on-a-chip systems.[41]
- **The gut microbiome**. The microbial xenobiotic metabolism influences the development of a large range of diseases, some locally at the intestinal level but also at a distance in other organs.[42] Many herbicides are considered safe for humans because their mode of action is not present in mammalian cells. However, when a herbicide affects gut bacteria, there can be follow-on effects not predicted by the current battery of toxicity assays, as in the case of glyphosate.[43]
- **Rodent diet**. Rodent diets are frequently contaminated by a variety of pesticide residues.[44, 45] This can mask toxic effects when exposed experimental animals are compared to control animals falsely assumed to be unexposed.[46]
- **Analytical methods**. The current battery of tests run according to OECD guidelines relies on standard histological and biochemical assays, the sensitivity of which has been outpaced by recently developed biomarkers. For instance, the measurement of kidney injury molecule-1 (KIM-1) can detect early renal damage that is not detectable with traditional biomarkers.[47] In farming communities with high rates of chronic kidney diseases suspected to be caused by high exposure to pesticides, KIM-1 was found to be elevated and a good predictor of early renal damage.[48] Recent studies also showed that using high-throughput assays can predict chronic diseases in short-term (5 day) toxicity tests.[49]

Predicting long-term effects with high-throughput and short-term assays

Moving away from animal tests: Adverse outcome pathways

Concerns have been raised regarding the reproducibility of preclinical research studies.[50] In a comparative analysis of the results of skin sensitization studies, a reproducibility of only 77% was found.[51] A comparison of 121 replicate rodent carcinogenicity tests found that they were reproducible only 57% of the time.[52] Interspecies concordance for cancer bioassays in mice and rats is poor, and these tests predict each other only 50% of the time.[53] Animal bioassays are increasingly criticized, and some scientists agree with the statement, "It is time to say goodbye to the standard 2-year rodent bioassay."[54]

Although they mostly relied on animal testing, OECD guidelines are progressively switching to mechanistically based in chemico and in vitro tests. This transition is reflected by the implementation of the adverse outcome pathway (AOP) framework, which aims at using mechanistic knowledge to support chemical safety assessment.[55]

AOPs consists of a molecular initiating event, the adverse outcome, and a variable of intermediate key events. Examples include:

- **Skin sensitization.** This was one of the earliest toxicological processes to be described by an AOP, in OECD testing guideline 168.[56] It is based on a common toxicological property of skin sensitizers: they are or can become electrophiles, which confers on them the property of reacting covalently with proteins. Key events for the skin sensitization AOP involve the production of inflammatory cytokines and the induction and proliferation of T-cells.
- **Pesticide toxicity to bees.** The prediction of pesticide toxicity in honeybees is done via the activation of the nicotinic acetylcholine receptor.[57]
- **Nonalcoholic fatty liver disease.** The AOP recapitulates the toxicological process leading to chronic conditions such as nonalcoholic fatty liver disease. An AOP-based study using HepaRG liver cells found that the activation of the receptors RARα and PXR was concomitant with lipid accumulation and mitochondrial dysfunction after exposure to the pesticide cyproconazole, a nonalcoholic fatty liver disease inducer.[58]
- **Endocrine-disrupting effects.** The use of in vitro assays to test the binding of chemicals to a hormone receptor involved in the development of endocrine-related diseases is a common basis to create AOPs. In tier 1 of the US Environmental Protection Agency (EPA) Endocrine Disruptor Screening Program, 52 pesticides were tested and 18 were found to interact with the thyroid, androgen, or estrogen pathway.[59]
- **Parkinson's disease.** The effect of the pesticide rotenone (Fig. 9.2) on the inhibition of mitochondrial respiratory chain activity in neurons leads to Parkinsonian motor deficits.[60, 61]

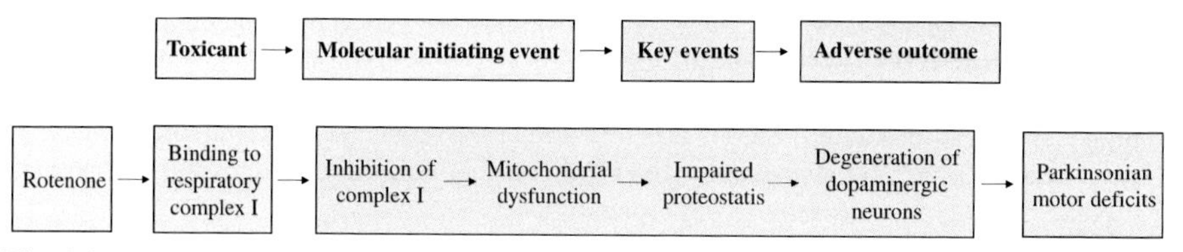

Fig. 9.2 The adverse outcome pathway framework, with the example of rotenone causing Parkinsonian motor deficits.

Toxicity forecasting

High-throughput screening assays such as those implemented by the US Tox21 collaborative program can uncover metabolic effects.[62] In the initial ToxCast program, the EPA used 700 high-throughput assay endpoints on approximately 1800 chemicals to rank and prioritize them on their potential to disturb various cellular pathways. It was estimated that the Tox21 program generated more than 100 million data points for more than 8000 chemicals. This can support mechanistic evaluation. In a recent study of the endocrine-disrupting properties of seven neonicotinoid insecticides, imidacloprid stimulated lipid accumulation in adipocytes.[63] Data mining of ToxCast high-throughput screening assays showed that this adipogenic effect was probably mediated by activation of the pregnane X receptor. In another study, ToxCast data suggested that the herbicide isoxaflutole activated the PPAR gamma receptor and pregnane X responsive elements in reporter gene assays, providing a mechanistic understanding of alterations in gene expression pointing to disruption in the lipid metabolism in liver cells.[64] Comparisons of the similarity of biological responses from ToxCast assays can also elucidate the mode of action of herbicides. In profiling 976 ToxCast chemicals across 331 assays, two triazine herbicides (cyanazine and prometryn) clustered with caffeine and theophylline for a selective effect on human adenosine receptors, suggesting that these compounds could affect adenosine signalling.[65] Scoring responses across assays can also help with comparing the relative toxic potential of different chemicals, as was done for perturbations of the estrogenic effects.[66] However, as with all assays, interpretation and reliability are sometimes questionable, such as in the identification of potential obesogens.[67, 68]

High-throughput omics approaches

Another recent development is the implementation of high-throughput methods allowing the measurement of a large number of biomarkers in a single assay[69] to determine gene expression across the whole genome for transcriptomics, the determination of protein profiles for proteomics, and the determination of a large number of chemical compounds for metabolomics. Collectively known as omics methods, these tools can determine herbicide toxic effects,[70] and they can also be used to study the evolution and genetic basis of herbicide resistance[71] and to understand the mode of herbicidal action.[72]

Although omics methods have been progressively introduced in biomedical sciences since the early 2000s,[70] they found a place in toxicology when the US National Research Council published a vision and strategy statement on toxicity testing in 2007.[73] This started a transition toward high-throughput predictive approaches using mechanistic information (Fig. 9.3). One of the most important motivations for this transition was the need to reduce vertebral animal testing, reduce the cost of toxicity tests, and improve predictive ability by increasing knowledge about the impact on human physiology.

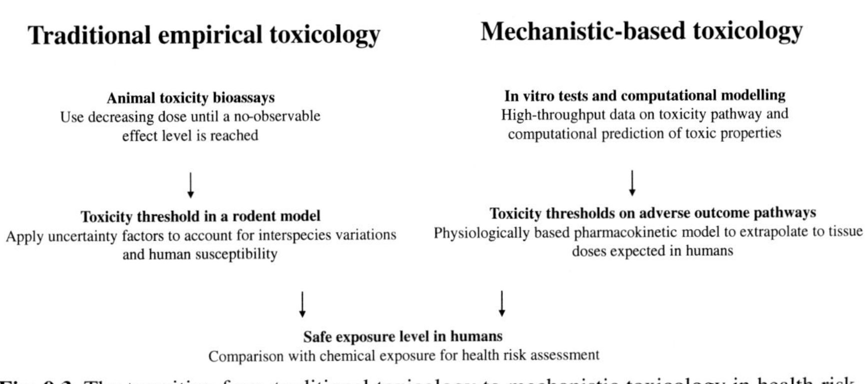

Fig. 9.3 The transition from traditional toxicology to mechanistic toxicology in health risk evaluation.

A limitation of omics is that they do not always provide a direct measure of adverse effects, but rather a snapshot of mechanisms of action. This is an important point because changes in physiology can occur without pathological consequences. The methods cannot be used in a traditional toxicology framework where the aim is to measure an observable toxic effect. However, they have their place in the most modern, mechanistic framework.

Most omics approaches are predictive. The study of gene expression with transcriptomics, or protein expression with proteomics, does not inform directly on biochemical changes. Gene expression profiling has been used to predict key events and outcomes in networks of adverse outcome pathways.[74] For instance, activation of the estrogen receptor, which could be an initiating event leading to the development of breast cancer, can be predicted by using a panel of 46 genes in human breast cancer cells in vitro.[75] We used this biomarker to assess the estrogenic effects of glyphosate (Fig. 9.4) and showed that activation occurs via a ligand-independent mechanism and only at relatively high concentrations.[76]

Metabolomics directly measures the levels of hundreds of biochemicals. Perturbations in the metabolome can thus be directly interpreted in terms of the toxicity process as well as providing mechanistic insights.[77,78] In a recent analysis of 130 human fecal

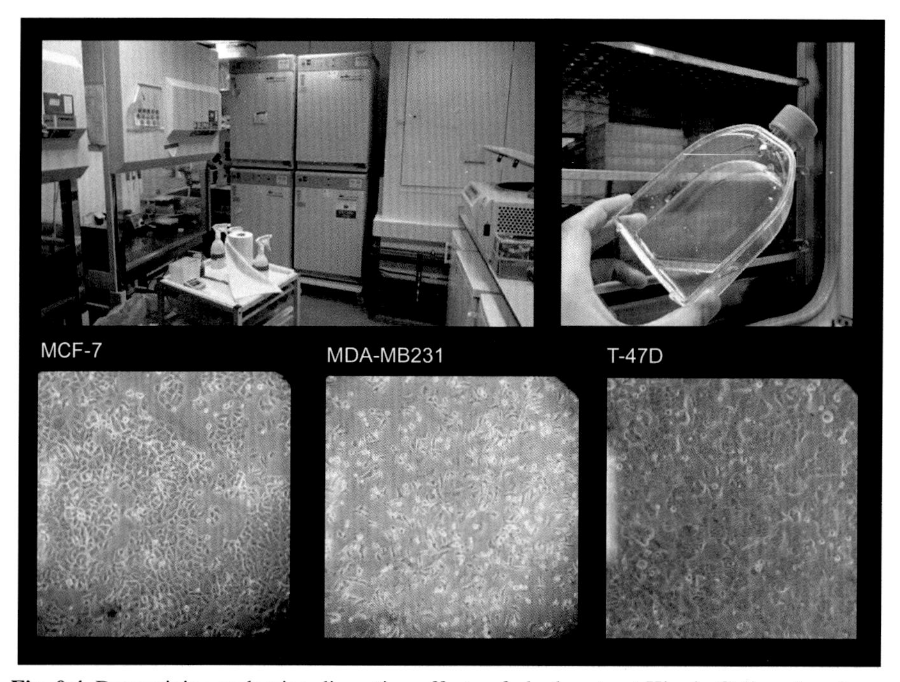

Fig. 9.4 Determining endocrine-disrupting effects of glyphosate at King's College London. Human cell-based assays using a panel of three breast cancer cell lines (MCF-7, MDA-MB231, T-47D) have been employed to study glyphosate estrogenic effects as described in Mesnage et al.[76] For this purpose, cells are cultivated in 75 cm^2 cell culture flasks (top right) in a tissue culture laboratory (top left).

samples by our group (not yet published), the metabolite profiles contained xenobiotics, including 84 food components and 46 drugs, together with a large number of endogenous compounds such as 197 amino acid derivatives, 30 carbohydrates, and 47 cofactors and vitamins as well as hundreds of lipids, steroids, corticosteroids, and endocannabinoids. A single metabolomics analysis can thus measure directly both pesticide levels and health consequences in a variety of biochemical networks (hormones, lipids, sugars, amino acids, metabolic intermediates). In our recently published study measuring host-gut microbiota interactions in rats exposed to glyphosate, we determined both glyphosate levels and the consequences of their inhibition of the shikimate pathway on microbial and blood metabolism.[43] Metabolomics have also been used to perform read-across evaluations of the toxic effects of phenoxy herbicides.[79] This suggested that a 28-day metabolomic study could predict toxic effects observed in a 90-day rat study. Untargeted metabolomics was also successfully used to discover early-response metabolic biomarkers predicting the reproductive fitness of *Daphnia magna* exposed to cadmium, 2,4-dinitrophenol, and propranolol.[80] Our group also showed that multi-omics phenotyping of the gut-liver axis allowed the health risk predictability of the effects of a mixture of six pesticide active ingredients frequently detected in foodstuffs with each ingredient at its regulatory permitted acceptable daily intake.[81]

Although omics technologies are sufficiently mature to inform health risk assessment, there is a need for standardization and for implementation of a consistent reporting framework for data collection, processing, interpretation, storage, and curation.[82] This is particularly true for emerging toxicity endpoints and tests such as the identification of microorganisms at species and strain levels using shotgun metagenomics. There are many experimental biases that could profoundly alter the results of a shotgun metagenomics study, including types of reagents, preservation methods, types of sequencing technologies, and the bioinformatics software used to extract biological information.

Herbicides in the era of big data

In the future, the toxicity of a new compound might be predictable from the knowledge accumulated from the study of other chemicals using artificial intelligence. Machine learning of big data from structure-activity relationship experiments is already suggested to be more accurate in predicting the effects of new chemicals than current animal-based approaches.[83] In a recent study of the six most common OECD test guidelines representing the use of 55% of the animals in toxicology in the EU (~600,000 rodents per year), machine learning achieved 87% accuracy and outperformed animal tests, which showed 81% accuracy.[84]

These computational methods allowing predictive modeling have a large range of applications. For instance, the training of a machine-learning algorithm on the gene expression alterations observed after the exposure of fish ovaries to different endocrine disruptors adequately classified the herbicide linuron, which has an antiandrogenic mode of action.[85] Besides toxicological studies, machine learning has also predicted the environmental impact of glyphosate as a function of soil variables from a regional soil survey,[86] and predicted the genes that could be involved in weed resistance.[87]

Remarks

The toxicological testing of herbicide effects is moving away from traditional animal testing. Modern methods allow quick, cheap, and effective evaluation of the effects on human physiology to set exposure limits. In many cases, high-throughput screening in a model coupled with a computational approach outperforms traditional animal testing. However, the adoption of new technologies is slow. Herbicide safety is re-evaluated every 15 years in the EU and the United States. Although testing and regulatory frameworks are available, they are not always fully implemented. The adoption of mechanistic-based toxicology as part of regulatory risk assessment can produce more sensitive and accurate prediction of long-term health risks. This should lead to better public health regulations.

Close-up: Endocrine-disrupting effects

Agrochemicals that are EDCs

This chapter describes some of the approaches as well as the challenges in evaluating the toxicity of chemicals such as herbicides. Endocrine disruption is a specific kind of toxicity that is governed by the principles of endocrinology,[88] and evaluating endocrine-disrupting chemicals (EDCs) requires a basic understanding of how hormones act.[89] These principles include the following: (1) hormones act at very low doses/concentrations; (2) hormones act via specific interactions with receptors; (3) hormones coordinate the health and function of all the tissues and organs of the body; (4) hormones have specific roles at different stages of life, from conception through aging; and (5) hormones often induce responses that are nonlinear, and even nonmonotonic, relative to dose/concentration.

In the EU, herbicides are regulated by two legislative actions, the 2009 *Plant Protection Products Regulation*[90] and the 2012 *Biocidal Products Regulation*,[91] which allow the use of an EDC on crops only if exposures are negligible. If herbicides (or other plant-protecting pesticides) are found to be EDCs, they must be listed as Substances of Very High Concern, which means that they will be phased out when safer alternatives are identified.[92]

In the United States, the 1996 *Food Quality Protection Act* requires the EPA to test all pesticides to determine if they are estrogenic.[93] To address this concern, the EPA developed a two-tiered suite of assays to be implemented in the Endocrine Disruptor Screening Program (EDSP).[94,95] The tier 1 EDSP assays include OECD tests for estrogenicity, androgenicity, antiandrogenicity, and alterations to steroidogenesis.

OECD test guidelines are commonly used to identify, characterize, and ultimately quantify the effects of chemicals on a range of outcomes. For the identification of EDCs, a few of these guideline assays are considered sufficient by regulators to evaluate (anti)estrogenic, (anti)androgenic, or thyroid-disrupting chemicals.[96]

In first-pass screening tests, chemicals can be evaluated for estrogenic (and antiestrogenic) behavior using a standard uterotrophic assay[97]: young female rodents, or female rodents that have had their ovaries removed, are administered a test chemical over a period of several days, and the weight of the uterus is measured at the end of the administration period. Chemicals can be evaluated for antiandrogenic (or androgenic) properties using the Hershberger assay[98]: young male rodents are administered the test chemical with or without a testosterone propionate, a potent androgen, over a period of several days, and five androgen-sensitive tissues are weighed at the end of the administration period.

In 2020, we reviewed the evidence for three agrochemicals that have been described as EDCs: DDT, endosulfan, and atrazine.[99] All had been extensively studied for endocrine-disrupting properties using OECD test guidelines, EDSP tier 1 screening tests, and similar assays. All three have been described as estrogens, but we found that all three have more than one mechanism of action contributing to their endocrine-disrupting properties.

A new approach, the Key Characteristics, is now available, which describes 10 key features of EDCs.[21] These offer a new way for scientists and regulators to organize information about chemicals when determining if they should be considered EDCs. The Key Characteristics identify chemicals that interact with or activate hormone receptors; antagonize hormone receptors; alter hormone receptor expression; alter signal transduction in hormone-responsive cells; induce epigenetic modifications in hormone-producing or hormone-responsive cells; alter hormone synthesis; alter hormone transport across cell membranes; alter hormone distribution or circulating hormone levels; alter hormone metabolism or clearance; and alter the fate of hormone-producing or hormone-responsive cells. This list was built based on the knowledge and studies of known EDCs, and it can be used to evaluate herbicides and other agrochemicals. Although OECD test guidelines are not yet established for several of these characteristics,[100] other well-developed methods are becoming available for use.[101–105]

Many debates around EDCs continue to affect the study, prioritization, and regulation of these chemicals. Unfortunately, some debates appear to be manufactured by individuals and organizations with vested interests in continuing to produce and use agrochemicals without restrictions.[106,107] Endocrinologists continue to press for the development of sensitive methods to identify EDCs,[108] the use of appropriate positive and negative controls in EDC studies,[89] the use of appropriate routes of exposure in animal tests,[29] and the use of health data when considering the cost of EDC usage.[109] Addressing these and other issues will enable better public health-protective decisions about the use of agrochemicals.

Laura N. Vandenberg
University of Massachusetts Amherst

References

1. EEA. *European Environment Agency. Late lessons from early warnings: science, precaution, innovation.* ISBN 978-92-9213-349-8; 2013. https://doi.org/10.2800/70069.
2. Leon ME, Schinasi LH, Lebailly P, et al. Pesticide use and risk of non-Hodgkin lymphoid malignancies in agricultural cohorts from France, Norway and the USA: a pooled analysis from the AGRICOH consortium. *Int J Epidemiol* 2019;**48**(5):1519–35.
3. Lerro CC, Hofmann JN, Andreotti G, et al. Dicamba use and cancer incidence in the agricultural health study: an updated analysis. *Int J Epidemiol* 2020.
4. Espandiari P, Glauert H, Lee E, Robertson L. Promoting activity of the herbicide dicamba (2-methoxy-3, 6-dichlorobenzoic acid) in two stage hepatocarcinogenesis. *Int J Oncol* 1999;**14**(1):79–163.

5. Robinson C, Portier CJ, ČAvoŠKi A, et al. Achieving a high level of protection from pesticides in Europe: problems with the current risk assessment procedure and solutions. *Eur J Risk Regul* 2020;1–31.
6. Benbrook C. Shining a light on glyphosate-based herbicide hazard, exposures and risk: role of non-Hodgkin lymphoma litigation in the USA. *Eur J Risk Regul* 2020;**11**(3): 498–519. https://doi.org/10.1017/err.2020.16.
7. Norwood FB, Oltenacu PA, Calvo-Lorenzo MS, Lancaster S. *Agricultural and food controversies: what everyone needs to know*. Oxford University Press; 2015.
8. Okumura T, Seto Y, Fuse A. Countermeasures against chemical terrorism in Japan. *Forensic Sci Int* 2013;**227**(1):2–6.
9. Gad SC. *Animal models in toxicology*. CRC Press; 2016.
10. Hill J. In: Baldwin R, Jackson J, editors. *Cautions against the immoderate use of snuff. Founded on the known qualities of the tobacco plant; and the effects it must produce when this way taken into the body*; 1761.
11. Yamagiwa K, Ichikawa K. Experimental study of the pathogenesis of carcinoma. *J Cancer Res* 1918;**3**(1):1–29.
12. Trevan JW. The error of determination of toxicity. *Proc R Soc Lond Ser B Biol Char* 1927;**101**(712):483–514.
13. Wax PM. Elixirs, diluents, and the passage of the 1938 Federal Food, Drug and Cosmetic Act. *Ann Intern Med* 1995;**122**(6):456–61.
14. Weideman M. Toxicity tests in animals: historical perspectives and new opportunities. *Environ Health Perspect* 1993;**101**(3):222–5.
15. Conner JD, McKenna C, Cuneo. *Pesticide regulation handbook*. Executive Enterprises Publications; 1983.
16. Cooperation of E Development. *Decision of the council concerning the mutual acceptance of data in the assessment of chemicals*. Paris, France: Organization for Economic Cooperation and Development; 1981.
17. Schneider K. *Faking it*; 1983.
18. Organization WH. *International code of conduct on the distribution and use of pesticides: guidelines for the registration of pesticides*. World Health Organization; 2010.
19. Manibusan MK, Touart LW. A comprehensive review of regulatory test methods for endocrine adverse health effects. *Crit Rev Toxicol* 2017;**47**(6):440–88.
20. Kassotis CD, Vandenberg LN, Demeneix BA, Porta M, Slama R, Trasande L. Endocrine-disrupting chemicals: economic, regulatory, and policy implications. *Lancet Diabetes Endocrinol* 2020;**8**(8):719–30.
21. La Merrill MA, Vandenberg LN, Smith MT, et al. Consensus on the key characteristics of endocrine-disrupting chemicals as a basis for hazard identification. *Nat Rev Endocrinol* 2020;**16**(1):45–57.
22. Margina D, Nițulescu GM, Ungurianu A, et al. Overview of the effects of chemical mixtures with endocrine disrupting activity in the context of real-life risk simulation (RLRS): an integrative approach. *World Acad Sci J* 2019;**1**(4):157–64.
23. Vandenberg LN, Colborn T, Hayes TB, et al. Hormones and endocrine-disrupting chemicals: low-dose effects and nonmonotonic dose responses. *Endocr Rev* 2012;**33**(3):378–455.
24. Hayes TB, Collins A, Lee M, et al. Hermaphroditic, demasculinized frogs after exposure to the herbicide atrazine at low ecologically relevant doses. *Proc Natl Acad Sci U S A* 2002;**99**(8):5476–80.
25. Heindel JJ, Vandenberg LN. Developmental origins of health and disease: a paradigm for understanding disease cause and prevention. *Curr Opin Pediatr* 2015;**27**(2):248–53.

26. Soffritti M, Belpoggi F, Esposti DD, Falcioni L, Bua L. Consequences of exposure to carcinogens beginning during developmental life. *Basic Clin Pharmacol Toxicol* 2008;**102**(2):118–24.
27. Kelsey KT, Rytel M, Dere E, et al. Serum dioxin and DNA methylation in the sperm of operation ranch hand veterans exposed to agent Orange. *Environ Health* 2019;**18**(1):91.
28. Anway MD, Cupp AS, Uzumcu M, Skinner MK. Epigenetic transgenerational actions of endocrine disruptors and male fertility. *Science* 2005;**308**(5727):1466–9.
29. Vandenberg LN, Welshons WV, Vom Saal FS, Toutain P-L, Myers JP. Should oral gavage be abandoned in toxicity testing of endocrine disruptors? *Environ Health* 2014;**13**(1):46.
30. Zeliger HI. Toxic effects of chemical mixtures. *Arch Environ Health: Int J* 2003;**58**(1):23–9.
31. Thiruchelvam M, McCormack A, Richfield EK, et al. Age-related irreversible progressive nigrostriatal dopaminergic neurotoxicity in the paraquat and maneb model of the Parkinson's disease phenotype. *Eur J Neurosci* 2003;**18**(3):589–600.
32. Zhan J, Liang Y, Liu D, et al. Antibiotics may increase triazine herbicide exposure risk via disturbing gut microbiota. *Microbiome* 2018;**6**(1):224.
33. Kalantzi OI, Hewitt R, Ford KJ, et al. Inter-individual differences in the ability of human milk-fat extracts to enhance the genotoxic potential of the procarcinogen benzo[a]pyrene in MCF-7 breast cells. *Environ Sci Technol* 2004;**38**(13):3614–22.
34. Bopp SK, Barouki R, Brack W, et al. Current EU research activities on combined exposure to multiple chemicals. *Environ Int* 2018;**120**:544–62.
35. Garner JP. The significance of meaning: why do over 90% of behavioral neuroscience results fail to translate to humans, and what can we do to fix it? *ILAR J* 2014;**55**(3):438–56.
36. Patisaul HB, Fenton SE, Aylor D. Animal models of endocrine disruption. *Best Pract Res Clin Endocrinol Metab* 2018;**32**(3):283–97.
37. Kim J, Koo B-K, Knoblich JA. Human organoids: model systems for human biology and medicine. *Nat Rev Mol Cell Biol* 2020.
38. Jang K-J, Otieno MA, Ronxhi J, et al. Reproducing human and cross-species drug toxicities using a liver-chip. *Sci Transl Med* 2019;**11**(517), eaax5516.
39. Jang KJ, Mehr AP, Hamilton GA, et al. Human kidney proximal tubule-on-a-chip for drug transport and nephrotoxicity assessment. *Integr Biol (Camb)* 2013;**5**(9):1119–29.
40. Kasendra M, Luc R, Yin J, et al. Duodenum intestine-chip for preclinical drug assessment in a human relevant model. *elife* 2020;**9**.
41. Dehne E-M, Marx U. Human body-on-a-chip systems. In: Hoeng J, Bovard D, Peitsch MC, editors. *Organ-on-a-chip*. Academic Press; 2020. p. 429–39 [chapter 13].
42. Koppel N, Maini Rekdal V, Balskus EP. Chemical transformation of xenobiotics by the human gut microbiota. *Science* 2017;**356**(6344).
43. Mesnage R, Teixeira M, Mandrioli D, et al. Use of shotgun metagenomics and metabolomics to evaluate the impact of glyphosate or Roundup MON 52276 on the gut microbiota and serum metabolome of Sprague-Dawley rats. *Environ Health Perspect* 2021;**129**(1):17005.
44. Mesnage R, Defarge N, Rocque LM, Spiroux de Vendomois J, Seralini GE. Laboratory rodent diets contain toxic levels of environmental contaminants: implications for regulatory tests. *PLoS One* 2015;**10**(7):e0128429.
45. Mesnage R, Defarge N, Spiroux de Vendomois J, Seralini GE. Letter to the editor regarding "Delaney et al., 2014": uncontrolled GMOs and their associated pesticides make the conclusions unreliable. *Food Chem Toxicol* 2014;**72**:322.

46. Kozul CD, Nomikos AP, Hampton TH, et al. Laboratory diet profoundly alters gene expression and confounds genomic analysis in mouse liver and lung. *Chem Biol Interact* 2008;**173**(2):129–40.
47. Han WK, Bailly V, Abichandani R, Thadhani R, Bonventre JV. Kidney injury molecule-1 (KIM-1): a novel biomarker for human renal proximal tubule injury. *Kidney Int* 2002;**62**(1):237–44.
48. De Silva PMCS, Mohammed Abdul KS, Eakanayake EMDV, et al. Urinary biomarkers KIM-1 and NGAL for detection of chronic kidney disease of uncertain etiology (CKDu) among agricultural communities in Sri Lanka. *PLoS Negl Trop Dis* 2016;**10**(9), e0004979.
49. Gwinn WM, Auerbach SS, Parham F, et al. Evaluation of 5-day in vivo rat liver and kidney with high-throughput transcriptomics for estimating benchmark doses of apical outcomes. *Toxicol Sci* 2020;**176**(2):343–54.
50. Freedman LP, Cockburn IM, Simcoe TS. The economics of reproducibility in preclinical research. *PLoS Biol* 2015;**13**(6), e1002165.
51. Kolle SN, Basketter DA, Casati S, et al. Performance standards and alternative assays: practical insights from skin sensitization. *Regul Toxicol Pharmacol* 2013;**65**(2):278–85.
52. Gottmann E, Kramer S, Pfahringer B, Helma C. Data quality in predictive toxicology: reproducibility of rodent carcinogenicity experiments. *Environ Health Perspect* 2001;**109**(5):509–14.
53. Hartung T. Opinion versus evidence for the need to move away from animal testing. *ALTEX* 2017;**34**(2):193–200.
54. Goodman JI. Goodbye to the bioassay. *Toxicol Res (Camb)* 2018;**7**(4):558–64.
55. Coady K, Browne P, Embry M, et al. When are adverse outcome pathways and associated assays "fit for purpose" for regulatory decision-making and Management of Chemicals? *Integr Environ Assess Manag* 2019;**15**(4):633–47.
56. Willett C. The use of adverse outcome pathways (AOPs) to support chemical safety decisions within the context of integrated approaches to testing and assessment (IATA). In: *Paper presented at: alternatives to animal testing; 2019; Singapore*; 2019.
57. LaLone CA, Villeneuve DL, Wu-Smart J, et al. Weight of evidence evaluation of a network of adverse outcome pathways linking activation of the nicotinic acetylcholine receptor in honey bees to colony death. *Sci Total Environ* 2017;**584-585**:751–75.
58. Luckert C, Braeuning A, de Sousa G, et al. Adverse outcome pathway-driven analysis of liver steatosis in vitro: a case study with cyproconazole. *Chem Res Toxicol* 2018;**31**(8): 784–98.
59. EPA U. *Endocrine disruptor screening program tier 1 screening determinations and associated data evaluation records*; 2015.
60. Segura-Aguilar J, Kostrzewa RM. Neurotoxin mechanisms and processes relevant to Parkinson's disease: an update. *Neurotox Res* 2015;**27**(3):328–54.
61. Greenamyre JT, Cannon JR, Drolet R, Mastroberardino PG. Lessons from the rotenone model of Parkinson's disease. *Trends Pharmacol Sci* 2010;**31**(4):141–2 [author reply 142-143].
62. Shukla SJ, Huang R, Austin CP, Xia M. The future of toxicity testing: a focus on in vitro methods using a quantitative high-throughput screening platform. *Drug Discov Today* 2010;**15**(23–24):997–1007.
63. Mesnage R, Biserni M, Genkova D, Wesolowski L, Antoniou MN. Evaluation of neonicotinoid insecticides for oestrogenic, thyroidogenic and adipogenic activity reveals imidacloprid causes lipid accumulation. *J Appl Toxicol* 2018;**38**(12):1483–91.

64. Mesnage R, Biserni M, Wozniak E, Xenakis T, Mein CA, Antoniou MN. Comparison of transcriptome responses to glyphosate, isoxaflutole, quizalofop-p-ethyl and mesotrione in the HepaRG cell line. *Toxicol Rep* 2018;**5**:819–26.
65. Sipes NS, Martin MT, Kothiya P, et al. Profiling 976 ToxCast chemicals across 331 enzymatic and receptor signaling assays. *Chem Res Toxicol* 2013;**26**(6):878–95.
66. Judson RS, Magpantay FM, Chickarmane V, et al. Integrated model of chemical perturbations of a biological pathway using 18 in vitro high-throughput screening assays for the estrogen receptor. *Toxicol Sci* 2015;**148**(1):137–54.
67. Houck KA, Judson RS, Knudsen TB, et al. Comment on "on the utility of ToxCast™ and ToxPi as methods for identifying new obesogens". *Environ Health Perspect* 2017;**125**(1): A8–A11.
68. Janesick AS, Dimastrogiovanni G, Vanek L, et al. On the utility of ToxCast™ and ToxPi as methods for identifying new obesogens. *Environ Health Perspect* 2016;**124**(8): 1214–26.
69. Hasin Y, Seldin M, Lusis A. Multi-omics approaches to disease. *Genome Biol* 2017;**18**(1):83.
70. Lederberg J, McCray AT. Ome SweetOmics—a genealogical treasury of words. *Scientist* 2001;**15**(7):8.
71. Patterson EL, Saski C, Küpper A, Beffa R, Gaines TA. Omics potential in herbicide-resistant weed management. *Plants (Basel, Switzerland)* 2019;**8**(12):607.
72. Grossmann K. What it takes to get a herbicide's mode of action. Physionomics, a classical approach in a new complexion. *Pest Manag Sci* 2005;**61**(5):423–31.
73. Krewski D, Acosta D, Andersen M, et al. Toxicity testing in the 21st century: a vision and a strategy. *J Toxicol Environ Health B* 2010;**13**(2–4):51–138.
74. Christopher CJ. Integrating gene expression biomarker predictions into networks of adverse outcome pathways. *Curr Opin Toxicol* 2019;**18**:54–61.
75. Ryan N, Chorley B, Tice RR, Judson R, Corton JC. Moving toward integrating gene expression profiling into high-throughput testing: a gene expression biomarker accurately predicts estrogen receptor α modulation in a microarray compendium. *Toxicol Sci* 2016;**151**(1):88–103.
76. Mesnage R, Phedonos A, Biserni M, et al. Evaluation of estrogen receptor alpha activation by glyphosate-based herbicide constituents. *Food Chem Toxicol* 2017;**108**(Pt A):30–42.
77. Tachibana C. What's next in'omics: the metabolome. *Science* 2014;**345**(6203):1519–21.
78. Mesnage R, Biserni M, Balu S, Frainay C, Poupin N, Jourdan F, et al. Integrated transcriptomics and metabolomics reveal signatures of lipid metabolism dysregulation in HepaRG liver cells exposed to PCB 126. *Arch Toxicol* 2018;**92**(8):2533–47. https://doi.org/10.1007/s00204-018-2235-7.
79. van Ravenzwaay B, Sperber S, Lemke O, et al. Metabolomics as read-across tool: a case study with phenoxy herbicides. *Regul Toxicol Pharmacol* 2016;**81**:288–304.
80. Taylor NS, Gavin A, Viant MR. Metabolomics discovers early-response metabolic biomarkers that can predict chronic reproductive fitness in individual Daphnia magna. *Metabolites* 2018;**8**(3):42.
81. Mesnage R, Teixeira M, Mandrioli D, et al. Multi-omics phenotyping of the gut-liver axis allows health risk predictability from in vivo subchronic toxicity tests of a low-dose pesticide mixture. *bioRxiv* 2020. 2020.2008.2025.266528.
82. Authority EFS, Aguilera J, Aguilera-Gomez M, et al. EFSA scientific colloquium 24–'omics in risk assessment: state of the art and next steps. *EFSA Support Publ* 2018;**15**(11), 1512E.

83. Luechtefeld T, Marsh D, Rowlands C, Hartung T. Machine learning of toxicological big data enables read-across structure activity relationships (RASAR) outperforming animal test reproducibility. *Toxicol Sci* 2018;**165**(1):198–212.
84. Hartung T. Predicting toxicity of chemicals: software beats animal testing. *EFSA J* 2019;**17**(S1):e170710.
85. Ornostay A, Cowie AM, Hindle M, Baker CJO, Martyniuk CJ. Classifying chemical mode of action using gene networks and machine learning: a case study with the herbicide linuron. *Comp Biochem Physiol D: Genomics Proteomics* 2013;**8**(4):263–74.
86. Giannini Kurina F, Hang S, Macchiavelli R, Balzarini M. Spatial predictive modelling essential to assess the environmental impacts of herbicides. *Geoderma* 2019;**354**:113874.
87. Meher PK, Sahu TK, Raghunandan K, Gahoi S, Choudhury NK, Rao AR. HRGPred: prediction of herbicide resistant genes with k-mer nucleotide compositional features and support vector machine. *Sci Rep* 2019;**9**(1):778.
88. Zoeller RT, Brown TR, Doan LL, et al. Endocrine-disrupting chemicals and public health protection: a statement of principles from the endocrine society. *Endocrinology* 2012;**153**:4097–110.
89. Vandenberg LN, Colborn T, Hayes TB. Regulatory decisions on endocrine disrupting chemicals should be based on the principles of endocrinology. *Reprod Toxicol* 2013;**38**:1–15.
90. European Parliament. *Regulation (EC) No 1107/2009 of 21 October 2009 concerning the placing of plant protection products on the market and repealing Council Directives 79/117/EEC and 91/414/EEC*; 2009.
91. European Parliament. *Regulation (EU) No 528/2012 of 22 May 2012 concerning the making available on the market and use of biocidal products (BPR)*; 2012.
92. Agency EC, Centre EFSAwttsotJR, Andersson N, et al. Guidance for the identification of endocrine disruptors in the context of Regulations (EU) No 528/2012 and (EC) No 1107/2009. *EFSA J* 2018;**16**(6):e05311.
93. Daston GP, Cook JC, Kavlock RJ. Uncertainties for endocrine disrupters: our view on progress. *Toxicol Sci* 2003;**74**.
94. The Endocrine Disruptor Screening and Testing Advisory Committee. *Endocrine Disruptor Screening and Testing Advisory Committee (EDSTAC) final report*. US Environmental Protection Agency; 1998.
95. Borgert CJ, Mihaich EM, Quill TF, Marty MS, Levine SL, Becker RA. Evaluation of EPA's tier 1 endocrine screening battery and recommendations for improving the interpretation of screening results. *Regul Toxicol Pharmacol* 2011;**59**:397–411.
96. Vandenberg LN, Bowler AG. Non-monotonic dose responses in EDSP tier 1 guideline assays. *Endocr Disruptors* 2014;**2**(1):e964530.
97. Kanno J, Onyon L, Haseman J, Fenner-Crisp P, Ashby J, Owens W. The OECD program to validate the rat uterotrophic bioassay to screen compounds for in vivo estrogenic responses: phase 1. *Environ Health Perspect* 2001;**109**:785–94.
98. Owens W, Zeiger E, Walker M, Ashby J, Onyon L, Gray Jr LE. The OECD program to validate the rat Hershberger bioassay to screen compounds for in vivo androgen and antiandrogen responses. Phase 1: use of a potent agonist and a potent antagonist to test the standardized protocol. *Environ Health Perspect* 2006;**114**(8):1259–65.
99. Vandenberg LN, Najmi A, Mogus JP. Agrochemicals with estrogenic endocrine disrupting properties: lessons learned? *Mol Cell Endocrinol* 2020;**110860**.
100. La Merrill MA, Vandenberg LN, Smith MT, et al. Consensus on the key characteristics of endocrine-disrupting chemicals as a basis for hazard identification. *Nat Rev Endocrinol* 2020;**518**, 110860.

101. Vandenberg LN, Hunt PA, Gore AC. Endocrine disruptors and the future of toxicology testing—lessons from CLARITY-BPA. *Nat Rev Endocrinol* 2019;**15**(6):366–74.
102. Myers JP, Zoeller RT, Vom Saal FS. A clash of old and new scientific concepts in toxicity, with important implications for public health. *Environ Health Perspect* 2009;**117**:1652–5.
103. Zoeller RT. Challenges confronting risk analysis of potential thyroid toxicants. *Risk Anal* 2003;**23**(1):143–62.
104. Zoeller RT. Regulation of endocrine-disrupting chemicals insufficient to safeguard public health. *J Clin Endocrinol Metab* 2014;**99**(6):1993–4.
105. Zoeller RT, Bergman A, Becher G, et al. A path forward in the debate over health impacts of endocrine disrupting chemicals. *Environ Health* 2014;**13**(1):118.
106. Autrup H, Barile FA, Berry SC, et al. Human exposure to synthetic endocrine disrupting chemicals (S-EDCs) is generally negligible as compared to natural compounds with higher or comparable endocrine activity. How to evaluate the risk of the S-EDCs? *Chem Biol Interact* 2020;**326**, 109099.
107. Bergman A, Becher G, Blumberg B, et al. Manufacturing doubt about endocrine disrupter science—a rebuttal of industry-sponsored critical comments on the UNEP/WHO report "state of the science of endocrine disrupting chemicals 2012". *Regul Toxicol Pharmacol* 2015;**73**(3):1007–17.
108. Vom Saal FS, Vandenberg LN. Update on the health effects of bisphenol A: overwhelming evidence of harm. *Endocrinology* 2020.
109. Trasande L, Zoeller RT, Hass U, et al. Estimating burden and disease costs of exposure to endocrine-disrupting chemicals in the European Union. *J Clin Endocrinol Metab* 2015;**100**(4):1245–55.

Glyphosate as an active substance authorized under EU pesticide regulations: Regulatory principles and procedures

Nicolas de Sadeleer
Université Saint-Louis, Brussels, Belgium

Chapter outline

Herbicides. https://doi.org/10.1016/B978-0-12-823674-1.00009-2

Introduction

Aiming at reducing health and environmental risks, the chemical policy has historically been related to a general preference for a certainty-seeking regulatory style in which a formal, science-based, and standardized risk assessment has been singled out as the predominant tool for decision-making. However, while risk assessments draw extensively on science, data are often incomplete and results may be unclear or contradictory. Indeed, as it is difficult to establish causal links between exposure to chemicals and health or environmental effects, there is generally a significant degree of uncertainty in estimates of the probability and magnitude of adverse effects associated with a chemical agent. The variety and complexity of the pathways of dispersion in the environment compound these uncertainties. As the result of limitations on knowledge, it is difficult to provide conclusive evidence of a threat to human health or to the environment. Nature does not reveal its secrets quickly: long latency periods may conceal hazards for decades.[1]

The implementation of Council directive 91/414/EEC of 15 July 1991, concerning the placing of plant protection products on the market proved so problematic that the European Union (EU) institutions decided to abrogate and replace it with a regulation. The adoption in 2009 of the EU pesticide regulations brought about a radical change in terms of legal bases, goals, risk assessment, and risk management. Moreover, this regulation was complemented by Directive 2009/128/EC of 21 October 2009, that requires the member states to achieve a reduction in pesticide-related risks.

Glyphosate is found in Roundup, the most widely used herbicide in the world. Its impacts on health and the environment have sparked much controversy, particularly in the EU. The active substance was approved under Council directive 91/414/EEC and later under the EU pesticide regulations. In spite of the opposition of several member states and parts of civil society, the European Commission (hereafter EC) renewed the approval of glyphosate as an active substance in 2017. To understand the scope of this approval, which will last until 2022, this chapter explores the intricacies of the 2009 EU pesticide regulations and attempts to shed light on their significance in the context of analogous chemical regulatory regimes.

The chapter is structured as follows. The EU pesticide regulations principles and procedures must be understood against the broader EU constitutional background, where the authors of EU treaties constantly strike a balance between economic integration (the functioning of the internal market) and nonmercantile interests (health and environmental protection in the case of glyphosate). The procedure for the approval of active substances must be distinguished from the procedure for authorization of plant protection products. The procedure under which the approval of glyphosate has been renewed is examined in detail. The core of the chapter discusses the consistency of the EU pesticide regulations with the precautionary principle.

In accordance with the recent European Court of Justice (hereafter CJEU) *Blaise* judgment, applicants are obligated to submit to the public authorities, in accordance with the precautionary principle, tests on the cocktail effects of active substances and the long-term carcinogenicity and toxicity of pesticide products.

Last but not least, the chapter is written for a broad scientific audience.

The pesticide regulations embedded in an economic common playing field

Glyphosate is authorized under Regulation (EC) No. 1107/2009 concerning the placing of plant protection products on the market (PPPR),[a] an EU act that is at the juncture of different and sometimes apposite policies.

It is settled case law that each piece of EU legislation must be founded on one or more legal bases set out in the founding EU treaties. The byzantine structure of EU treaty law, with its diversification of legal bases likely to provide for specific competences to address issues as complex as the marketing and the use of pesticides, remains the subject of an ongoing debate.[2] The choice of the relevant legal base for legislation aimed at enhancing the free circulation of pesticides while achieving a high level of environmental and health protection represents a critical juncture in relations between institutions as well as the relations between member states and the EU. The choice of legal base is not a purely formal question, but rather one of substance, being a matter of "constitutional significance"[b] that is regularly ruled on by the CJEU. It is settled case law that "the choice of the legal base for a measure may not depend simply on an institution's conviction as to the object pursued."[c] Instead, the determination of the legal base is amenable to judicial review, which includes in particular the aim and the content of the measure.[d]

If it is established that the act simultaneously pursues different objectives or has several components that are indissociably linked (e.g., agricultural production, internal market, environmental protection), and if one of these is identifiable as the main or predominant purpose or component whereas the others are merely incidental, it will have to be founded on a single legal base, namely, that required by the main or predominant purpose or component–the center of gravity of the act–rather than its effects.[e] Accordingly, the act concerned should be adopted in principle on a sole legal base, namely that required by the main or predominant purpose or component.

[a] Regulation (EC) No 1107/2009 concerning the placing of plant protection products on the market, *OJ L 309, 1*.

[b] Opinion 2/00 [2001] ECR I-9713, para. 5.

[c] Case C-300/89 *Commission v. Council* (Titanium dioxide) [1991] ECR I-2867, para. 10.

[d] See, *inter alia*, Case C-300/89 *Titanium Dioxide*, cited above, para. 10; Case C-269/97 *Commission v. Council* [2000] ECRI-2257, para. 43; and Case C-211/01 *Commission v. Council* [2003] ECR I-3651, para. 38; and Case C-338/01 *Commission v Council* [2004] ECR I-4829, para. 54.

[e] See, *inter alia*, Case C-155/91 *Commission v. Council* [1993] ECR I-939, paras. 19 and 21, Case C-36/98 *Spain v. Council* [2001] ECR I-779, para. 59; and Case C-211/01 *Commission v. Council*, cited above, para. 39; and Case C-281/01 *Commission v. Council* [2002] ECR I-12049, para. 57; Case C-338/01 *Commission v. Council*, cited above, para. 55; and Case C-91/05 *Commission v. Council* [2008] [illegible], para. [illegible]

However, it may be the case that the twin objectives and the two constituent parts of the act are "inseparably" or inextricably linked without one being secondary and indirect in relation to the other. In such a case, it is impossible to apply the predominant aim and content test. Exceptionally, the CJEU accepts that such a measure must be founded on the corresponding legal bases and the applicable legislative procedures respected.[f] In this connection, the 2009 EU pesticide regulations are a case in point.

Given that the marketing of pesticides intermingles with a swath of policy issues, the EU lawmaker (the European Parliament and the Council) based this regulation on three provisions of the Treaty on the Functioning of the European Union (hereafter TFEU):

- Article 43(2) with respect to the Common Agricultural Policy.
- Article 114 with respect to the internal market.
- Article 168(4)(b) with respect to health protection.

This choice amounted to a significant departure from the previous regulatory approach, as Council directive 91/414/EEC of 15 July 1991 concerning the placing of plant protection products on the market was related to the Common Agricultural Policy as it was adopted in virtue of former Article 43 EEC. In other words, the lawmaker abandoned the traditional agricultural base in favor of a mixed approach combining agriculture, the internal market, and health protection. Being based on the internal market legal base, the PPPR came closer to the other regulatory regimes related to chemical substances, which are mainly based on Article 114 TFEU, a treaty provision fostering the functioning of the internal market. The importance afforded by the EU lawmaker in 2009 to internal market integration requires a few words of explanation.

The EU policy for the placing on the market of chemical substances was established in the early days of the environmental debate. It consists of a complex regulatory system made up of an intricate network of regulations. Several features of this risk regulatory framework need to be explained before focusing on the legal status of glyphosate in EU law.

The EU institutions have been producing a web of varied, fragmented, and complex regulations that harmonize the procedures related to the placing on the market of chemical substances. Given that all these sectors are product-related, the EU institutions have favored regulations adopted pursuant to Article 114 TFEU. In sharp contrast with other environmental sectors (air, water, waste, nature, listed installations), these regulations increase the centralization of the decision-making process. The preference for regulations based on Article 114 could be explained by the fact that the more flexible nature of a directive entails a genuine risk of market fragmentation.

[f] Case C-300/89 "*Titanium dioxide*," cited above, para. 13; Case C-336/00 *Huber* [2002] ECR I-7699, para. 31; Case C-281/01 *Commission v. Council* [2002] ECR I-12049, para. 35, Case C-211/01 *Commission v. Council*, cited above, para. 40; Case C-211/01 *Commission v. Council* [2003] ECR I-8913, paragraph 40, Case C-91/05 *Commission v. Council*, cited above, para. 75; and Opinion 2/00 [2001] ECR I-9713, para. 23.

For instance, in spite of the significant environmental and health concerns they give rise to, both Regulation 528/2012 concerning making available on the market and the use of biocidal products and the PPPR have been based on Article 114 TFEU. What is more, this harmonization confers an exclusive competence on the EU authorities for the assessment of the active substances found in these products. Given the completeness of their procedures,[g] this regulation led to a complete harmonization that constrains the member states' room for maneuvering.[3]

Although it was not mentioned in the 1957 Treaty of Rome, environmental and health concerns have, through the various treaty reforms, gradually been able to establish themselves among the greatest values enshrined in the EU treaties. Needless to say, internal market and environmental and health policies have traditionally focused on opposite, albeit entangled, objectives: the abolition of unilateral national measures hindering free trade, and protection of individuals and vulnerable populations as well as natural resources. In other words, whereas the internal market is about facilitating the free circulation of products and substances, environmental and health policies encourage the adoption of regulatory measures that are likely to jeopardize free trade. Besides, the internal market favors economic integration through total harmonization while environmental law allows for differentiation.[4]

Does it mean that the environmental and health concerns are absent from the internal market harmonization process? Several TFEU provisions oblige EU lawmakers to take into account, as part of the decision-making process, not only the full range of the interests affected by the harmonization process (freedom of enterprise, confidentiality of private data, etc.) but also a number of interests that received a lesser degree of priority at the earlier stage of the European integration.[4]

First, pursuant to Article 114(3) TFEU, the EC's proposals, which have as their object the establishment and functioning of the internal market, must pursue a high level of protection while addressing the concerns about health, safety, environmental protection, and consumer protection. These concerns have to be fully integrated into the internal market harmonization process.

Second, Article 168(1) TFEU, read in combination with Article 35 EUCHR, requires that the protection of human health be integrated into the definition and implementation of EU policies and activities. In accordance with this integration clause, the PPPR is also based on Article 168(4)(b).

Third, by the same token, Article 11 TFEU, read in combination with Article 37 EUCHR, requires that environmental protection requirements be integrated into the definition and implementation of EU policies and activities. However, the environmental legal base–Article 192TFEU–has not been added to the other legal bases of the PPPR.

In light of these treaty provisions, the EU institutions must display particular sensitivity to nonmercantile concerns and must reconcile the various objectives laid down in the TFEU. The internal market's goals are thus not solely economic, but also social

[g] Case T-31/07, *Du Pont de Nemours* [2013] T:2013:167, para. 203.

and environmental. Prioritizing one objective should not render the achievement of the other objectives impossible.[h]

These various integration clauses are not devoid of legal consequences. In accordance with treaty law, the internal market regulations of hazardous substances seek to strike a balance between a high level of protection of human health and the environment and the free circulation of substances in the internal market. The objectives of the PPPR are testament to such a conciliatory approach. Article 1(3) is worded as follows[i]:

> *The purpose of this Regulation is to ensure a high level of protection of both human and animal health and the environment and to improve the functioning of the internal market through the harmonisation of the rules on the placing on the market of plant protection products, while improving agricultural production.*

The combination of these different objectives has legal effects. It was settled case law that former Directive 91/414 (replaced by the PPPR) sought to enhance the internal market (removal of barriers to intra-EU trade in plant products) while maintaining a high level of protection of the environment and of human and animal health. The CJEU held in this respect that the EC was enjoying broad discretion to effectively pursue the objective assigned to it.[j] However, the exercise of that discretion does not escape judicial review. As a result, the EU judicature must determine whether the relevant procedural rules have been complied with, whether the facts established by the Commission are correct, and whether there has been a manifest error of appraisal of those facts or a misuse of powers.[k]

It is also settled case law that health requirements take precedence over economic interests.[l]

The pesticide regulations epitomizing specific features of EU regulatory governance

The regulations on hazardous substances display specific features of EU governance. On the one hand, they empower the Commission to adopt implementing acts in accordance with the comitology procedure; on the other hand, they delegate significant

[h] See in particular the CJEU case law on CAP objectives. Joined Cases 197-200/80 *Ludwigshafener Walzmühle v. Council and Commission* [1981] ECR 3211, para. 41; Case 59/83 *Biovilac v. Commission* [1984] ECR 4057, para. 16; Case C-280/93 *Germany v Council* [1994] ECR I-4973, paras. 47 and 51; Case C-122/94 *Commission v Council* [1996] ECR I-881, para. 24.

[i] Along the same lines, see Regulation (EC) No. 1907/2006 concerning the Registration, Evaluation, Authorisation and Restriction of Chemicals [2006] OJ L396/1, Art 1(3); and the Biocide Regulation, Art 1. It must be noted that the European Commission is not empowered to undermine the equilibrium sought by the EU lawmaker. See Case T-521/14 *Sweden v Commission* [2015] T:2015:976, para. 72.

[j] Case C-326/05 P *Industrias Químicas del Vallés v Commission* [2007] ECR I-6557, paras. 74 and 75; Case T-31/07, *Du Pont de Nemours* [2013] T:2013:167, para. 155.

[k] Case C-326/05 P *Industrias Químicas del Vallés v Commission* [2007] ECR I-6557, para.106; Case T-326/07 *Cheminova* [2009] II-2685, para 107.

[l] Joined Cases T-74, 76, & 83/00 to T-85, 132, & 137/00 and T-141/00 *Artegodan* [2002] ECR II-4945, para. 184.

administrative tasks, in particular in the realm of risk assessment, to two EU agencies. In effect, the regulatory decisions on chemicals policy, such as those relating to registration, authorization, restriction, classification, and labelling under the Registration, Evaluation, Authorization, and Restriction of Chemicals (REACH), are backed by the opinions of the European Chemicals Agency (ECHA), whereas the assessment of the active substances in pesticides is subject to the opinions of the European Food Safety Agency (EFSA). The interaction among these two agencies (risk assessment), the regulatory committees, and the Commission (risk management) is testament to the principle of institutional balance enshrined in Article 4(3) of the treaty on the EU. In this connection, the PPPR straddles risk assessment and risk management alike.

Additional precautionary features of the pesticide regulations: Cut-off hazard-based criteria and substitution

In contrast to former Directive 91/414/EEC, the PPPR represents a real step forward for health and environmental protection. In shifting the risk-based approach to the hazard-based approach, in providing new regulatory mechanisms such as substitution, the PPPR not only fleshed out the precautionary principle but also allocated more clearly the responsibilities for safety and improved the risk assessment requirements.[5]

Lately, the EU regulatory approach has been shifting from risk to hazard: substances that do not meet the EU's predetermined cut-off hazard-based criteria (persistent bioaccumulative and toxic, persistent organic pollutants, very persistent very bioaccumulative, or endocrine disruptive) cannot receive approval or renewal of approval.[m] Whenever one of these properties is ascertained, the substance is deemed to be intrinsically dangerous and its use must be forbidden.[7] These cut-off criteria do not require any additional risk exposure assessment. It follows that the EC cannot list an active substance if it displays some hazardous properties, regardless of the likelihood of the hazard causing actual harm (i.e., the risk).

In avoiding the need to perform an entire risk assessment on a case-by-case basis, which can be time and resource consuming, relying on cut-off hazard-based criteria reduces considerably the administrative burden entailed by full risk-assessment procedures. As a result, it is faster and less expensive.[8] In practical terms, this means that experts are not called on to fully perform the additional steps of the assessment procedure (hazard characterization, risk identification, and risk characterization). It comes as no surprise that this regulatory approach has been championed by different EU institutions and several member states and strongly resisted by others. So far, few substances have been regulated in relation to their hazard.

Furthermore, the adoption in 2009 of the PPPR represented a watershed in the development of the substitution principle, according to which the mere existence

[m] PPPR, Annex II, 3.6.2 to 3.6.5; Biocides Regulation, preamble recital 12. However, in 2017, the Commission took on board "potency" and transformed the hazard-based for the listing of EDCs (PPPR, Annex II, 3.6.5) into a risk based one. See Commission Delegated Regulation (EU) 2017/2100 (OJ L 301, 17.11.17, p. 1) and Commission Regulation (EU) 2018/605 (OJ L 101, p. 33). See also Kurai.[6]

of an alternative substance that appears to be less dangerous than the substance in question constitutes a sufficient basis for restriction or prohibition. This principle has been enshrined in the PPPR (Article 50). The EC is required to define a list of active substances in pesticides considered to be "Candidates for Substitution" (CfS) that go through a comparative assessment.

Procedures for approval of active substances and for authorization of plant protection products

Plant protection products containing active substances[n] can be formulated in many ways and used on a variety of plants and plant products under different agricultural, plant health, and environmental (including climatic) conditions.[o]

The approval procedure for active substances (Articles 7 to 13 PPPR) such as glyphosate is separated from the authorization procedure for the formulated plant protection product, such as Roundup, that is then used by operators (Articles 33 to 39 PPPR). In so doing, the EU lawmakers have been endorsing a two-pronged approach:

- In the first stage, the active substance such as glyphosate must be assessed and approved by the EC "in order to achieve the same level of protection in all member states" (Preamble, recital 10; Articles 7 to 13 PPPR).
- In the second stage, the pesticide containing the active substances approved by the EU cannot be placed on the market or used unless it has been assessed and authorized in one of the member states (Articles 28 to 39 PPPR). Consequently, the applicant who wishes to place a pesticide on the market is obligated to apply for an authorization to each member state where the pesticide is intended to be placed on the market (Article 33 PPPR). As a result, the national authorities and not the EC have to authorize the placing on the market of products such as Roundup containing the active substance glyphosate.

The procedure for the approval of active substances and the procedure for authorization of plant protection products are not waterproof. In this connection, two examples will suffice to highlight how the two procedures are closely linked.[p]

- As stressed above, the authorization of a plant protection product presupposes that its active substances have previously been approved by the EC.
- The EU lawmakers have obliged the authorities to take into account the potential effects of a combination of the various constituents of a plant protection product both in the procedure for the approval of the active substances and in the procedure for the authorization of the plant protection products.

[n] Although the PPPR does not contain any definition of the expression "active substance," it is clear from Article 2(2) of Regulation No 1107/2009 that substances, including micro-organisms, having general or specific action against harmful organisms or on plants, parts of plants or plant products are to be regarded as active substances, for the purposes of that regulation. See Case C-616/17 *Blaise* [2019] C:2019:800, paras. 53-54.

[o] Case T-545/11 RENV *Stichting Greenpeace Nederland* [2016] EU:T:2018:817, para. 74.

[p] Case C-616/17 *Blaise* [2019] C:2019:800, para. 64. See case note Bailleux A. Common Market Law Review; 2020; 57: 861–876.

Level of harm

"It is for the [EU] institutions to determine the level of protection (that) they deem appropriate for society."[q] Accordingly, it is by reference to that level of protection that the institutions may require preventive measures in spite of any existing scientific uncertainty. Therefore, determining the level of risk deemed unacceptable involves "the [EU] institutions in defining the political objectives to be pursued under the powers conferred on them by the treaty."[r] Is it possible for EU lawmakers to pursue an absolute level of protection? It is settled case law that precautionary measures "must not aim at zero risk," for they may be deemed disproportionate.[s] That being said, nothing precludes the EU institutions from endorsing a zero tolerance policy with regard to certain risk factors for which the producer of the substance cannot adduce proof that they are acceptable.[t] In particular, the concept of zero tolerance may through the precautionary principle result in the total ban of a substance provided that its potential risk is supported by scientific data.[u]

Against this background, the determination of the illicit level of harm varies significantly from one regulation to another.[9, 10] Whereas several chemical regulations prohibit squarely the use of chemical substances such as persistent organic pollutants,[v] companies may under the PPPR seek authorization for placing on the market hazardous active substances if their risks can be "adequately controlled."[w]

The preamble of the PPPR stresses that "when granting authorisations of plant protection products, the objective of protecting human and animal health and the environment should take priority over the objective of improving plant production. Therefore, it should be demonstrated, before plant protection products are placed on the market, that they present a clear benefit for plant production and do not have any harmful effect on human or animal health, including that of vulnerable groups, or any unacceptable effects on the environment" (Preamble, recital 24). Against this background, the PPPR requires that an active substance can only be included in a pesticide product if it has been demonstrated to present a clear benefit for plant production and is not expected to have any harmful effect on human or animal health or any unacceptable effects on the environment [Preamble, recitals 10, 24, Article 4(2)b, Article 4 (3)c and e]. This requirement is not devoid of legal consequences. "Determining the

[q] Case T-13/99 *Pfizer, cited above*, para. 151.

[r] Case T-13/99 *Pfizer, cited above*, para. 151.

[s] Communication of the European Commission, n° 6.3.1, para. 18. See Case T-13/99 *Pfizer,* cited above, para. 145.

[t] Case C-121/00 *Hahn* [2002] ECR I-9193, para. 93; Case T-392/02 *Solvay Chemicals*, cited above, para. 97.

[u] Taking account of the genuine risk that the intake of fluoride in food supplements will exceed the upper safe limit established for that mineral, a Member State may set the maximum amount of fluoride which may be used in the manufacture of food supplements at a zero level. Case C-446/08 *Solgar Vitamin's France* [2010] ECR I-03973, para. 47.

[v] Regulation (EU) 2019/1021 of the European Parliament and of the Council of 20 June 2019 on persistent organic pollutants [2019] *JO L 169, 45.*

[w] See also REACH, Art. 57(2).

level of risk deemed unacceptable involves the [EU] institutions in defining the political objectives to be pursued under the powers conferred on them by the Treaty."[x] It is by reference to that level of protection–absence of harmful effect on human or animal health–that the EU institutions and national authorities may be required to take preventive measures.

The General Court and the CJEU alike have been endorsing lately a harder look at the Commission's attempts to relax somewhat the level of safety requirements in the area of active substances found in plant protection products and chemicals. In this respect, the *Paraquat* judgment handed down by the General Court on 11 July 2007 is a case in point. Under the former Directive 91/414,[y] the Commission could list such a substance under Annex I inasmuch as the use of the products "in the light of current scientific and technical knowledge" will not have any harmful effects on animal health. Adjudicating an action for annulment brought by Sweden against the decision listing paraquat under Annex I, the General Court stressed that the safety requirement had to be interpreted "in combination with the precautionary principle." It follows that "in the domain of human health, the existence of solid evidence which, while not resolving scientific uncertainty, may reasonably raise doubts as to the safety of a substance, justifies, in principle, the refusal to include that substance in Annex I to Directive 91/414."[z]

However, in *Blaise* the CJEU stresses the "need to strike a balance between several objectives and principles."[aa] This paragraph conveys the feeling that the competing interests have been placed on equal footing. But a closer reading of the PPPR indicates that the EU lawmakers clearly departed from a traditional weighing of interests in highlighting the prevalence of health and environmental protection over plant production (Preamble, recital 24). The high level of health and environmental protection pursued by the EU institutions limits their room for maneuvering in implementing the regulation.[ab]

Burden of proof

While in the past regulators had to prove that particular substances were hazardous,[ac] under EU secondary law the applicants who wish to place a hazardous substance on the market must provide evidence that it is safe for health and the environment.[11] The

[x] Case T-13/99 *Pfizer Animal Health v Council* [2002] ECR II-3305, para. 151.

[y] Directive 91/414/EEC concerning the placing of plant protection products on the market [1991] OJ L 230/1. This directive has been replaced by Regulation (EC) 1107/2009.

[z] Case T-229/04 *Sweden v Commission* [2007] ECR I-2437, paras. 161 and 224.

[aa] Case C-616/17 *Blaise* [2019] C:2019:800, para. 50.

[ab] In sharp contrast, the choice of the Commission to authorize the chemical substance DEHP under the REACH "socio-economic procedure" (Article 60(4)) does not constitute a breach of the precautionary principle on the ground that this authorization procedure 'was precisely conceived to enable undertakings to place on the market substances which pose in particular a risk to human health but whose socio-economic advantages prevail'. See Case T-108/17 *ClientEarth* [2019] EU:T:2019:215, para. 287. Under such a procedure, the weighing of interest tilts in favour of the socio-economic interests.

[ac] By way of illustration, Council Regulation (EEC) 793/93 required the Member States to carry out the risk assessment of priority substances because of their potential impacts on man or the environment. The implementation of that regulation has been so laborious that it has paralyzed regulatory action.

PPPR is testament to this shift. In effect, it is for the applicant and not for the authorities to prove that the active substance or the plant protection product that is the subject of an application for approval or authorization fulfills the relevant criteria laid down by the PPPR. As a result, the tests, studies, and analyses must be provided by the applicant to permit approval of the active substance [Article 7(1) and from Article 8(1) and (2)] and authorization of the product [Article 33(3)(a) and (b)]. Placing the obligation on the applicant "contributes to achieving compliance with the precautionary principle by ensuring that there is no presumption that active substances and plant protection products have no harmful effects."[ad]

In *Blaise*, the French criminal referring court was uncertain whether the tests, studies, and analyses required in the procedures for the approval of glyphosate submitted by the applicant, with no independent counteranalysis, were contrary to the precautionary principle. As the studies performed by the applicant are confidential, there is a risk that the applicant submits only studies showing no potential health and environmental risks. In particular, the absence of counteranalysis could imply that the applicant's tests, studies, and analyses might be biased.

The CJEU held that the EU legislature provided a number of safeguards to control the quality of the tests, studies, and analyses submitted by the applicant.[ae] To enhance transparency, the PPPR requires, for instance, that the applicant submit a summary dossier that contains, in respect to each point of the data requirements that apply to the active substances and the plant protection products, the summaries and results of tests and studies and the name of their owner and of the person or institute that carried out the tests and studies [Article 8(1)].[af]

However, the authorities are not totally dependent upon the scientific assessment submitted by the applicant. Their duty is "to take account of the most reliable scientific data available and the most recent results of international research and not to give in all cases preponderant weight to the studies provided by the applicant."[ag] Accordingly, the authorities can rely on information other than merely the tests, analyses, and studies submitted by the applicant.[ah] Nothing precludes these studies contradicting the data gathered by the applicant.[ai] In acknowledging the possibility for the authorities to take into consideration dissenting opinions, the CJEU *Blaise* judgment takes full stock of the precautionary principle.

In addition, the authorities are allowed to organize a consultation of experts and to ask the Commission to consult a community reference laboratory, to which the applicant may be required to submit samples and analytical standards.[aj] So far, the reference laboratories appointed by the EC in 2019 have played no role in the assessment of glyphosate.[ak]

[ad] Case C-616/17 *Blaise* [2019] C:2019:800, paras. 79-80.

[ae] Case C-616/17 *Blaise* [2019] C:2019:800, para. 82.

[af] Case C-616/17 *Blaise* [2019] C:2019:800, para. 83.

[ag] Case C-616/17 *Blaise* [2019] C:2019:800, para. 94.

[ah] Case C-616/17 *Blaise* [2019] C:2019:800, para. 96.

[ai] Case C-616/17 *Blaise* [2019] C:2019:800, para. 93.

[aj] Case C-616/17 *Blaise* [2019] C:2019:800, para. 98.

[ak] [illegible]

These various safeguards preclude biased risk assessments breaching the precautionary principle. Accordingly, the PPPR is vitiated by a manifest error of assessment.[al]

Inclusion of glyphosate as an active substance in EU pesticide legislation

As discussed above, glyphosate had to be approved at the EU level before being included in various pesticide products. The EC adopted Directive 2001/99/EC on 20 November 2001, with a view to amending Annex I to Council Directive 91/414/EEC concerning the placing of plant protection products on the market, to include glyphosate as an active substance.

Pursuant to Directive 2001/99, glyphosate was approved from 1 July 2002 to 30 June 2012. Since 2002, it has been authorized in member states and is one of the most widely used herbicides in the EU.[am]

The EC received a renewal request for glyphosate before the expiration of the approval. As the information necessary for the renewal of active substances had yet to be adopted at that time, the inclusion of glyphosate was extended until 31 December 2015.[an]

Subsequently, in 2009, Directive 91/414 was repealed, with effect from 14 June 2011, by the PPPR. As active substances included in Annex I to Directive 91/414/EEC are deemed to have been approved under Regulation (EC) No. 1107/2009, glyphosate was listed in the annex to the Implementing Regulation No. 540/2011.[ao] Under that regulation, the expiry date of the approval period was fixed at 31 December 2015.

Renewal of the authorization of glyphosate as an active substance

On 20 December 2013, the Federal Republic of Germany, as the rapporteur member state, submitted, in collaboration with the Slovak Republic as the corapporteur, the renewal assessment report for glyphosate. The EFSA forwarded it to the applicant and to the other member states for comments. Assessment of the carcinogenicity of glyphosate has been a matter of controversy ever since. In March 2015, the International Agency for Research on Cancer (IARC) of the World Health Organization published its monograph on glyphosate, concluding that it should be classified as "probably carcinogenic to humans."[ap] In the wake of this international assessment,

[al] Bailleux A. Case note on Blaise, *cited above*, 874. para. 100.

[am] Implementing Regulation 2017/2324, recital 19.

[an] Commission Directive 2010/77/EU of 10 November 2010 amending Directive 91/414 as regards the expiry dates for inclusion in Annex I of certain active substances (OJ 2010 L 293, p. 48).

[ao] Commission Implementing Regulation (EU) No 540/2011 of 25 May 2011 implementing Regulation No 1107/2009 as regards the list of approved active substances (OJ 2011 L 153, p. 1).

[ap] In *Pilliod et al. c Monsanto Company et al.*, the California Superior Court held that Roundup's alleged risk of NHL was "known or knowable in light of the generally recognized and prevailing scientific and medical knowledge." *Alva and Alberta Pilliod v. Monsanto Co.* (Case No. RG17862702, JCCP No. 4953) Cal.1d.

the Commission mandated EFSA to review the information in the IARC's findings. The EFSA stated that "glyphosate is unlikely to pose a carcinogenic hazard to humans and the evidence [did] not support classification [of that active substance] with regard to its carcinogenic potential" according to REACH.

Doubtful of the EFSA's risk assessment, several member states of the Standing Committee on Plants, Animals, Food and Feed decided to seek the opinion of the Committee for Risk Assessment of the ECHA. In 2017, ECHA concluded by consensus that, on the basis of the information currently available, no hazard classification for carcinogenicity was justified for glyphosate.

The opposing views of the IARC and the two EU agencies can be explained by their diverging methodologies. First, while the IARC looked at both glyphosate–the active substance–and the plant protection products (e.g., Roundup), the EU risk assessments considered only glyphosate, on the grounds that member states are responsible for authorizing each plant protection product that is marketed in their territories. Second, while the IARC considered only published studies, the EU agencies also took into consideration studies submitted by applicants as part of their dossiers that were not in the public domain. These divergent methodologies explain the differences in how EFSA/ECHA and IARC weighed the available data.[aq]

In light of the diverging view between EFSA/ECHA and IARC, the Commission extended the period of the validity of the approval of glyphosate in amending the Implementing Regulation No. 540/2011. Since then, the approval period for glyphosate has been extended several times.

In 2017, the EFSA communicated to the Commission its report on the potential endocrine-disrupting properties of glyphosate, confirming that the weight of evidence indicates that glyphosate does not have endocrine-disrupting properties.

In 2016 and 2017, the European Parliament adopted several resolutions on the different draft Commission Implementing Regulations renewing the approval of the active substance glyphosate.

Considering that the approval of glyphosate was going to expire on 15 December 2017, the Commission presented in 2016 a draft implementing regulation to renew the approval of the substance for a period of 10 years to the Standing Committee on Plants, Animals, Food and Feed. As the Standing Committee did not manage to deliver an opinion within the time limit (Comitology Regulation 182/2011), the draft implementing act was submitted to the Appeal Committee. Finally, the Appeal Committee delivered a favorable opinion by qualified majority (Article 4 Rules of Procedure 182/2012), which enabled the EC to adopt its implementing regulation (EU) 2017/2324 on 12 December 2017.

The quorum of 65% provided for in Article 278 TFEU was narrowly reached (65.71%). The Commission's success was due to Germany's about-face in the Appeal Committee. During the previous vote on 6 October 2007, this quorum was not

[aq] Communication from the Commission on the European Citizens' Initiative "Ban glyphosate and protect people and the environment from toxic pesticides" C(2017) 8414 final.

reached, as the 16 member states in favor of the measure represented only 47.46% of the EU population.

The 2017 Commission Implementing Regulation renewed the approval of the active substance until 15 December 2022.[ar] Shorter than the maximum duration of 15 years provided for in Article 14.2 of Regulation No.1107/2009, the 5-year renewal period was justified by the EC on the grounds that it had to take into account the opinions of the European Parliament and "future scientific and technological developments." It is important to note that the implementing regulation itself contains the maintenance of the marketing authorizations for the pesticides containing the active substance "glyphosate."

European citizens' initiative

In allowing one million citizens residing in one-quarter of the member states to invite the Commission to submit a proposal for a legal act to implement the EU treaties, the European Citizens' Initiative (ECI) mirrors participatory democracy in the EU.[as] It allows the citizens to apply directly to the Commission to submit a request inviting it to submit a proposal for a legal act of the EU, for the purposes of the application of the treaties.[at]

On 6 October 2017, the EC officially received a successful ECI referring specifically to glyphosate. On 23 October, 2017, the Commission responded to the ECI stating that "as regards the first aim seeking to ban glyphosate-based herbicides it [took] the view that there [was] no scientific or legal grounds for a ban on glyphosate and [did] not intend to introduce legislative proposals to that effect." It added that "in particular, the scientific evidence [did] not support the conclusion that glyphosate could cause cancer" and "therefore, the decision adopted ... to renew the approval of glyphosate (for a period of five years) [was] completely justified."

Actions for annulment of Commission implementing regulation (EU) 2017/23241

The CJEU guarantees rule of law[au] in reviewing the conformity of EU legislative acts, regulatory acts, and individual acts against the superior rules of the EU legal order. An action can be brought within 2 months of the publication. Applicants are divided into three categories: privileged, semiprivileged, and nonprivileged. Although they are considered privileged applicants, neither the member states nor the European

[ar] Commission Implementing Regulation (EU) 2017/2324 of 12 December 2017 renewing the approval of the active substance glyphosate [2017] *JO* L 333, 10.

[as] Art. 11(4) and Art. 24(1) TFEU. See Regulation (EU) No 2019/788.

[at] Case C-589/15 P *Anagnostakis/Commission* [2017] EU:C:2017:663, para. 24

[au] Art. 2 TEU.

Parliament lodged within 2 months of the publication of the Commission implementing regulation an action for annulment.

Nonprivileged applicants, including regional and local governments, may bring an action for annulment only if they prove that the contested regulation infringes upon their interests. In particular, they may bring an action against an act provided that it is of direct and individual concern to them as well as against a regulatory act that is of direct concern to them and does not entail implementing measures.[12]

The first action for annulment of the Commission implementing regulation was lodged by an Italian association of wheat producers. The General Court held that the renewal of the approval of glyphosate "is not, in itself, the confirmation, extension, or renewal of the marketing authorisation granted by the Member States for a phytopharmaceutical product containing that active substance."[av] It concluded that the renewals of marketing authorizations constitute implementing measures of the contested act, within the meaning of the fourth paragraph of Article 263 TFEU.[aw] The existence of implementing measures precluded the claimants from demonstrating that they were "directly" affected by the contested implementing regulation.[ax] Accordingly, the applicants also had to demonstrate that they were "individually" affected. Although the applicants claim that the continued use of glyphosate gives rise to material damage for some of its members who are wheat producers, the General Court held that the harm relied on by the applicants was "no different from that which could be relied on by any farmer who, for his own reasons, abstains from using glyphosate in favour of other solutions which give rise to certain costs for him."[ay] The action was thus dismissed as inadmissible.

Another case must be mentioned. With a view to protecting its own regulatory prerogatives, the Brussels Capital Region brought an action for annulment of the Commission implementing regulation. The region alleged an infringement of the obligation to ensure a high level of protection of human health and of the environment. It argued that the implementing regulation is based on a scientific assessment of risks to health and the environment that does not meet the requirements of the precautionary principle. The region also alleged an infringement of the general principle of sound administration. With respect to standing, the region argued that the Commission implementing regulation prevents it from exercising its regulatory competences as it sees fit.

The General Court did not rule on the merit of the case as, on 28 February 2019, it declared the action inadmissible on the ground of lack of standing to bring proceedings.[az] Specifically, the General Court held that the Brussels Capital Region was not "directly concerned" within the meaning of Article 263 fourth paragraph TFUE by the regulation at issue. On appeal, Advocate General Michal Bobek found

[av] Case T-125/18, *Associazione GranoSalus* [2019] T:2019:92, para. 83.

[aw] Case T-125/18, *Associazione GranoSalus* [2019] T:2019:92, para. 85.

[ax] Case T-125/18, *Associazione GranoSalus* [2019] T:2019:92, para. 96.

[ay] Case T-125/18, *Associazione GranoSalus* [2019] T:2019:92, para. 59.

[az] Case T-178/18 *Région de Bruxelles-Capitale v Commission* [2019] EU.T.2019:130.

recently that by denying standing to the Brussels Capital Region, the General Court erred in law, misinterpreting the fourth paragraph of article 263 TFEU as well as a number of provisions of applicable secondary law.[ba]

Validity of the pesticide regulations in light of the precautionary principle

Enshrined in Article 192(2) TFEU–a provision declaring the principles underpinning EU action in the field of environmental protection–the precautionary principle was not defined by the treaty framers. The CJEU filled this gap in 1998 by asserting that "where there is uncertainty as to the existence or extent of risks to human health, protective measures may be taken without having to wait until the reality and seriousness of those risks become fully apparent."[bb] The precautionary principle quickly developed into one of the foundations of the high level of environmental protection in the EU.[bc] While the CJEU has been more careful in speculating about the nature of that principle, the General Court has classified precaution as a general principle of EU law. According to the General Court, the precautionary principle empowers the EU institutions "to take appropriate measures to prevent specific potential risks to public health and safety."[bd]

The precautionary principle is explicitly mentioned in the PPPR [Article 1(4)]. In this connection, EU institutions as well as national authorities are called on to conduct risk assessments of pesticide products and their substances considering the extent of lingering uncertainties.

Because the principle is binding on the EU institutions and on the member states when their measures fall within the scope of secondary law, EU courts may be called on to review the consistency of measures on hazardous substances with the principle. This happened in *Blaise*, when the CJEU was called on to review the consistency of the PPPR substances with the environmental principle.[be] All the questions referred by the French criminal court inquire as to the conformity of the EU pesticide regulations with the principle. In answering these questions, the CJEU stressed that there is "an obligation"[bf] on the EU legislature, when it adopts rules governing the placing on the market of pesticides, to comply with the principle in order to ensure a high level of

[ba] Advocate General's Opinion in Case C-352/19 P *Région de Bruxelles-Capitale v Commission.*

[bb] See Case C-157/96 *NFU* [1998] ECR I-2211, para. 63; Case C-180/96 *UK v Commission* [1998] ECR I-2265, para. 99. This interpretation of the PP has become settled case law: Case C-236/01 *Monsanto Agricoltura Italia* [2003] ECR I-8105, para. 111; Case C-77/09 *Gowan* [2010] C:2010:803, para. 73; Case C-333/08 *Commission v France* [2010] ECR I-757, para. 91; Case C-343/09 *Afton* [2010] C:2010:419, para. 62. See also Case T-13/99 *Pfizer* [2002] ECR II-3305, para. 139.

[bc] Case C-127/02 *Waddenzee* [2004] ECR I-7405, para. 44; Case T-125/17, *BASF Grenzach GmbH* [2019] T:2019:638, para. 272. See de Sadeleer.[13, 14]

[bd] Cases T-429/13 and T-451/13 *Bayer* [2018] T:2018:624, para. 109.

[be] Case C-616/17 *Blaise*, cited above, C:2019:800.

[bf] Case C-616/17 *Blaise*, cited above, C:2019:800 para. 42.

protection of human health.[bg] As a result, the Court ruled on the validity of the PPPR in light of the precautionary principle.

At the outset, the CJEU stresses that the normative framework stemming from the PPPR "ensures that the competent authorities ..., when they decide on that authorisation and that approval, sufficient information in order adequately to assess, ..., the risks to health resulting from the use of those active substances and those plant protection products."[bh]

However, "in view of the need to strike a balance between several objectives and principles, and of the complexity of the application of the relevant criteria," judicial review was limited to whether the EU legislature, in adopting the PPPR, committed a manifest error of assessment.[bi] As a result of the rather limited scope of review, the Article 191(2) TFEU principle grants wide discretion as to the measures that can be taken by the EU institutions.

This restrained judicial review of the internal legality of EU legislation is testament to a rather classical reasoning. Traditionally, the EU judiciary has shown judicial restraint, as it is not entitled to substitute its assessment of the facts for that of the EU institutions on which the treaty confers sole responsibility for that duty.[bj] Where the EU institutions are called upon to make "complex assessments," they enjoy a wide measure of discretion when they adopt risk management measures.[bk] In this respect, when invoking the principle of precaution, the CJEU[bl] and the General Court[bm] have on various occasions in the past rejected lawsuits founded on manifest errors of appraisal committed by the institutions when taking decisions that were not fully justified in the light of prevailing scientific knowledge. In any case, a court is no substitute for legislative power.

Although the CJEU confirmed the validity of the PPPR, it interpreted that regulation in light of the precautionary principle. In so doing, the Court enhanced the precautionary obligations placed on the authorities when approving an active substance.

In particular, it must be noted that in the course of risk assessment, the authorities have to take into account the "known cumulative and synergistic effects" of the residues having a harmful effect on human or animal health [Articles 4(2)(a)–4(3)(b)]. This entails that the cocktail effects caused by the interaction between glyphosate

[bg] CFR, Art 35 and TFEU, Art 9 and Art 168(1).

[bh] Case C-616/17 *Blaise*, cited above, para. 47.

[bi] Case C-616/17 *Blaise*, cited above, para. 50.

[bj] See Case T-13/99 *Pfizer*, above, para. 169.

[bk] Case C-180/96 *UK v Commission* [1998] ECR I-2269, para. 97; Case T-74/00 *Artegodan* [2002] ECR II-4945, para. 201; Case T-392/02 *Solvay Chemicals* [2003] para. 126; Case C-77/09 *Gowan* [2010] ECR I-13533, paras 55 and 82; Case C-343/09 *Afton* [2010] ECR I-7027, para. 28.

[bl] See Case 174/82 *Sandoz* [1983] ECR 2445, para. 17; Case C-331/88 *Fedesa* [1990] ECR I-4023, para. 9; Case C-180/96 *UK v Commission* [1998] ECR I-2269, paras 99 and 100; and Case C-127/95 *Norbrook Laboratories Ltd* [1998] ECR I-1531.

[bm] See Case T-199/96 *Laboratoires pharmaceutiques Bergaderm S.A.* [1998] ECR II-2805, paras. 66 and 67; cases T-13/99 *Pfizer Animal Health v Council* [2002] ECR II 3305 and T-70/99 *Alpharma v Council* [2002] ECR II-3495

and, inter alia, the other constituents of the product must be taken into consideration.[bn] However, EFSA has found providing such a methodology difficult. The CJEU judgment should boost the agency's effort in that respect.[bo]

Furthermore, in order to obtain the authorization of plant protection products, the applicants must submit to the national authorities, in addition to the "known cumulative and synergistic effects," "any information on potentially harmful effects of the plant protection product on human and animal health or on the environment."[bp] In other words, an information duty applies whenever the impacts are potential and not yet fully demonstrated.

By the same token, the CJEU ruled that the applicant is not exempted from submitting tests of long-term carcinogenicity and toxicity relating to the plant protection product that is the subject of an application for authorization.[bq] In effect, a plant protection product cannot be considered to satisfy the safety requirement laid down by the EU lawmaker "where it exhibits any long-term carcinogenicity and toxicity."[br] As a result, the applicants would be required to submit tests of long-term carcinogenicity and toxicity relating to their plant protection products. Whether such tests will be submitted with respect to the renewal of the approval of pesticides containing glyphosate remains to be seen.

Validity of the restrictions placed by the member states on the use of glyphosate

At the outset, the interactions between the free movement of goods, enshrined in the TFEU, and the national restrictive measures placed on the marketing and use of pesticides are at odds with one another. Thus far, scant attention has been paid to the scope of these restrictive measures.

Total harmonization

As far as pesticides are concerned, until the entry into force of former directive 91/414/EEC, there were no common harmonized rules relating to the production and marketing of plant protection. In the absence of harmonization, it was therefore "for the Member States to decide what degree of protection of the health and life of humans they intended to assure… having regard to the fact that their freedom of action is itself restricted by the Treaty."[bs]

[bn] Advocate General Sharpston' Opinion in Case C-616/17 *Blaise*, cited above, para. 58.

[bo] Bailleux A. Case note on Blaise, cited above, 872.

[bp] Case C-616/17 *Blaise*, cited above, para. 73.

[bq] Case C-616/17 *Blaise*, cited above, para. 113.

[br] Case C-616/17 *Blaise*, cited above, para. 115.

[bs] Case 104/75 *De Peijper* [1981] ECR 613; Case 272/80 *Biologische Produkten* [1981] [illegible], para. 12.

Since the entry into force of directive 91/414/EEC (replaced by the PPPR), this field is fully harmonized. As a result, member states may thus no longer rely upon Article 36 TFEU or a mandatory requirement.[bt]

The PPPR enshrines the principle of mutual recognition with the aim of ensuring the free movement of pesticides within the EU (PPPR Preamble, recital 2), albeit imperfectly.[15] To avoid the duplication of administrative burden for the application, the authorization to place the pesticide on the market granted by one member state has to be accepted by other member states where agricultural, plant health, and environmental (including climatic) conditions are comparable. To facilitate such mutual recognition, the EU has been divided into zones.

The renewal of glyphosate approval had the immediate effect of preserving the validity of existing authorizations to place pesticides containing glyphosate on the market.[bu] In the absence of a renewal, those authorizations would ipso facto have lapsed.

However, state authorities still keep room for maneuvering. First, harmonization does not equate with uniformity. By way of illustration, Directive 91/414 concerning the placing of plant protection products on the market did not contain any provision that specifically governed the conditions for the granting of marketing authorization for plant protection products in the context of parallel imports.[bv] As this area was not harmonized, the CJEU ruled that a member state is entitled pursuant to Article 36 TFEU to subject a farmer who imports a plant protection product as a parallel import solely for the needs of his farm to a simplified authorization procedure.[bw]

The PPPR safeguard clauses

The TFEU empowers the EU lawmaker to allow the EC and member states, subject to an EU control procedure, to adopt temporary measures in the event of a sudden and unforeseen danger to health, life, etc. [Article 114(10) TFEU]. Along the same lines of the majority of regulations founded on Article 114 that contain safeguard clauses encompassing health and environmental risks, the PPPR provides for such safeguard clauses. A detailed examination of the manner in which such mechanisms are implemented should be made.

[bt] See among others Case 5/77 *Tedeschi* [1977] ECR 1555, para. 35; Case 148/78 *Ratti* [1979] ECR 1629, para. 36; Case 251/78 *Denkavit Futtermittel* [1979] ECR 3369, para. 14; Case 190/87 *Moormann* [1988] ECR 4689, para. 10; Case 215/87 *Schumacher* [1989] ECR 617, para. 15; Case C-369/88 *Delattre* [1991] ECR I-1487, para. 48; Case C-62/90 *Commission v Germany* [1992] ECR I-2575, para. 10; Case C-323/93 *Centre d'insémination de la Crespelle* [1994] ECR I-5077, para. 30; and Case C-320/93 *Ortscheit* [1994] ECR I-5243, para. 14. Regarding the non-exhaustive character of food additives see Case C-121/00 *Walter Hahn* ECR I-9193.

[bu] Advocate General Bolbek's Opinion in Case C-352/19 P *Région de Bruxelles-Capitale v Commission*, para. 38.

[bv] Advocate General Trstenjak's Opinion delivered on 10 July 2007 in Joined Cases C-260/06 and C-261/06 *Escalier and Bonnarel* [2007] ECR I-9717, para. 8.

[bw] Joined Cases C-260/06 and C-261/06 *Escalier and Bonnarel*, above, paras. 31 and 36

At the outset, a member state may request the Commission "to restrict or prohibit the use and/or sale" of a substance or product (that) has been authorized in accordance with the Regulation (Article 69), provided that the following conditions are fulfilled:

- It is "likely to constitute a serious risk to human or animal health or the environment."
- This risk cannot be contained satisfactorily by means of measures taken by the member state (s) concerned.
- The evidence must be examined by the Commission with the possibility to request the EFSA's opinion.

Moreover, in cases of extreme urgency, the Commission may provisionally adopt emergency measures after consulting the member states concerned and informing the other member states (Article 70).

However, the Commission has full discretion to adopt either the restrictive measures or the provisional measures requested by the member state.

In case no action has been taken in accordance with Article 69 or 70, the member state may adopt interim protective measures without obtaining the prior authorization of the Commission. In this event, it shall immediately inform the other member states and the Commission [Article 71(1)]. In such a case, within 30 working days of the notification, the Commission shall put the matter before the regulatory Committee "with a view to the extension, amendment, or repeal of the national interim protective measure" [Article 71(2)].

Meanwhile, the member state may maintain its national interim protective measures. When the EU measures have been adopted, it must comply with them. Accordingly, if the Commission backed up by the regulatory committee decides to repeal the unilateral domestic measure, the member state is obliged to repeal it.

This analysis prompts several observations.

- First, the Article 71 safeguard clause allows the member states to deal with exceptional situations of limited duration. In other words, the measures of prohibition or limitation taken by national authorities are only authorized for as long as is necessary for a new decision to be taken by the EU authorities.
- Second, precautionary provisions of the PPPR [Article 1(4)] do not prevent the member states from applying the precautionary principle where there is scientific uncertainty as to the risks with regard to human or animal health or the environment posed by the pesticides to be authorized in their territory.[bx] Accordingly, the condition that the substance or the product is "likely to constitute a serious risk to human or animal health or the environment" (Article 69) must be interpreted in light of the precautionary principle.
- Third, the member state must justify the ban or restriction in the light of the noneconomic reasons mentioned in Article 69, such as human or animal health or the environment.
- Fourth, regarding the burden of proof, the member state must provide the relevant evidence demonstrating the existence of a risk. In this connection, the CJEU ruled in *Monsanto Agricultura Italia* that "protective measures, notwithstanding their temporary character and even if they are preventive in nature, can be adopted only if they are based on a risk assessment which is as complete as possible in the particular circumstances of an individual case" [Case C-236/01, *Monsanto Agricoltura Italia* (2003) C:2003:431, para 107].

[bx] *Blaise*, para. 44.

- Finally, in accordance with principles traditionally applicable to safeguard clauses, the application of a safeguard clause is subject to a "control procedure" undertaken by the Commission. The member state is thus obligated to notify the Commission of the derogating measure taken in order to enable the latter to ascertain whether it is consistent with the PPPR. Generally speaking, the Commission shall either authorize the provisional measure for a time period defined or require the member state to revoke the provisional measure. As a result, the interim national measure is temporary.

The restrictive measures authorized by Directive 2009/128/EC

In implementing Directive 2009/128/EC of the European Parliament and of the Council of 21 October 2009, establishing a framework for Community action to achieve the sustainable use of pesticides (hereafter Directive 2009/128/EC), member states are empowered to regulate the use of pesticides containing glyphosate. In this connection, the following provisions are illustrative of the preventive measures that member states should enact.

- Integrated Pest Management: "professional users of pesticides [ought to] switch to practices and products with the lowest risk to human health and the environment among those available for the same pest problem" (Article 14.1).
- Use nonchemical alternatives: "Member states shall take all necessary measures to promote low pesticide-input pest management, giving wherever possible priority to non-chemical methods" (Article 14).
- Minimization or prohibition of pesticide use in specific areas (Article 12) and the establishment of appropriately sized buffer zones to protect nontarget aquatic organisms and safeguard zones for surface and groundwater used for the abstraction of drinking water (Article 11).

Review by national courts of restrictive measures placed on the use of glyphosate

The following judgments exemplify to some extent the room for maneuvering left to member states.

In its judgment of 15 January 2019, the Administrative Court of Lyon struck the authorization granted by the French Agency for Food, Environmental and Occupational Health & Safety (ANSES) with respect to the marketing of Roundup Pro 360.[by] It criticized ANSES for failing to produce a risk assessment making it possible to establish that Roundup Pro 360 was neither carcinogenic nor toxic to reproduction, even though the EFSA considered that glyphosate preparations may not be carcinogenic. Indeed, the court stressed that Roundup Pro 360 is a preparation that is more toxic than glyphosate. In other words, the product at issue is likely to prove to be carcinogenic without the active substance it contains being carcinogenic as such. Despite the restrictions placed on the use of Roundup Pro 360, the marketing authorization granted by ANSES is "likely to cause serious damage to health." On the basis of that

[by] Administrative Court of Lyon, 15 January 2019.

conclusion, the Court annulled the marketing authorization on the ground that, by authorizing that herbicide, ANSES had committed a manifest error of appraisal in the light of the precautionary principle, which is enshrined in Article 5 of the Constitutional Charter on the Environment. That principle is to be implemented by the public authorities where there is a risk of serious and irreversible damage to the environment or damage to the environment likely to cause serious harm to health. By omitting to take into consideration serious health risks, ANSES could not grant such a marketing authorization. This reasoning seems to us to be in line with EU law because the PPPR only allows the marketing of safe plant protection products.

On 28 February 2019, the Belgian Constitutional Court dismissed a claim lodged by the Belgian association of the plant protection products industry against a decree of the Flemish region restricting the use of glyphosate (judgment No. 38/2019). The Court held that the Flemish decree implements Directive 2009/128/EC that allows member states to regulate the use of pesticides. Furthermore, the Court emphasized that the restrictions placed on the use of pesticides containing glyphosate are authorized in virtue of Article 12 of that directive.[bz]

The Swedish Supreme Administrative Court reversed a decision taken by the Swedish Chemicals Agency (KemI) and, on appeal by the government, restricting the use of the active substance glyphosate, on the ground that the substance was authorized under former Directive 91/414 on pesticides.[ca] The Supreme Court held that a concrete risk assessment of the impact of the substance on ground water was missing.

Access to information

The solitary exercise of power linked to the administrative tradition of secrecy has long been reflected in the considerable inertia that arises when it comes to disclosing information about technical choices relating to environmental issues. Yet information constitutes the core of the struggle to protect the environment because ignorance renders rights to participation and access to justice ineffective. The right to information is therefore central among procedural rights.[16] Furthermore, access to environmental information plays an important role as a procedural aspect of a substantive right such as a right to a clean environment.[17]

Information held by EU institutions

Access to information held by EU institutions is covered by two regulations, the second of which implements the Convention on Access to Information, Public Participation in Decision-Making, and Access to Justice in Environmental Matters to community institutions and bodies (hereafter the Aarhus Convention):

[bz] B.5.

[ca] Case Raa 2005.

- Regulation (EC) No. 1049/2001 of the European Parliament and of the Council of 30 May 2001, regarding public access to European Parliament, Council, and Commission documents.[cb]
- Regulation (EC) No. 1367/2006 of the European Parliament and of the Council of 6 September 2006, on the application of the provisions of the Aarhus Convention on Access to Information, Public Participation in Decision-Making, and Access to Justice in Environmental Matters to community institutions and bodies (hereafter the Aarhus Regulation).[cc]

At the outset, the EU institutions could refuse to grant access to the information on glyphosate by invoking the exception in Article 4(2), first indent, of Regulation No. 1049/2001, namely the protection of the commercial interests of the company producing glyphosate.

However, the Aarhus Regulation is a *lex specialis* that derogates from Regulation No. 1049/2001. Indeed, the aim of the Aarhus Regulation is to ensure the widest possible systematic availability and dissemination of the environmental information held by the institutions and bodies of the EU.[cd] As a result, exceptions to that principle must be interpreted and applied strictly. Furthermore, account must be taken of the Aarhus Convention for the purposes of interpreting the Aarhus Regulation.[ce]

The fact that the EU institutions are obliged under the Aarhus Regulation to ensure the dissemination of environmental information has significant consequences. The rule laid down in Article 4(2) of Regulation 1049/2001 requires the weighing of the interests. In contrast, the Aarhus Regulation derogates from that rule by establishing a presumption in favor of the disclosure of information that "relates to emissions into the environment" [Article 6(1) first sentence]. It follows that Article 6(1) of the Aarhus Regulation requires the disclosure of a document where the information requested relates to "emissions into the environment," even if there is a risk of undermining the protection of the commercial interests of the company.

The question arose as to whether the concept of "information relating to emissions into the environment" must be interpreted broadly or restrictively.

Greenpeace and the Pesticide Action Network Europe (PAN Europe) have been attempting to gain access to the records concerning the authorization of glyphosate for use in pesticides. The EC disclosed some of the documents in question, but withheld others on the grounds of protection of the commercial interests of the undertakings concerned. The General Court, on the other hand, ruled that the withheld documents also relate to emissions into the environment, and an overriding interest in their disclosure must therefore be presumed to exist. On appeal, the CJEU held that the concept of "information [which] relates to emissions into the environment" must not be interpreted strictly.[cf] Consequently, an EU institution cannot justify its refusal

[cb] OJ 2001 L 145, p. 43.

[cc] OJ 2006 L 264, p. 13.

[cd] Art. 1. See Case C-673/13 P, *Commission v Stichting Greenpeace Nederland and PAN Europe* [2016] EU:C:2016:889, para. 52.

[ce] Case C-673/13 P *Commission v Stichting Greenpeace Nederland and PAN Europe* [2016] EU:C:2016:889, para. 61.

[cf] C-673/13 P *Commission v Stichting Greenpeace Nederland and PAN Europe* [2016] EU:C:2016:889, paras 49 and 53.

to divulge it on the basis of the exception relating to the protection of the commercial interests of a particular natural or legal person for the purposes of Article 4(2), first indent, of Regulation No. 1049/2001, where the information contained in that document constitutes information that "relates to emissions into the environment" within the meaning of Article 6(1) of the Aarhus Regulation.[cg]

The CJEU concluded that it was necessary to include in the concept of information that "relates to emissions into the environment" information enabling the public to check whether the assessment of actual or foreseeable emissions, on the basis of which the competent authority authorized the product or substance in question, was correct, and the data relating to the effects of those emissions on the environment.[ch] Accordingly, the Court endorsed a broad interpretation of the notion of the concept of "emissions."

Overriding public interest in disclosing information relating to emissions into the environment

The CJUE judgment in *Commission* v *Stichting Greenpeace Nederland and PAN Europe* left a number of questions unanswered on the account that an active substance such as glyphosate is inevitably released into the environment at some stage of its life cycle. They referred the case back to the General Court. The parties disagree on whether that information is covered by the concept of "information relating to emissions into the environment" as defined by the CJUE in the judgment on appeal.

The nongovernmental organizations requesting the information [Stichting Greenpeace Nederland and Pesticide Action Network Europe (PAN Europe)], supported by Sweden, argued that the information concerns all the substances released into the environment when the authorized substance glyphosate is used and applied in pesticides. In particular, the nongovernmental organizations sought access to information relating to the "identity" and quantity of impurities present in the glyphosate; the analytical profile of the batches, in particular their composition; the "identity" and quantity of chemical substances added during the tests; the duration of those tests; and the actual effects on the active substance.[ci] In their view, that information could allow the determination of the level of emission of those impurities into the environment.[cj]

In contrast, the EC argued that this information relates to the manufacturing processes used by the various operators that notified glyphosate for the purpose of its inclusion in Annex I to Directive 91/414 and was thus not directly linked to emission into the environment. In particular, the disclosure of such information would make it possible to reconstitute the manufacturing process of glyphosate and the related business secrets.[ck]

[cg] Case T-716/14, *Tweedale* [2019] T:2019:141, para. 58.

[ch] C-673/13 P *Commission v Stichting Greenpeace Nederland and PAN Europe*, cited above, para. 80.

[ci] C-673/13 P *Commission v Stichting Greenpeace Nederland and PAN Europe*, cited above, para. 60.

[cj] C-673/13 P *Commission v Stichting Greenpeace Nederland and PAN Europe*, cited above, paras. 62–64.

[ck] C-673/13 P *Commission v Stichting Greenpeace Nederland and PAN Europe*, cited above, para. 65.

The General Court held that

> *while it is not necessary to apply a restrictive interpretation of the concept of 'information [which] relates to emissions into the environment,' that concept may not, in any event, include information containing any kind of link, even direct, to emissions into the environment. If that concept were interpreted as covering such information, it would to a large extent deprive the concept of 'environmental information' as defined in Article 2(1)(d) of Regulation No. 1367/2006 of any meaning. Such an interpretation would deprive of any practical effect the possibility, laid down in the first indent of Article 4(2) of Regulation No. 1049/2001, for the institutions to refuse to disclose environmental information on the ground, inter alia, that such disclosure would have an adverse effect on the protection of the commercial interests of a particular natural or legal person and would jeopardise the balance which the EU legislature intended to maintain between the objective of transparency and the protection of those interests. It would also constitute a disproportionate interference with the protection of business secrecy ensured by Article 339 TFEU (judgment on appeal, paragraph 81).*[cl]

The General Court drew a distinction between

- On the one hand, "the use and conditions of use of the plant protection product covered by authorisation in a Member State" that "may be very different from those which have been subject to the theoretical assessment at EU level."[cm]
- On the other, the plant protection product for which authorization is requested that is often produced by a different undertaking than that which requested approval for the active substance at the EU level.[cn]

It concluded that

> *it is only at the stage of the national authorisation procedure to place a specific plant protection product on the market that the Member State assesses any emissions into the environment and that specific information emerges concerning the nature, composition, quantity, date, and place of the actual or foreseeable emissions, under such conditions, from the active substance and the specific plant protection product containing it.*[co]

and further that the Commission did not commit an error of assessment in considering that the draft report, drawn up in the course of the approval procedure at the EU level, does not contain information relating to emissions into the environment.

This narrow interpretation has been criticized by E. Brosset, who argues that the CJEU judgment allowed a case-by-case approach while the General Court endorses a global approach (*une approche d'ensemble*).[18]

In 2009, in *Tweedale* and *Hautala*, the General Court endorsed an interpretation that was more consistent with the *Stichting Greenpeace Nederland and PAN Europe*

[cl] Case T-545/11 *RENV Stichting Greenpeace Nederland* [2016] EU:T:2018:817, para. 58.

[cm] Case T-545/11 *RENV Stichting Greenpeace Nederland* [2016] EU:T:2018:817, para. 83.

[cn] Case T-545/11 *RENV Stichting Greenpeace Nederland* [2016] EU:T:2018:817, para. 84.

[co] Case T-545/11 *RENV Stichting Greenpeace Nederland* [2016] EU:T:2018:817, para. 88.

CJEU judgment. The General Court ruled that key studies intended to determine the effects of exposure to glyphosate on human health [determining, inter alia, the acceptable daily intake (ADI) and acute reference dose (ARfD) for glyphosate] and used in the renewal dossier amount to information on foreseeable emissions into the environment. What is more, "an active substance contained in plant protection products, such as glyphosate, in the course of normal use, is intended to be discharged into the environment by virtue of its function, and its foreseeable emissions cannot, therefore, be regarded as purely hypothetical."[cp]

Environmental information held by the member states

In adopting the PPPR, the EU lawmaker sought to balance the confidential nature of the information submitted by the applicant and the right of the public to access to environmental information. This led to a political compromise, as Article 63 of the PPPR is worded as follows:

(1) A person requesting that information submitted under this regulation is to be treated as confidential shall provide verifiable evidence to show that the disclosure of the information might undermine his commercial interests, or the protection of privacy and the integrity of the individual.

(2) Disclosure of the following information shall normally be deemed to undermine the protection of the commercial interests or of (the) privacy and the integrity of the individuals concerned:

(a) the method of manufacture;

...

(f) information on the complete composition of a plant protection product;

...

The third paragraph of this provision states expressly that the right to keep some information confidential is to be without prejudice to the application of Directive 2003/4/EC, which means that requests for access by third parties to the information contained in authorization application dossiers are subject to the general provisions of that directive.[cq]

In virtue of Directive 2003/4/EC on public access to environmental information, member states have to ensure that public authorities make the environmental information they hold available to any legal or natural person on request. Account must be taken of the Aarhus Convention for the purposes of interpreting Directive 2003/4/EC.[cr]

The CJEU has provided guidelines in its interpretation of the concept of "information on emissions into the environment" for the purposes of the second subparagraph of Article 4(2) of Directive 2003/4/EC.[cs]

[cp] Case T-716/14 *Tweedale* [2019] T:2019:141; Case T-329/17 *Hautala* [2019] T:2019:142.

[cq] Blaise, para. 106.

[cr] Case C-442/14 *Bayer CropScience and Stichting De Bijenstichting* [2016] EU:C:2016:890, para. 54.

[cs] Directive 2003/4/EC of the European Parliament and of the Council of 28 January 2003 on public access to environmental information.

In *Bayer CropScience*, the CJEU had to assess whether the information regarding the foreseeable emissions into the environment of the residues of the active substance glyphosate could be disclosed in accordance with Directive 2003/4/EC. Studies of residues and reports of field trials submitted in connection with a procedure for extending the authorization of a product in accordance with the legislation of plant protection products are deemed to be "environmental information" for the purpose of Article 2 of Directive 2003/4/EC on access to environmental information. In effect, this information "concerns elements of the environment which may affect human health if excess levels of those residues are present."[ct]

The Court took the view that the information to be communicated encompasses "studies which seek to establish the toxicity, effects, and other aspects of a product or substance under the most unfavourable realistic conditions which could possibly occur, and studies carried out in conditions as close as possible to normal agricultural practice and conditions which prevail in the area where that product or substance is to be used."[cu]

Conclusions

The EC has always claimed that EU environmental laws, and in particular the PPPR, are among the most ambitious environmental regulations. This level of ambition should be buttressed by the constitutional obligations laid down in the EU founding treaties (high level of environmental and health protection). In particular, the PPPR has to be interpreted in light of the key objectives set forth under treaty law as well as the general principles of EU law, such as precaution. Moreover, in adopting the PPPR, the EU lawmaker has been favoring environmental and health interests over plant production. Therefore, one has to take seriously the place occupied by environmental and health interests in both primary and secondary law.

As it is underpinned by the principles of hazard identification, precaution, and substitution, the PPPR has been paving new ways in the reduction of health and environmental risks. As a result, the EU's goals are not solely economic, but also social and environmental. The proper functioning of the internal market must be accommodated with these nonmarket values, the legal protection of which is nonetheless essential.

The approval of glyphosate as an active substance has been subject to significant tension between the member states and the EU institutions. To understand this tension, this chapter has explored the key features of the regulatory framework empowering the EC to approve glyphosate and the member states' restrictions placed on its use. The understanding of this framework is indispensable, as a new approval procedure has just commenced.

[ct] Case C-266/09 *Stichting Natuur en Milieu* [2010] C:2010:779, paras 42-43.

[cu] Case C-442/14 *Bayer CropScience and Stichting De Bijenstichting* [2016] C:2016:890, para. 81.

References

1. Cranor C. *Toxic torts: science, law, and the possibility of justice*. Cambridge University Press; 2006. p. 216.
2. de Sadeleer N. Environmental governance and the legal bases conundrum. *Oxford Yearbook Eur Law* 2012;373–401.
3. de Sadeleer N. *EU environmental law and the internal market. Oxford University* Press; 2014. p. 157–61. 291;304;353;358–382.
4. de Sadeleer N. *EU environmental law and the internal market. Oxford University* Press; 2014. p. 21–33.
5. Bozzini E. *Pesticides policy and politics in the EU*. 58. Mannheim: Springer; 2017.
6. Kuraj N. Complexities and conflicts in controlling dangerous chemicals. In: Maitre-Kern E, et al., editors. *Preventing environmental damage from products*. Cambridge University Press; 2018. p. 299.
7. Bozzini E. *Pesticides policy and politics in the EU*. 30. Mannheim: Springer; 2017.
8. Bozzini E. *Pesticides policy and politics in the EU*. 32. Mannheim: Springer; 2017. p. 67.
9. de Sadeleer N. *EU environmental law and the internal market. Oxford University Press*; 2014. p. 50–1.
10. de Sadeleer N. *Environmental principles*. 2nd ed. *Oxford University Press*; 2020. 93, 94, 107.
11. de Sadeleer N. *Environmental principles*. 2nd ed. *Oxford University Press*; 2020. p. 337–9.
12. de Sadeleer N, Poncelet C. Protection against acts harmful to human health and the environment adopted by the EU institutions. *Cambridge Yearbook of EU Law* 2011-2012;**14**:177–208.
13. de Sadeleer N. *EU environmental law and the internal market. Oxford University Press*; 2014. p. 45–56.
14. de Sadeleer N. *Environmental principles*. 2nd ed. *Oxford University Press*; 2020. p. 140–5.
15. Bozzini E. *Pesticides policy and politics in the EU*. Mannheim: Springer; 2017. p. 43–5.
16. de Sadeleer N. *Environmental principles*. 2nd ed. *Oxford University Press*; 2020. p. 425.
17. de Sadeleer N. *Environmental principles*. 2nd ed. *Oxford University Press*; 2020. p. 418.
18. Brosset E. Le glyphosate devant la Cour: quels enseignements sur le droit d'accès au documents et à la justice? *Revue Trimestrielle de Droit Européen* 2019;629–50.

Case law on the validity of glyphosate in the EU legal order

Court of Justice of the European Union

Joined Cases C-260/06 and C-261/06 *Escalier and Bonnarel* [2007] ECR I-9717.
Case C-442/14 *Bayer CropScience and Stichting De Bijenstichting* [2015] C:2016:890.
Case C-616/17 Blaise [2019] C:2019:800.

General Court

Case T-125/18 *Associazione GranoSalus* [2019] T:2019:92.
Case T-716/14 *Tweedale* [2019] T:2019:141.
Case T-329/17 *Hautala* [2019] T:2019:142.
Case T-178/18 *Région de Bruxelles-Capitale v Commission* [2019] EU:T:2019:130.

Herbicides: A necessary evil? An integrative overview

Robin Mesnage[a] *and Johann G. Zaller*[b]
[a]King's College London, London, United Kingdom, [b]Institute of Zoology, University of Natural Resources and Life Sciences (BOKU), Vienna, Austria

Chapter outline

Introduction

The rapid expansion of the human population in the last century has prompted an ever more intensive agricultural production system. Among a plethora of technologies, herbicides are used to eliminate weeds—or, to put it simply, plants present at the wrong place at the wrong time, such as a sunflower in a sorghum field (Chapter 1).

Herbicides are used mostly in agriculture to support the intensification of food, animal feed, and biofuel production. In large-scale monocultures, or when plowed fields are left unsown, some plants can grow without competition. These weeds have been estimated to cause a crop yield loss of 43% worldwide.[1] Herbicides are also applied in urban environments mainly for cosmetic purposes, along roadsides and railway tracks, in landscape turf management, and even in nature conservation areas to kill invasive neophytes (Chapter 1).

The use of herbicides increases when countries get richer. In a recent analysis, a 1% increase in gross domestic product per capita was shown to result in a 1.2% annual increase in herbicide use between 1990 and 2014.[2] With the beneficial influence of pesticides on the yields and production costs of a farming system, there is a downside: environmental degradation such as water contamination and loss of biodiversity as

Herbicides. https://doi.org/10.1016/B978-0-12-823674-1.00003-1

well as public health impact such as acute poisoning and chronic diseases from exposure.[34]

In this book, authors from various disciplines have summarized the knowledge of the chemistry, toxicology, and legal aspects as well as the practical aspects of application in conventional agricultural systems and their documented nontarget effects in human populations and ecosystems. Fig. 11.1 gives a quantitative idea of some of the matters covered.

Avoiding a technological dead end in weed management

From the first agricultural settlements to modern agricultural systems, weed control has been central to food production. While nonchemical and chemical methods were already known in ancient times, control accelerated massively after World War II with the development of synthetic herbicides. These products allowed the intensification of food production by implementing large-scale monocultures in the Green Revolution. Cereals can receive up to three herbicide treatments per cropping season, with dozens of active ingredients approved for a single crop species (Chapter 1). Herbicides are

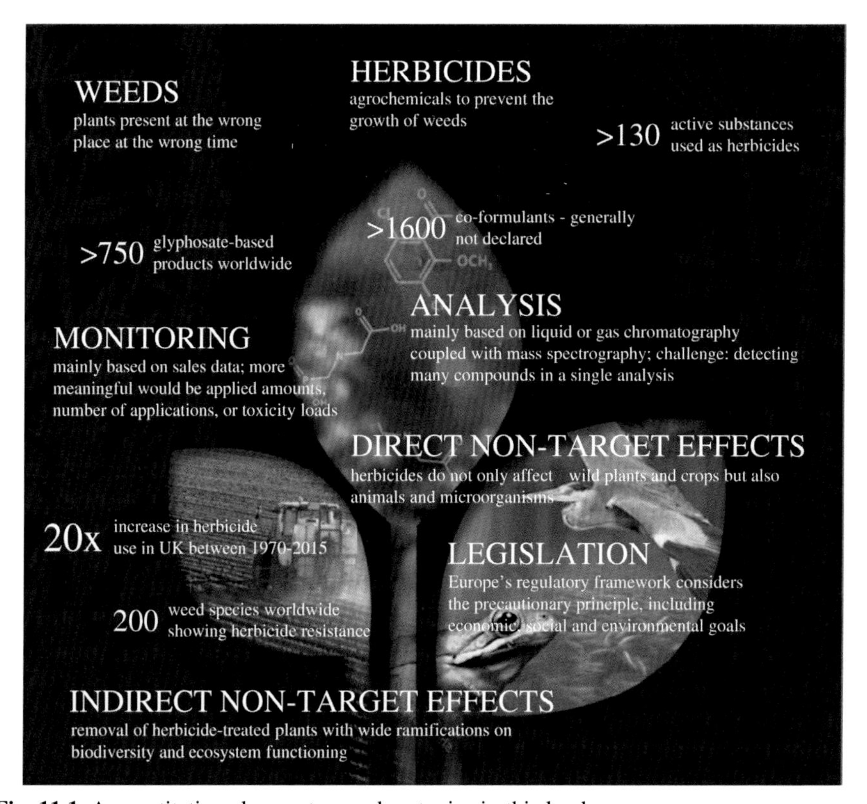

Fig. 11.1 A quantitative glance at some key topics in this book.

also widely used to facilitate harvest by killing both crop species and weeds, for instance in lentils, cotton, and palm oil.

Herbicide-based management systems rely on constant innovation. The steady creation of new mechanisms of action is needed to prevent or mitigate weed resistance (Chapters 3 and 8). During the 1970s and 1980s, the number of chemical solutions available to farmers increased rapidly, and there was always a new herbicide to cope with weed resistance or to diversify management practice. However, while herbicides with novel modes of action were introduced at the rate of 5.5 annually until the 2000s, that average declined to about 2 in the next decade.[3] Product introduction was slowed by the increase in research, development, and registration costs: the average cost of developing a new pesticide product rose from $152 million in 1995 to $286 million in 2014.

To diversify weed management practices and maintain a competitive advantage, pesticide companies started to acquire seed companies. In the 1980s, they introduced crops engineered to tolerate herbicides, initially glyphosate-based herbicides (GBHs). This was a double-edged success. The technology was widely adopted by farmers in North and South America, but weed management became highly dependent on glyphosate. Over the past two decades, glyphosate accounted for half the total herbicide volume (Chapter 1). Reliance on glyphosate favored the development of weed resistance.

One response was to engineer herbicide tolerance into crop seeds so that plants would not be damaged by multiple herbicides such as dicamba, 2,4-D, isoxaflutole, glyphosate, and glufosinate. All major agrochemical companies have now developed stacked herbicide-tolerant genetically modified (GM) crops so that higher doses and multiple agents can be used on resistant weeds (Chapter 6). However, the escalation of weed resistance continues. To maintain food production sufficient for a growing world population while dealing with ever-widening herbicide resistance, new weed management methods must be found.

Good weed control ultimately depends on cultural knowledge—what a farmer knows. Programs should not be focused on yield and production costs alone but must integrate long-term sustainability, environmental impact, and human health. More research and education are needed to define and implement future weed management. A farmer's freedom of choice for a weed management system is becoming increasingly difficult. This is because of the economic and license-based connection between the cropping of GM herbicide-tolerant seeds and the obligation to apply a certain herbicide with it.

Toward a better measure of herbicide use across time and space

Evaluating the costs and benefits of herbicide use requires a reliable assessment of the quantity used. To the question whether use is going up or down or staying the same, there is no gold standard: the answer depends on the metric chosen. To provide the most accurate estimation and allow comparisons across time and space, several metrics have been defined (Chapter 2). They describe changes in land area treated (hectares planted in a given crop, percent hectares treated with a given herbicide, and number of hectare treatments) or changes in the intensity of herbicide use on a given hectare (rate of application, number of applications, rate per crop year).

GBH use in the United States skyrocketed toward the end of the 1990s after the introduction of glyphosate-tolerant (Roundup Ready) soybeans (Chapter 2). The changes were first driven by an increase in the area treated as the technology was progressively adopted by farmers. An analysis of herbicide volume suggests that weed control was done with a single application in the early years. Area-based treatments of soybeans even decreased toward the end of the 1990s. Since 2010, when the number of resistant weeds started to increase, farmers have used agents with other modes of action and have increased the rate of GBH application, resulting in further elevation of the total volume.

Another approach calculates pesticide load indicators to monitor nontarget effects on human health, ecotoxicology, and environmental fate at the farm level, and provides detailed information on the use pattern of the various pesticide modes of action.[4]

Monitoring exposure to herbicide mixtures in food and human matrices

Statistics on the quantity of herbicides sold and used, the agricultural area treated, the average one-time rate of application, or the number of applications can be used to estimate use. However, to avoid or assess the risk of exposure, analytical chemistry methods are necessary to directly measure concentrations in food, animal feed, human matrices, and the environment.

Although measurement methods are available, a challenge for the future will be the integration of use data with other information in large databases accessible to herbicide regulators and researchers. Cross-linking use data with meteorological, geographic, and public health information can help with monitoring effects.[5] This allows the retrospective assessment of exposure to herbicide drift from agricultural use in rural areas near intensively managed apple orchards and vineyards[6] and even in urban areas.[7] In a recent study, the risk of autism spectrum disorder was shown to be increased for women exposed prenatally or in infancy to glyphosate and other pesticides applied within 2000 m of their residences.[8] There are still important challenges to be addressed in getting accurate estimates of use over time and space in most non-occupationally exposed populations. Toxicologists should appreciate the potential of big data in understanding the effects on human populations and ecosystems; weed managers can use it and digitalization to develop precision weed control, which will reduced herbicide use.[9]

The challenges in the measurement of herbicide analytes in food and human samples are discussed in Chapter 5. The most frequent method is liquid chromatography coupled with tandem mass spectrometry (LC-MS/MS) and gas chromatography coupled with single (GC-MS) or tandem mass spectrometry (GC-MS/MS). While some other methods such as ELISA (enzyme-linked immunosorbent assay) have been applied, they are efficient for water analysis but failed to fulfill the initial promise for urine.

One of the main challenges for the future is to develop methods allowing the fast generation of quantitative data for a large number of compounds in a single analysis. This is a challenge for herbicides because they have very different biochemical structures (Chapter 3). The pH and ionic strength are factors affecting the development of reliable analytical methods, as most herbicides are weak acids and bases. For instance, in the case of glyphosate, the development of methods compatible with regulatory requirements is problematic because this compound and its metabolite AMPA have short retention times and low molecular weights, and interferences are frequent. This enhances the probability of false positives. In addition, the measurement of glyphosate can hardly be multiplexed with those of other herbicides because of its structural particularities.

Many controversies arise from the results of studies reporting the presence of glyphosate in human matrices such as breast milk, blood, and hair, or in tissues of animals fed herbicide-tolerant GM crops. There is no consensually recommended analytical method for the measurement of glyphosate and AMPA in human matrices. Given the controversies associated with the presence of herbicides in food, there is an urgent need to develop sensitive and robust analytical methods for the determination of herbicide levels, including metabolites, in biological fluids, and to improve the performances of current methods for analyses of feed and food to better understand the sources of exposure.

What are the health effects of herbicides in humans?

All pesticides carry risks. Exposure to spray dilutions is implicated in the development of cancer, diabetes, infertility, and neurodegenerative disorders in agricultural workers. Whether current exposure to environmental levels of herbicide residues causes adverse health effects is less clear.

The use of herbicides has become safer in the last decade because the most toxic ingredients have been progressively banned. In addition, the use of the most toxic agents was reduced when glyphosate-tolerant crops were introduced in the late 1990s. However, the development of glyphosate-resistant weeds prompted farmers to mix glyphosate with other herbicides (Chapter 6). The toxicological characteristics and impacts of these multiherbicide tank mixtures are not known.

The literature review in Chapter 6 also showed that the apparent increase in safety of herbicide use patterns could be misleading because recently introduced compounds remain poorly tested. There is an inescapable time gap between the introduction of a new herbicide and the detection of potential health effects.

Inadequacies of current toxicity testing strategies have been revealed and a potential way forward suggested. Analytical methods may not be sufficiently sensitive to detect the ability of ingredients to promote chronic disease. This can be attributed to the lack of suitable animal models to detect hormonal or metabolic disruptions, insufficiencies in experimental design, and the use of artificial conditions that do not reflect real-life exposure. Although toxicity tests in laboratory animals have always been the first choice to evaluate human health risks, they are being progressively replaced by new in vitro approaches. Artificial intelligence methods have even been shown to predict acute toxic effects better than some short-term animal studies.

In addition to conducting toxicity studies with combinations of herbicides and other types of pesticides using advanced high-throughput methods,[10] it is important to conduct epidemiological studies to ascertain the distribution and body burden of herbicide residues (Chapter 9). A combination of laboratory toxicity studies and epidemiological surveys will contribute to more appropriate regulations to better protect public health.

Toxicologists distinguish herbicides from other pesticides and evaluate the safety of each herbicide individually. Insecticides and herbicides are very different types of chemicals. Many insecticides are nerve toxins, and these and others are broadly toxic to vertebrates, including humans, at field rates of application. In contrast, herbicides are designed to act on biochemical pathways that are unique to plants. Many inhibit enzymes that plants use in photosynthesis or in the synthesis of amino acids, which animals do not possess. Thus, most herbicides appear to have little or no acute toxicity to humans and wildlife at environmental levels.

However, the real unknown about herbicides (and other pesticides as well) is the health effects of low-level chronic exposure. Most research on long-term effects has focused on cancer risk, with far less attention and low funding priority to neurological, immunological, developmental, and reproductive effects. The bottom line is that we do not presently have the tools and comprehensive herbicide application data to accurately evaluate the human and environmental risk of long-term exposure to herbicides.

Underneath herbicide product labels: The role of coformulants

The success of weed management is dependent not only on the mechanisms of action of active substances (Chapter 3) but also the physicochemical properties of the spray mixture. The solubility, volatilization, adherence, penetration, rainfastness, foaming, and drift of a herbicide can be controlled by the inclusion of compounds other than the active principles (Chapter 4). They are called coformulants in formulated herbicides, or adjuvants when they are added to tank mixes. These compounds are considered chemically inert and are commonly not tested in environmental risk assessments, despite their supportive effect of the active substance. In Germany, more than 1600 coformulants for pesticides are approved.[11]

Adjuvant technologies have become cornerstones of chemical weed control. It is also a way for manufacturers to differentiate products containing the same active ingredients, to target specific market segments, and to compete for market share. For instance, about 750 GBH products were on the market in 2015. They differ not only by containing different glyphosate forms (e.g., isopropylammonium, diammonium, and potassium salts) in various concentrations but also in their composition and coformulants.

Coformulants can also have toxic effects, which are ignored in most laboratory animal and epidemiological studies.[12] When surfactants and oils were used at the beginning of the 20th century as weed killers, they were recognized as an important source of toxicity. The first regulations considered that each component of a herbicide should have the same safety requirements. Nowadays, substances used as coformulants are mostly considered nontoxic within the regulatory frameworks in Europe and the United States.

If assessed at all, safety information on coformulants is held as confidential for business reasons. Confidentiality can serve an important purpose for manufacturers to maintain competitive advantages, but it also causes a gap in knowledge because some toxic effects can remain unexplained if only the active ingredient is considered in assessing the effects of spraying formulated products.[12] This is the case for the debate on glyphosate carcinogenicity, which would be very different if pesticide product labelling requirements included a description of the full chemical composition. Improving transparency is key to building public trust in herbicides and other pesticides.

Toxicity on other species, biodiversity loss, and the environment

Contrary to a widespread assumption, herbicides not only affect weeds but have direct and indirect impacts on a wide variety of nontarget organisms. The direct effects of course concern weed diversity, but also crop plants, invertebrates, vertebrates, and microorganisms (Chapter 7). To what extent they contaminate soil depends on the mode of action (e.g., leaching potential, volatility), soil characteristics, and environmental parameters. Various soil organisms and processes can then be directly affected by residues in soil, and those residues can be more or less prone to leaching into water bodies. Although environmental risk assessments are required for regulatory purposes, they are done on a small selection of surrogate species—mammals (mice, rats), birds (quail, ducks), and invertebrates (honeybees, compost worms).[35] So our knowledge of the effects on most native species is still quite scarce. Some animal groups, such as amphibians and bats, are not considered at all. Herbicides can affect nontarget organisms at all levels, altering physiology, reproduction, and behavior.

Because of the great variety in modes of action, deriving consistent response patterns is impossible. Effects can differ even within the same herbicide class. Importantly, most studies have tested a single active ingredient on a single nontarget species without target plants present, under controlled conditions, for a short duration. It is difficult to extrapolate the findings to the complex multispecies agroecosystem level where impacts of herbicides interact with other management practices, other pesticides, and environmental conditions over a long period.

Indirect effects are not as obvious as direct ones but can still have consequences on the performance and fitness of organisms and overall ecosystem functioning (Chapter 8). Spray drift and erosion by wind or water distribute herbicides widely within the agricultural landscape, so that not only crop fields but also noncrop areas are affected. The indirect effects alter overall biodiversity and food web interactions that further trigger ecosystem functions and services. An increasing body of evidence shows that herbicides affect crop disease development.[13] As a consequence, herbicide effects could trigger further fungicide or insecticide use because crop plants have been weakened in their natural potential to defend against disease.

Interactions of herbicide effects with climatic factors and other environmental stressors are becoming increasingly important and will continue to be so in the future as climate change accelerates. Because of the delayed effects of previous herbicide applications (legacy effects), there is a knowledge gap in assessing exposure under real-world conditions. This is compounded by multiple applications of herbicides and interactions with other pesticide classes within a season and especially long term. When addressing indirect impact, it is not often easy to separate the effects of other agricultural management practices such as tillage and crop rotation from herbicide effects. The sheer number of available herbicide formulations and mixtures, and inadequate information on their ingredients, make an analysis of environmental effects generally difficult.

When research on herbicide effects is ripped apart by corporate interests

Perspectives from political economy are also important in understanding why the commercialization of a herbicide such as glyphosate is at the heart of a heated debate. It is common sense, but one should bear in mind that herbicides are commercial products, and the primary purpose of their commercialization is to maximize profits for owners and stakeholders of pesticide-producing companies. The agrochemical industry spends several million dollars every year on lobbying decision makers to oppose bills seeking to ban their products, and sometimes via campaign contributions to political parties.[14] The scientific debate can become secondary when a substance like glyphosate is a strategic piece of a billion dollar market of a company such as the US-based Monsanto, with ramifications for food sovereignty and international trade.

Maximizing profits and maintaining corporate social responsibility have been difficult exercises for pesticide companies. This is illustrated by a series of public health scandals in which the stewardship of Monsanto undermined the fulfilment of the aims of US public health regulations.[15] The large quantity of internal Monsanto emails released during cancer litigation revealed the strategies by which the company influenced the reporting of scientific studies and colluded with regulatory agencies to misrepresent the safety profile of Roundup herbicides (Chapter 10). In 2012, a rat study assessed the health effects of a Roundup-tolerant GM maize cultivated with or without Roundup and Roundup alone (from 0.1 ppb in water) over 2 years.[16] In females, all treated groups died at a rate 2–3 times that of controls, and more rapidly. Females also developed large mammary tumors almost always earlier and more often than controls. The authors explained these results by the nonlinear endocrine-disrupting effects of Roundup, and by the overexpression of the transgene in the GM maize and its metabolic consequences. However, the study was not conceived to form conclusions on long-term effects because the pathologies of normal aging introduce background noise that limits statistical power with 10 rats per group. When the "Séralini study" was published in September 2012, a letter was sent to the editor-in-chief of the journal by a former Monsanto researcher stating that "The implications and the impacts of this uncontrolled study is having HUGE impacts, in international trade, in consumer confidence in all aspects of food safety, and certainly in US state referendums on labelling."[17] Roundup cancer litigation in the United States later revealed that this was part of a strategy deployed by Monsanto to undermine the scientific study, which led to its retraction after the same researcher was hired as an editor to reconsider the publication of the study. Even the editor-in-chief of the journal was a consultant for Monsanto and directly in contact with the company at the time of the publication of the article.[15] The Séralini publication and resultant media attention raised the profile of fundamental challenges faced by science in a world increasingly dominated by corporate influence. The issue was addressed in an open letter signed by hundreds of scientists around the world.[18] These challenges are important for all of science but are rarely discussed in scientific venues.

One should bear in mind that the influence of advocate groups is not unilateral. While companies develop strategies to preserve the commercialization of their pesticides, other advocate groups deploy strategies to ban pesticides (Chapter 10). This was the case in the Séralini affair. The network of activists involved in the publication of this study orchestrated a communication operation causing an explosion of media coverage by spreading graphic images of rats with large tumors.[19] This affected public opinion about GM foods, with political consequences on international trade. Kenya and Peru placed a moratorium on GM crops.[20] Although we do not pass judgment on the soundness of the retraction, we highlight that the series of controversies that became the Séralini affair is a well-documented case of the influence of advocate groups on academic science.

In some cases, the safety evaluation of herbicides has been compromised by scientific misconduct, including the selective use and omission of published data besides undisclosed industry data, the dismissal of adverse effects, the misuse of historical data and statistical methods, and plagiarism.[21] In the current approval framework, the industry seeking commercial approval performs the risk assessment of their own results, which is then used by a regulatory agency for decision making. Also, industry studies are not registered in advance, and unfavorable findings can be discarded. Improving the rigor, transparency, and independence of the process is urgently needed and could have far-reaching consequences, including the promotion of public trust in agricultural technologies. Solutions have been proposed.[21]

Weed management for a resilient Earth system

The ecological pressure of human societies has become unsustainable and threatens the resilience and integrity of the Earth system.[22] The population is expected to continue increasing and reach a plateau of about 9 billion in the next few decades while agricultural land area cannot increase at the same rate. The use of herbicides has increased the yield,[23] but meeting the global food demand of 9 billion or more is complex and goes beyond a race toward yield improvement. Especially with climate change and other environmental stressors, emphasis should be on long-term environmental sustainability rather than short-term yield.[24] As the planet warms, crop failure may become more frequent due to climate instability and increased rates of natural disasters. Sudden losses in food production have become more frequent because of geopolitical (e.g., conflicts across sub Saharan Africa and the Middle East) and extreme weather events (e.g., droughts and floods in South Asia).[25]

There is no one-size-fits-all approach for weed control. Like our economies, it will have to become more localized by the inclusion of regional farmers' knowledge. This will help in tackling issues caused by climate change, as the development of local solutions will be facilitated. Each country will account for the particularities of its environment. The globalization of agriculture is not compatible with the reality of the environment. Agriculture is linked to climate change in three ways. First, it contributes to global CO_2 emissions. Second, it is detrimentally affected by global warming, weather extremes, and changes in diseases and pests. Third, it can contribute to climate change mitigation by capturing CO_2 in soil, thereby increasing soil carbon content.

Industrialized agriculture is a major contributor to Earth system degradation. Modern agricultural methods are highly dependent on coal, natural gas, and oil. Human labor has been converted into machine labor, improving efficiency but creating dependence on fossil energy. Petroleum is also essential to synthesize herbicides, which may become increasingly expensive as resources are exhausted. It is estimated that per kilogram of herbicide produced, 1.7–12.6 kg of carbon are emitted; in contrast, for each kg of nitrogen fertilizer produced, 0.9–1.8 kg of carbon are emitted.[26] The energy returned to energy invested index of petroleum is declining as more energy becomes necessary to obtain it than the amount of usable energy extracted from it.[27]

The heavy use of herbicides also has human health consequences. Generally, these agents are assumed to be safe because they are designed to specifically target plant metabolism. However, evidence suggesting a role in the etiology of neurological, immunological, metabolic, developmental, and reproductive diseases is increasing. These effects are not always fully predicted from bioassays before the products are placed on the market. This amplifies the gap between the introduction of new herbicides and the detection of their toxic effects in humans.

The challenge of global food security has two aspects: quantity and quality. An estimated 2 billion people lack access to one or more essential nutrients.[28] While almost a billion people still suffer from inadequate diet and insecure food supplies, the global transition toward diets high in processed foods has contributed to 2.1 billion people becoming overweight or obese.[29] Nutritional deficiencies are the result of not only poor food quality but a lack of food diversity. Increasing diversity is therefore an important strategy to ensure health sustainability. It implies that agricultural production needs to be diversified because intensive agricultural systems have contributed to narrowing global production patterns to a limited number of major crop plants.[30]

Two other important issues are food waste and biofuels. An estimated 30% of the world's agricultural land is lost to produced but uneaten food.[31] Food wastage reduction would not only avoid pressure on scarce natural resources but also decrease herbicide waste and the need to raise food production by 60% to meet the estimated 2050 population demand. About 4% of the world's agricultural land is used to grow biofuels; about 280 million of the malnourished people in the world could be fed by using these resources for food production.[32]

The diversification of weed management practices will be key to reducing the selection pressure for herbicide-resistant weed evolution. This can be accomplished through diversification of herbicide modes of action, ideally in combination as mixtures but also in rotation as part of a weed management program. Most importantly, combinations of mechanical, cultural, biological, and other nonchemical methods have recently increased the scope of available techniques for many crops. These herbicide-free management programs using such techniques as the reintroduction of crop rotation, cultivation designed to minimize erosion, timing of planting, high-density planting, cover cropping, crop row orientation, allelopathy, specific irrigation, intercropping, and biological control have been shown to produce yields equivalent to those with conventional herbicide use in multiple long-term field experiments. Recently, bioeconomic models revealed that a ban of glyphosate and all other herbicides in wheat production in Switzerland would lead to only moderate economic loss and to only about 6% yield reduction.[33] Possible economic loss would be outweighed by existing agri-environmental subsidies aiming to reduce risks for the environment and human health.

The use of herbicides for weed control is not only an issue of human and environmental health but also has important politic, legal, and economical implications. Providing 9 billion people with healthy food through sustainable farming is a great challenge humanity has to face. It goes beyond the enhancement of farm performance and implies a global revolution of food systems. Socioeconomic factors such as food availability, disparity in wealth, waste management, and dietary choice are equally important for ensuring global food security. It is up to scientifically informed policymakers to decide whether food production without synthetic herbicides is fostered. This book is meant to provide insights from multiple scientific disciplines to reach this decision.

References

1. Oerke EC. Crop losses due to pests. *J Agric Sci (Camb)* 2006;**144**(1):31–43.
2. Scarrow R. Ever-increasing use. *Nat Plants* 2019;**5**(12):1199.
3. Duke SO. Why have no new herbicide modes of action appeared in recent years? *Pest Manag Sci* 2012;**68**(4):505–12.
4. Kudsk P, Jørgensen LN, Ørum JE. Pesticide load—a new Danish pesticide risk indicator with multiple applications. *Land Use Policy* 2018;**70**:384–93.
5. Lyseen AK, Nøhr C, Sørensen EM, et al. A review and framework for categorizing current research and development in health related geographical information systems (GIS) studies. *Yearb Med Inform* 2014;**9**(1):110–24.
6. Linhart C, Niedrist GH, Nagler M, et al. Pesticide contamination and associated risk factors at public playgrounds near intensively managed apple and wine orchards. *Environ Sci Eur* 2019;**31**(1):28.

7. Sapcanin A, Cakal M, Imamovic B, et al. Herbicide and pesticide occurrence in the soils of children's playgrounds in Sarajevo, Bosnia and Herzegovina. *Environ Monit Assess* 2016;**188**(8).
8. von Ehrenstein OS, Ling C, Cui X, et al. Prenatal and infant exposure to ambient pesticides and autism spectrum disorder in children: population based case-control study. *BMJ* 2019;**364**:l962.
9. Partel V, Charan Kakarla S, Ampatzidis Y. Development and evaluation of a low-cost and smart technology for precision weed management utilizing artificial intelligence. *Comput Electron Agric* 2019;**157**:339–50.
10. Mesnage R, Teixeira M, Mandrioli D, Falcioni L, Ducarmon QR, Zwittink RD, et al. Use of shotgun metagenomics and metabolomics to evaluate the impact of glyphosate or roundup MON 52276 on the gut microbiota and serum metabolome of Sprague-Dawley rats. *Environ Health Perspect* 2021;**129**(1):17005. https://doi.org/10.1289/EHP6990.
11. BVL. *Beistoffe in zugelassenen Pflanzenschutzmitteln*; 2018. https://www.bvl.bund.de/SharedDocs/Downloads/04_Pflanzenschutzmittel/zul_info_liste_beistoffe.html. [Accessed 29 September 2019].
12. Mesnage R, Benbrook C, Antoniou MN. Insight into the confusion over surfactant co-formulants in glyphosate-based herbicides. *Food Chem Toxicol* 2019;**128**:137–45. https://doi.org/10.1016/j.fct.2019.03.053.
13. Kortekamp A. Unexpected side effects of herbicides: modulation of plant-pathogen interactions. In: Kortekamp A, editor. *Herbicides and environment*. Rijeka: InTech; 2011. p. 85–104.
14. Wouters OJ. Lobbying expenditures and campaign contributions by the pharmaceutical and health product industry in the United States, 1999-2018. *JAMA Intern Med* 2020;**180**(5):688–97.
15. McHenry LB. The Monsanto Papers: poisoning the scientific well. *Int J Risk Saf Med* 2018;**29**:193–205.
16. Séralini G-E, Clair E, Mesnage R, et al. RETRACTED: long term toxicity of a Roundup herbicide and a Roundup-tolerant genetically modified maize. *Food Chem Toxicol* 2012;**50**(11):4221–31.
17. Séralini G-E, Mesnage R, Defarge N, Spiroux de Vendômois J. Conflicts of interests, confidentiality and censorship in health risk assessment: the example of an herbicide and a GMO. *Environ Sci Eur* 2014;**26**(1):13.
18. Bardocz S, Clark A, Ewen S, et al. Seralini and science: an open letter. *Indep Sci News* 2012. https://www.independentsciencenews.org/health/seralini-and-science-nk603-rat-study-roundup/.
19. Butler D. Rat study sparks GM furore. *Nature* 2012;**489**(7417):484.
20. Gerasimova K. Advocacy science: explaining the term with case studies from biotechnology. *Sci Eng Ethics* 2018;**24**(2):455–77.
21. Robinson C, Portier CJ, ČAvoŠKi A, et al. Achieving a high level of protection from pesticides in Europe: problems with the current risk assessment procedure and solutions. *Eur J Risk Regul* 2020;1–31.
22. Steffen W, Richardson K, Rockstrom J, et al. Sustainability. Planetary boundaries: guiding human development on a changing planet. *Science* 2015;**347**(6223):1259855.
23. Gianessi LP. The increasing importance of herbicides in worldwide crop production. *Pest Manag Sci* 2013;**69**(10):1099–105.
24. Ehrlich PR, Harte J. Opinion: to feed the world in 2050 will require a global revolution. *Proc Natl Acad Sci U S A* 2015;**112**(48):14743–4.

25. Cottrell RS, Nash KL, Halpern BS, et al. Food production shocks across land and sea. *Nat Sustain* 2019;**2**(2):130–7.
26. Lal R. Carbon emission from farm operations. *Environ Int* 2004;**30**(7):981–90.
27. Hall CAS, Lambert JG, Balogh SB. EROI of different fuels and the implications for society. *Energy Policy* 2014;**64**:141–52.
28. Institute IFPR. *Global nutrition report 2014: Actions and accountability to accelerate the world's progress on nutrition.* Washington, DC: International Food Policy Research Institute (IFPRI); 2014.
29. Tilman D, Clark M. Global diets link environmental sustainability and human health. *Nature* 2014;**515**(7528):518–22.
30. Khoury CK, Bjorkman AD, Dempewolf H, et al. Increasing homogeneity in global food supplies and the implications for food security. *Proc Natl Acad Sci U S A* 2014;**111**(11):4001–6.
31. FAO. *Food wastage footprint. Impacts on natural resources*; 2013 http://wwwfaoorg/3/i3347e/i3347epdf.
32. Rulli MC, Bellomi D, Cazzoli A, De Carolis G, D'Odorico P. The water-land-food nexus of first-generation biofuels. *Sci Rep* 2016;**6**(1):22521.
33. Böcker T, Möhring N, Finger R. Herbicide free agriculture? A bio-economic modelling application to Swiss wheat production. *Agric Syst* 2019;**173**:378–92.
34. Zaller JG. *Daily Poison. Pesticides – an underestimated danger.* Springer International; 2020.
35. Brühl CA, Zaller JG. Biodiversity decline as a consequence of an inappropriate environmental risk assessment of pesticides. *Front Environ Sci* 2019;**7**:177. https://doi.org/10.3389/fenvs.2019.00177.

Index

Note: Page numbers followed by *f* indicate figures, *t* indicate tables, and *b* indicate boxes.

M

N

O

P

Printed in the United States
by Baker & Taylor Publisher Services